Dissertation

Adaptive Wavelet Frame
Domain Decomposition Methods
for Elliptic Operator Equations

Manuel Werner

2009

Bibliografische Information der Deutschen Nationalbibliothek

Die Deutsche Nationalbibliothek verzeichnet diese Publikation in der
Deutschen Nationalbibliografie; detaillierte bibliografische Daten sind
im Internet über http://dnb.d-nb.de abrufbar.

ISBN 978-3-8325-2286-5

Logos Verlag Berlin GmbH
Comeniushof, Gubener Str. 47,
10243 Berlin
Tel.: +49 030 42 85 10 90
Fax: +49 030 42 85 10 92
INTERNET: http://www.logos-verlag.de

Adaptive Wavelet Frame Domain Decomposition Methods for Elliptic Operator Equations

Dissertation
zur
Erlangung des Doktorgrades
der Naturwissenschaften
(Dr. rer. nat.)

dem

Fachbereich Mathematik und Informatik
der Philipps–Universität Marburg

vorgelegt von

Manuel Werner
aus Marburg/Lahn

Marburg/Lahn Mai 2009

Vom Fachbereich Mathematik und Informatik
der Philipps–Universität Marburg als Dissertation
angenommen am: 2. Juli 2009

Erstgutachter: Prof. Dr. Stephan Dahlke, Philipps–Universität Marburg

Zweitgutachter: Prof. Dr. Rob Stevenson, University of Amsterdam

Tag der mündlichen Prüfung: 8. Juli 2009

Acknowledgments

First of all, my sincere gratitude goes to my supervisor Professor Stephan Dahlke for giving me the opportunity to work in his group. I thank him for his willingness to invest a lot of work and time to organize new positions for young unexperienced mathematicians and his belief in their abilities. During my time as a PhD student, I have been given the chance to present my results at various conferences, workshops, and seminars and to cooperate with other mathematicians working in the field of adaptive wavelet methods, for which I am particularly grateful. I also thank Professor Dahlke for the many discussions and valuable comments and corrections during the preparation of this thesis.

Furthermore, I want to express my gratitude to Professor Rob Stevenson. I thank him for the invitation to Utrecht, his readiness to write the second referee report, and for the very kind cooperation.

I also want to thank Thorsten Raasch for countless mathematical discussions and private conversations. He has always had an open ear for any kinds of problems. Moreover, parts of the software used to produce the numerical results have been developed in association with him since I started this work. Also the second part of the proof of Proposition 2.2 stems from a nice collection of elementary proofs of some important results in frame theory worked out by Thorsten Raasch.

During his numerous visits in Marburg, Massimo Fornasier has certainly enriched the scientific life in our group. I want to thank him for this lively collaboration and the many productive conversations.

My thanks go to all other members of our group. I have particularly enjoyed the very friendly atmosphere.

I feel grateful to the Deutsche Forschungsgemeinschaft which has supported the first two years of my PhD position under Grant Da 360/4-2 and Da 360/4-3.

I want to thank my parents for always supporting my academic education, and my uncle Hans Effinger for encouraging my interest in mathematics and for many good advice. I also thank my parents-in-law for always being there when any kind of help is needed.

Most importantly, I am deeply indebted to my wife, Steffi, for her great support, patience, and understanding.

Zusammenfassung

Bei der mathematischen/physikalischen Modellierung vieler realer Probleme aus den verschiedensten Bereichen der Wissenschaft spielen *Partielle Differentialgleichungen und Integralgleichungen* eine zentrale Rolle. Diese werden in der Regel in einem funktionalanalytischen Rahmen als Operatorgleichungen

$$\mathcal{L}u = f$$

studiert. Im Mittelpunkt dieser Arbeit steht die numerische Behandlung der speziellen Klasse *elliptischer* Operatorgleichungen, wobei $\mathcal{L} : \mathcal{H} \rightarrow \mathcal{H}'$ einen linearen Operator darstellt, welcher von einem Hilbertraum $\mathcal{H}$ in dessen normierten Dualraum $\mathcal{H}'$ abbildet und die Elliptizitätsbedingung $c\|v\|_{\mathcal{H}} \leq \|\mathcal{L}v\|_{\mathcal{H}'} \leq C\|v\|_{\mathcal{H}}$, $v \in \mathcal{H}$, mit Konstanten $0 < c, C < \infty$ erfüllt. Hierbei ist $f \in \mathcal{H}'$ gegeben und $\mathcal{H}$ in der Regel ein Funktionenraum über einem Gebiet $\Omega \subset \mathbb{R}^n$ oder einer geschlossenen Mannigfaltigkeit.

Hintergrund und Motivation

Die numerische Lösung der in der Regel nicht explizit auflösbaren Gleichungen beruht typischerweise auf der Approximation des kontinuierlichen Problems mit Hilfe eines diskreten Modells, z.B. aufbauend auf einem endlichen Netz oder Gitter. Heute verfügbare Rechnerleistungen ermöglichen die Betrachtung immer feinerer Auflösungen mit Millionen von Freiheitsgraden. Hierbei ist es von entscheidender Bedeutung, dass der Rechenaufwand und die benötigte Anzahl von Speicherplätzen eines numerischen Verfahrens zur Lösung des Problems bis auf eine bestimmte Genauigkeit proportional zur Dimension des diskreten Modells bleibt, wenn das Gitter oder Netz verfeinert wird. Um dieses Ziel zu erreichen, muss sichergestellt werden, dass die Konditionszahlen der Systemmatrizen, der im diskreten Modell auftretenden linearen Gleichungssysteme, bei Verfeinerung der Diskretisierung, gleichmäßig beschränkt bleiben. Hierzu wird üblicherweise eine *Vorkonditionierung* der Systemmatrizen benötigt. In diesem Kontext besteht die zentrale Technik in der Einbeziehung verschiedener Auflösungsstufen und deren Interaktion in das diskrete Modell. Bekannte Verfahrensklassen sind *Multigrid-Methoden* (z.B. [15, 72]) und die eng verwandten *Multilevel-Verfahren* [16, 93, 121, 123]. Diese Arbeit befasst sich mit der Entwicklung von Lösungsverfahren, welche auf einer *Multiskalen-Diskretisierung* mit Hilfe von *Wavelet-Funktionen* aufbaut. Diese Funktionen erlauben die Diskretisierung elliptischer Operatorgleichungen, wobei die Kondition der diskreten Probleme mit wachsender Anzahl der Freiheitsgrade gleichmäßig beschränkt bleibt [42, 44].

In den letzten Jahren wurden Wavelet-Verfahren sehr erfolgreich z.B. in der Signal-Analyse/Kompression eingesetzt. Seit einiger Zeit wird aber auch deren Verwendung zur Lösung von Operatorgleichungen untersucht. Wavelets sind spezielle Multiskalen-Basen, zunächst in der Regel für $L_2(\mathbb{R}^n)$, welche durch Translation, Dilatation und Skalierung einer oder mehrerer *Mother-Wavelets* entstehen. Der Erfolg von Wavelets beruht auf der Tatsache, dass für den Fall $\Omega = \mathbb{R}^n$ mächtige analytische Eigenschaften garantiert werden können. Es ist möglich, Wavelets so zu konstruieren, dass

- sie *kompakten Träger* bei beliebig wählbarer Regularität besitzen,

- sie eine *Charakterisierung von Funktionenräumen* durch eine Äquivalenz der Funktionenraumnorm mit einer gewichteten Folgenraumnorm der Wavelet-Koeffizienten ermöglichen,

- sie *verschwindende Momente* besitzen, woraus sich ergibt, dass L_2-Innenprodukte glatter Funktionen mit Wavelets mit wachsender Waveletskala dem Betrage nach exponentiell abfallen.

Die Lösungen vieler Probleme besitzen lokale Charakteristika, wie z.B. Singularitäten, für deren effiziente Auflösung *adaptive Verfahren* unabdingbar sind. Ein wichtiges Beispiel stellen elliptische Operatorgleichungen auf einem Gebiet mit nicht glattem Rand (vgl. [53, 69, 70, 86]) oder Probleme mit singulärer rechter Seite f dar. Unter einem adaptiven Verfahren verstehen wir einen Algorithmus, der iterativ eine Folge von speziell an die unbekannte Lösung u angepasste Approximationen generiert, wobei, um von einer Iterierten zur nächsten zu gelangen, ausschließlich Information verwendet wird, welche bereits in der aktuellen Approximation enthalten ist oder aus der rechten Seite f und $\mathcal{L}$ gewonnen wird. Es wird hierbei zu keinem Zeitpunkt a priori-Information über die exakte Lösung u verwendet.

In einer Reihe von Arbeiten wurden adaptive Wavelet-Verfahren zur Lösung von Operatorgleichungen über einem beschränkten Gebiet oder einer geschlossenen Mannigfaltigkeit studiert [8, 27–29, 34, 35, 40, 42, 64, 65, 97, 102]. Hierbei sind die generierten Approximationen $\{u_{\Lambda_j}\}_{j \geq 0} \subset \mathcal{H}$ in der Regel als endliche Linearkombinationen der Elemente einer *Wavelet-Riesz-Basis* $\Psi = \{\psi_\lambda\}_{\lambda \in \Lambda}$ für $\mathcal{H}$ gegeben, das bedeutet $u_{\Lambda_j} = \sum_{\lambda \in \Lambda_j} \mathbf{u}_\lambda \psi_\lambda$, $\#\Lambda_j < \infty$, $\Lambda_j \subset \Lambda$. In [27] wurde für eine große Klasse symmetrischer, positiv definiter, elliptischer Operatoren ein konvergentes adaptives Wavelet-Galerkin-Verfahren entwickelt. Es konnte bewiesen werden, dass bei diesem Algorithmus

(O1) *das Verhältnis zwischen der erzielten Genauigkeit $\|u - u_{\Lambda_j}\|_\mathcal{H}$ und der Anzahl der Freiheitsgrade $\#\Lambda_j$ optimal ist, in dem Sinne, dass $\|u - u_{\Lambda_j}\|_\mathcal{H} \leq C_1(u)(\#\Lambda_j)^{-s}$, mit dem größtmöglichen $s > 0$, für eine von u abhängige Konstante $C_1 > 0$ gilt.*

Die optimale *Konvergenzrate* $s > 0$ ist hierbei stets gegeben durch die Abfallrate des Fehlers der besten N-Term-Approximation $\|u - u_N\|_{\mathcal{H}} = \inf\{\|u - v\|_{\mathcal{H}} : v = \sum_{\lambda \in \mathcal{J}} \mathbf{v}_\lambda \psi_\lambda, \#\mathcal{J} \leq N, \mathcal{J} \subset \Lambda\}$. Darüber hinaus benötigt das in [27] entwickelte Verfahren zur Berechnung von u_{Λ_j} im Prinzip größenordnungsmäßig $\mathcal{O}(\#\Lambda_j)$ Rechenoperationen und Speicherplätze. Es besitzt also

(O2) *lineare Komplexität.*

Die Konvergenzrate der besten N-Term-Wavelet-Approximation der Lösung u hängt in der Regel von deren Besov-Regularität ab. Für viele elliptische Probleme mit singulären Lösungen ist nun bekannt, dass diese Regularität wesentlich höher ist als deren Sobolev-Regularität, welche für die Konvergenzrate eines nicht-adaptiven Verfahrens, basierend auf einer uniformen Verfeinerungsstrategie, maßgebend ist. Optimale adaptive Verfahren lohnen sich in diesen Fällen also, da höhere Konvergenzraten erreicht werden können.

Eine wesentliche Schwierigkeit liegt nun jedoch in der Konstruktion von Wavelet-Basen mit den oben genannten vorteilhaften Eigenschaften für die entsprechenden Funktionenräume über dem in der Regel nicht trivialen, *beschränkten* Gebiet $\Omega \subset \mathbb{R}^n$. Die im translationsinvarianten Fall $\Omega = \mathbb{R}^n$ verwendbaren, mächtigen Fourier-Techniken stehen hier z.B. nicht zur Verfügung. Bisher bekannte Konstruktionen (z.B. [19,32,48,49,79,87,106]) sind sehr kompliziert und führen zum Teil auf schlecht konditionierte Basen. Diese oft unbefriedigenden, quantitativen Eigenschaften haben dann negative Auswirkungen auf die Effizienz der hierauf aufbauenden adaptiven Wavelet-Verfahren.

Aus diesem Grund wurden in [38,104] erste Verfahren basierend auf dem schwächeren Konzept der *Frames* (vgl. [23,50,51,80]) an Stelle von Riesz-Basen studiert, welches redundante Darstellungen zulässt. In der Tat kann ein geeigneter Wavelet-Frame mittels einer sehr einfachen Konstruktion erzeugt werden. Durch eine überlappende Überdeckung von Ω mit $m \in \mathbb{N}$ Teilgebieten Ω_i, $\Omega = \bigcup_{i=0}^{m-1} \Omega_i$, und Vereinigung von Wavelet-Riesz-Basen über den Teilgebieten der Überdeckung, entsteht ein *aggregierter Wavelet-Frame*. Mit diesem kann jedes Element des Funktionenraums über dem Gesamtgebiet dargestellt werden, wobei diese Darstellung nicht länger notwendig eindeutig ist, da in den Überlappungsbereichen Redundanzen entstehen. Hierbei werden die Teilgebiete als parametrisches Bild des n-dimensionalen Einheitskubus gewählt. Gut konditionierte Wavelet-Basen über dem Einheitskubus stehen mittlerweile zur Verfügung [95], so dass hierdurch eine einfache Implementierung eines Wavelet-Frames mit guten quantitativen Eigenschaften über einem beschränkten Gebiet erreicht werden kann. In [104] wurde für das im Zuge der Diskretisierung eines symmetrischen Problems entstehende unendlich-dimensionale Gleichungssystem ein auf einer *gedämpften Richardson-Iteration* basierendes, implementierbares adaptives Verfahren entwickelt, welches ebenfalls mit optimaler Rate und linearer Komplexität konvergiert. Mit dem in [38] eingeführten Begriff der *Gelfand-Frames* steht darüber hinaus ein allgemeines Konzept zur Verfügung, welches den üblichen analytischen

Rahmen bei der Konstruktion von Wavelet-Riesz-Basen umfasst und solche Funktionensysteme beschreibt, welche neben der Charakterisierung eines Hilbertraumes V (häufig $L_2(\Omega)$) auch jene eines in $\mathcal{H}$ stetig und dicht eingebetteten Raumes $\mathcal{H} \subset V$ (z.B. ein L_2-Sobolev-Raum) erlauben.

Nach den ersten grundlegenden Arbeiten [38, 104] (vgl. auch [97]) bestehen auch in diesem Kontext noch verschiedene prinzipielle Schwierigkeiten. Erste numerische Tests der in [104] vorgeschlagenen Verfahren in [97, 117] haben gezeigt, dass die praktisch beobachtbare quantitative Konvergenzgeschwindigkeit nicht befriedigend ist. Darüber hinaus beruhen die Beweise der optimalen Komplexität zum Teil auf einer theoretischen Annahme, deren Verifikation für die meisten praktisch relevanten Probleme bisher nicht erbracht werden konnte.

Ziele und Inhalt der Arbeit

Der wesentliche Beitrag dieser Arbeit besteht daher in der Entwicklung neuer, *effizienter*, adaptiver Frame-Verfahren für den Fall elliptischer, symmetrisch positiv definiter Operatoren, mit denen die erwähnten theoretischen und praktischen Schwierigkeiten überwunden werden können. Zum einen spielt jeweils der rigorose theoretische Nachweis von Konvergenz und asymptotischer Optimalität ((O1), (O2)) eine zentrale Rolle. Zum anderen steht ebenso die praktische Realisierung und die Analyse der numerischen Eigenschaften der Verfahren im Mittelpunkt. Es soll gezeigt werden, dass adaptive Wavelet-Frame-Methoden existieren, welche mit Standard-Verfahren, z.B. basierend auf einer Finite-Elemente-Diskretisierung, konkurrieren können und im Hinblick auf die benötigte Anzahl von Freiheitsgraden sogar effizienter sein können.

Es werden Operatorgleichungen, welche durch eine variationelle Formulierung einer elliptischen partiellen Differentialgleichung mit homogenen Dirichlet-Randdaten entstehen, betrachtet. In diesem Kontext ist $\mathcal{H} = H_0^t(\Omega)$, $t \in \mathbb{N}_0$, ein randangepasster L_2-Sobolev-Raum (vgl. Kapitel 1). Zwei verschiedene neue Verfahrensklassen werden behandelt. Der erste zentrale Aspekt dieser Arbeit lautet daher nun wie folgt.

(T1) Es soll ein auf der *Methode des steilsten Abstiegs* aufbauendes, konvergentes adaptives Wavelet-Frame-Verfahren entwickelt und dessen Optimalität im Sinne von (O1) und (O2) bewiesen werden. Es soll untersucht werden, inwieweit dieser Ansatz die bestehenden Probleme im Zusammenhang mit der bisher verwendeten adaptiven gedämpften Richardson-Iteration auflöst oder deren Nachteile verbessert. Die praktischen Eigenschaften der neuen Methode werden durch eine Reihe numerischer Tests analysiert.

Die Betrachtung der Methode des steilsten Abstiegs ist vor allem motiviert durch die Tatsache, dass die besonders im Kontext einer Frame-Diskretisierung schwierige Vorberechnung des optimalen Relaxationsparameters, wie sie für die in [38, 104] verwendete gedämpfte Richardson-Methode benötigt wird, so vermieden werden kann. In einer Reihe numerischer Experimente wird die Optimalität des neuen Verfahrens bestätigt. Des Weiteren zeigt ein Vergleich mit der Richardson-Methode die verbesserte Effizienz des Verfahrens für den Fall, dass der optimale Relaxationsparameter unterschätzt wird, womit in der Praxis zu rechnen ist.

Die zweite, zugleich vielversprechendste Verfahrensklasse, welche in dieser Arbeit entwickelt wird, basiert auf *überlappenden Schwarz-Gebietszerlegungsmethoden.* Die Grundidee dieser Ansätze geht auf eine frühe Arbeit von H. A. Schwarz [103] zurück. Hierbei wird das elliptische Problem in Ω durch iteriertes Lösen elliptischer Probleme in den Teilgebieten Ω_i gelöst. Hierdurch ergeben sich zwei wesentliche Vorteile. Zum einen kann durch einen solchen Ansatz eine zusätzliche Vorkonditionierung des Ausgangsproblems gewonnen werden. Man unterscheidet dabei zwischen *multiplikativen* und *additiven* Vorkonditionierern, wobei vor allem additive Methoden die *simultane* Lösung der lokalen Teilprobleme zulassen, so dass auf natürliche Weise effiziente *parallele* Algorithmen entwickelt werden können [18,22,96,110]. Dies stellt einen weiteren entscheidenden Vorteil dar. Die wiederholte Lösung der lokalen elliptischen Probleme kann dann wiederum mit Hilfe eines adaptiven Wavelet-Verfahrens in Bezug auf die jeweils über Ω_i zur Verfügung stehenden Wavelet-Basen erfolgen. Die zweite zentrale Thematik dieser Arbeit lautet daher wie folgt.

(T2) Es werden adaptive *multiplikative und additive Wavelet-Frame-Gebietszerlegungsverfahren* entwickelt und deren Konvergenz und Optimalität im Sinne von (O1) und (O2) bewiesen. Des Weiteren stehen ausführliche numerische Tests im Mittelpunkt. Neben der praktischen Verifikation der theoretisch verifizierten Optimalität spielt der Vergleich des multiplikativen Schwarz-Verfahrens mit einem standard adaptiven Löser basierend auf einer Finite-Elemente-Diskretisierung (vgl. [6,83]) und der zuvor entwickelten Methode des steilsten Abstiegs eine zentrale Rolle. Es werden darüber hinaus die Ergebnisse numerischer Experimente vorgestellt, die mit einer ersten parallelen Implementierung des entwickelten additiven Schwarz-Verfahrens erzielt werden.

Vor allem bei der multiplikativen Variante kann hierbei auf eine zuvor in [104] benötigte, schwer zu verifizierende theoretische Voraussetzung zum Nachweis der Optimalität verzichtet werden. Die numerischen Resultate verdeutlichen die Effizienz der Gebietszerlegungsverfahren. Für den Fall der Poisson-Gleichung mit homogenen Dirichlet Randbedingungen in einem zwei-dimensionalen L-förmigen Gebiet benötigt die multiplikative Schwarz-Methode zum Beispiel deutlich weniger Freiheitsgrade als die betrachtete Finite-Elemente-Methode. Die Optimalität dieses Verfahrens wird auch für die biharmonische Gleichung der Ordnung 4 praktisch belegt. Man beachte, dass dieses Problem mit den bisher oft verwendeten *composite wavelets* aus [48] auf Grund mangelnder Regularität dieser Ansatzfunktionen nicht diskretisiert werden konnte. Zusätzlich verifizieren die numerischen Experimente einerseits die Optimalität und Effizienz des additiven Algorithmus als sequentieller Löser. Andererseits zeigen die ersten mit einer *parallelen* Implementierung erzielten Ergebnisse das hohe Potential dieses Ansatzes. Denn für relevante Modellprobleme zeigt sich, dass sich bei fester Gebietszerlegung die Rechenzeit beim Übergang von einem zu m (Anzahl der Teilgebiete Ω_i) Prozessoren tatsächlich um einen Faktor m verringert. In diesem Fall benötigt die additive Schwarz-Methode tendenziell weniger Rechenzeit als das sequentielle multiplikative Verfahren.

Neben diesen zentralen Betrachtungen befasst sich diese Arbeit mit zwei weiteren wichtigen Problemstellungen im Kontext adaptiver Wavelet-Frame-Methoden.

Die Elemente der konstruierten Wavelet-Frames sind in der Regel stückweise polynomiale Funktionen bezüglich eines dyadischen Gitters auf Ω_i. Diese lokalen Gitter müssen im Überlappungsbereich der verschiedenen Teilgebiete nun nicht zusammenpassen. Die Berechnung der diskreten Darstellung eines elliptischen Operators bezüglich des Wavelet-Frames mittels numerischer Integration stellt somit ein praktisches Problem dar, vor allem im Hinblick auf den linearen Gesamtrechenaufwand der Verfahren, welcher garantiert werden soll. In der bisherigen Komplexitätsanalyse wurde vereinfachend angenommen, dass ein einzelner Eintrag der diskreten Darstellung eines Operators in konstanter Zeit berechnet werden kann. Dieses Problem wird in Kapitel 5 behandelt. Hier werden summierte Quadraturformeln entwickelt, welche in einem adaptiven Algorithmus zur Approximation der Einträge komprimierter Versionen der diskretisierten elliptischen Operatoren verwendet werden können, so dass der lineare Gesamtrechenaufwand erhalten bleibt. Diese Resultate zeigen also, dass die möglicherweise unpassenden dyadischen Gitter im Überlappungsbereich der Teilgebiete die lineare Komplexität nicht zerstören.

Das zentrale theoretische Hilfsmittel zum Nachweis der Frame-Eigenschaft eines aggregierten Frames ist eine *hinreichend glatte* Partition der Eins $\{\sigma_i\}_{i=0}^{m-1}$, so dass $\operatorname{supp}\sigma_i \subset \overline{\Omega_i}$, $\sum_{i=0}^{m-1}\sigma_i \equiv 1$. Speziell für praktisch relevante Gebietszerlegungen, welche "unpassende Gitter" vermeiden, existieren aber häufig keine beliebig glatten Funktionen σ_i mit diesen Eigenschaften. Als ein Beispiel kann hier ein zwei-dimensionales L-förmiges Gebiet genannt werden, welches mit zwei kongruenten Rechtecken überdeckt wird. Aus praktischer Sicht ist diese Zerlegung sinnvoll. Allerdings ist die Existenz einer ausreichend glatten Teilung der Eins nicht klar (siehe §2.7). Darüber hinaus wird eine solche Teilung auch im Rahmen des Studiums der Approximationseigenschaften eines aggregierten Frames benötigt. Um Kenntnis der optimalen Konvergenzrate eines adaptiven Frame-Verfahrens zu erlangen, müssen die Konvergenzraten der besten N-Term-Frame-Approximation der Lösung u studiert werden (vgl. (O1)). Dies stellt einen weiteren wichtigen Aspekt dieser Arbeit dar. Für den Fall einer Wavelet-Riesz-Basis hängt diese Rate in der Regel von der Besov-Regularität der Lösung ab. Die zentrale Technik, um diese Ergebnisse auf den Fall aggregierter Frames zu übertragen, besteht in der Zerlegung der Funktion u in m Teile $\sigma_i u$ und das Studium der besten N-Term-Approximation dieser Funktionen bzgl. der lokalen Wavelet-Basen über Ω_i. Die Regularität der Teilung der Eins spielt also auch in diesem Zusammenhang eine entscheidende Rolle. Für solche Gebietszerlegungen, welche die Konstruktion einer beliebig glatten Partition der Eins erlauben, lassen sich die klassischen Ergebnisse auf den Frame-Fall einfach übertragen. Als eine prototypische Situation, in der die Regularität der Teilung der Eins beschränkt ist, betrachten wir in dieser Arbeit den Fall des erwähnten L-Gebietes mit der angesprochenen Überdeckung. Es wird bewiesen, dass auch in diesem Fall die klassischen Resultate im wesentlichen gültig sind.

Die vorliegende Arbeit liefert eine geschlossene Darstellung der in [37, 39, 108, 109] veröffentlichten Ergebnisse. Diese werden durch eine erweiterte Theorie hinsichtlich additiver Schwarz-Methoden und eine Vielzahl zusätzlicher praktischer Tests komplementiert. Im Detail ist die vorliegende Arbeit nun wie folgt strukturiert.

Kapitel 1 stellt die benötigten theoretischen Grundlagen hinsichtlich elliptischer Operatorgleichungen bereit, liefert eine kurze Einführung in die Theorie der Sobolev-Räume und zeigt als elementares Lösungsverfahren die Funktionsweise des klassischen Galerkin-Ansatzes auf.

In **Kapitel 2** wird zunächst die klassische Theorie von Frames und Riesz-Basen in separablen Hilberträumen zusammengefasst. Nach einer Präsentation klassischer Konstruktionsprinzipien für Wavelet-Basen und der Einführung des Konzepts der aggregierten Frames wird deren explizite Konstruktion basierend auf lokalen Wavelet-Systemen behandelt. Danach wird aufgezeigt, wie hiermit die Diskretisierung einer Operatorgleichung erreicht werden kann. Im Anschluss steht die Diskussion der Verfügbarkeit einer geeigneten Partition der Eins im Mittelpunkt.

Das Studium der Konvergenzraten der besten N-Term-Frame-Approximation der Lösung u ist Gegenstand von **Kapitel 3**. Zu diesem Zweck werden zunächst die benötigten Zusammenhänge der Funktionenraumtheorie, der Interpolationstheorie von Funktionenräumen und der Approximationstheorie dargestellt. In diesem Kapitel stellen Theorem 3.6 und dessen Anwendungen auf die wichtigen Spezialfälle der Poisson und biharmonischen Gleichung in Korollar 3.5 und 3.6 die zentralen Ergebnisse dar.

Kapitel 4 umfasst die Entwicklung des ersten adaptiven Verfahrens dieser Arbeit (vgl. (T1)) und dessen praktische Analyse. Zunächst werden sämtliche zur Entwicklung eines optimalen adaptiven Verfahrens benötigten numerischen Bausteine vorgestellt. Danach wird das neue adaptive Verfahren präsentiert und dessen Konvergenz und Optimalität bewiesen. Hier stellt Theorem 4.2 ein zentrales Ergebnis dieser Arbeit dar. Abschließend werden die Resultate numerischer Experimente diskutiert, welche unter anderem den Vergleich mit der gedämpften Richardson-Iteration liefern.

In **Kapitel 5** werden geeignete Quadraturformeln zur Berechnung der Einträge der Matrixdarstellung von $\mathcal{L}$ entwickelt. Die in diesem Kontext hergeleiteten Fehlerabschätzungen werden durch die Ergebnisse numerischer Tests bestätigt.

Im zentralen Kapitel dieser Arbeit, nämlich **Kapitel 6**, erfolgt dann die Behandlung von (T2). Zunächst wird eine adaptive multiplikative Schwarz-Methode entwickelt. Die Konvergenz und Optimalität dieses Verfahrens wird in Proposition 6.1 und Theorem 6.3 formuliert und bewiesen. Dies sind zwei weitere zentrale, theoretische Resultate dieser Arbeit. Im Anschluss wird ein additives Schwarz-Verfahren entwickelt, dessen Konvergenz bewiesen (Proposition 6.7) und dessen Optimalitätsnachweis skizziert.

Kapitel 7 enthält die bereits oben erwähnten Ergebnisse der in diesem Zusammenhang durchgeführten numerischen Experimente.

Diese Arbeit schließt mit einer Zusammenfassung der erzielten Ergebnisse und einer Reihe von Anregungen für zukünftige Untersuchungen im Kontext adaptiver Frame-Methoden für Operatorgleichungen.

Contents

Introduction   1

1 Elliptic Operator Equations   13
  1.1 Elliptic boundary value problems 13
  1.2 Sobolev spaces 14
  1.3 Weak formulation 17
  1.4 Inhomogeneous boundary conditions 19
  1.5 The Galerkin approach 20
  1.6 Boundary integral equations 22

2 Wavelet Frames on Bounded Domains   23
  2.1 Basic frame theory 23
      2.1.1 Frames for Hilbert spaces 23
      2.1.2 Riesz bases 27
      2.1.3 Banach frames 28
      2.1.4 Hilbert frames revisited 28
  2.2 Classical wavelet bases 29
      2.2.1 Common construction principles 29
      2.2.2 Norm equivalences 34
      2.2.3 Cancellation properties 36
      2.2.4 Spline wavelet bases on the interval 36
      2.2.5 Wavelet bases on the unit cube 38
  2.3 Construction of wavelet frames 39
      2.3.1 Aggregated frames and stable space splittings 39
      2.3.2 Aggregated wavelet frames 43
      2.3.3 Gelfand frames 45
  2.4 Summary of fundamental wavelet properties 50
  2.5 Discretization of operator equations 51
  2.6 How to select the domain decomposition 53
  2.7 Partitions of unity 54

3 Nonlinear Approximation with Aggregated Wavelet Frames   61
  3.1 Function spaces 62
  3.2 Interpolation and approximation spaces 64
      3.2.1 Interpolation spaces 64

3.2.2   Approximation classes   65
3.3   The classical wavelet basis setting   66
     3.3.1   Linear approximation with wavelets in $H^t(\Omega)$ 66
     3.3.2 N-term wavelet approximation in $H^t(\Omega)$ 66
     3.3.3 N-term approximation in ℓ_2   69
     3.3.4   Spline wavelets and boundary conditions   71
3.4   Aggregated wavelet frames   72
     3.4.1   Basic assumptions   73
     3.4.2   Coverings permitting a smooth partition of unity   73
     3.4.3   The L-shaped domain   73

4   An Adaptive Steepest Descent Wavelet Frame Algorithm   85
4.1   Optimality of adaptive algorithms   86
4.2   The exact steepest descent scheme   87
4.3   Development of the adaptive solver   89
     4.3.1   The adaptive matrix-vector product   91
     4.3.2   Coarsening   93
     4.3.3   Adaptive approximation of the right-hand side   94
     4.3.4   Computation of an approximate residual   94
     4.3.5   The adaptive steepest descent method   97
     4.3.6   Assumption 4.2 and ways to circumvent it   101
4.4   Numerical examples   104
     4.4.1   The Poisson equation in an interval   105
     4.4.2   The Poisson equation in an L-shaped domain   111
     4.4.3   Conclusion and motivation for further improvement   113

5   Computation of Differential Operators in Frame Coordinates   119
5.1   Compressibility   119
5.2   Computability   123
     5.2.1   The quadrature paradigm   123
     5.2.2 s^*-computability of $\mathbf{A}^{(r)}$ 125
     5.2.3 s^*-computability of $\mathbf{A}^{(s)}$ 125
     5.2.4   A second look at the regular case   128
5.3   Numerical confirmation of the derived estimates   129

6   Adaptive Schwarz Domain Decomposition Solvers   135
6.1   A multiplicative Schwarz adaptive wavelet frame method   136
     6.1.1   The exact multiplicative Schwarz method   136
     6.1.2   Preparations for the design of the adaptive method   139
     6.1.3   The adaptive method and its convergence   141
     6.1.4   Optimality   144
     6.1.5   Construction of the limits on the subdomains: Proof of Theorem 6.2 (a)   151

6.1.6   Smoothness of the limits on the subdomains: Proof of Theorem 6.2 (b) 158

6.2   An additive Schwarz adaptive wavelet frame method 161

6.2.1   The exact additive Schwarz method 161

6.2.2   The adaptive method and its convergence 164

6.2.3   Optimality 168

6.2.4   Implementation of the projector **P** 172

6.2.5   Concluding remarks 174

7   Numerical Tests   **175**

7.1   The adaptive multiplicative Schwarz method 175

7.1.1   The Poisson equation in the unit interval 175

7.1.2   The Poisson equation in the L-shaped domain 180

7.1.3   Comparison with an adaptive finite element code 189

7.1.4   The biharmonic equation in the L-shaped domain 191

7.2   The adaptive additive Schwarz method 194

7.2.1   Parallelization strategy 194

7.2.2   The L-shaped domain 195

7.2.3   A ring-shaped domain 198

Conclusion and Outlook   **201**

List of Figures   **205**

List of Tables   **209**

Bibliography   **211**

Contents

Introduction

The scientific fields of Mathematics and Physics provide powerful techniques for the description of real-life problems in terms of *Partial Differential Equations* (PDEs) or *Integral Equations*. Such problems can be formulated in a functional analytic setting as an operator equation

$$\mathcal{L}u = f,$$

and they are then the subject of analytical and computational studies. This thesis will be concerned with the special class of linear operators $\mathcal{L} : \mathcal{H} \to \mathcal{H}'$, mapping from a Hilbert space $\mathcal{H}$ into its normed dual $\mathcal{H}'$, with $\mathcal{L}$ being *elliptic* in the sense that there exist constants $0 < c, C < \infty$ such that $c\|v\|_{\mathcal{H}} \leq \|\mathcal{L}v\|_{\mathcal{H}'} \leq C\|v\|_{\mathcal{H}}$, for all $v \in \mathcal{H}$. In this context, typically $\mathcal{H}$ is a function space over a domain $\Omega \subset \mathbb{R}^n$ or a closed manifold, and $f \in \mathcal{H}'$ is given. One prominent example fitting into this framework (as shown in Chapter 1) is the *Poisson equation*

$$-\Delta u = g \text{ in } \Omega, \quad u = 0 \text{ on } \partial\Omega,$$

in a bounded domain $\Omega \subset \mathbb{R}^n$. The computational solution of an operator equation is usually based on a discretization, for instance, with respect to some finite grid. The ongoing increase of computing power comes along with the desire to treat larger and larger problems with millions of unknowns, involving large-scale numerical simulation. In this context, for the feasibility of a sufficiently accurate computation of the true solution u, it is of particular importance to be equipped with *asymptotically optimal* algorithms, meaning that the number of floating point and storage operations, needed to approximate the solution up to a certain accuracy, stay *proportional* to the dimension of the problem when the discretization is refined. Therefore, one has to make sure that the condition numbers of the matrices that appear in the arising linear systems of equations, stay uniformly bounded. To ensure this, usually *preconditioning strategies* are mandatory. The most prominent examples in this direction and other standard techniques in large-scale computations are shortly summarized in the sequel.

Standard tools in large-scale numerical simulation

One successful way to realize asymptotically optimal methods, is to incorporate the interaction of multiple levels of resolution into the discrete model. The most prominent examples are *multigrid methods*, see [15,72], and *multilevel methods*, being similar in spirit [16,93,121,123].

1

Another important technique in large-scale computations are *domain decomposition methods* [22, 96, 110]. The domain decomposition paradigm consist in the splitting of a PDE into coupled problems on smaller subdomains, which form a partition of the domain Ω in which the original PDE is posed. The fundamental idea is that the iterated solution of subproblems on smaller domains is more convenient and more efficient than the treatment of one huge problem on the global domain. By a simultaneous treatment of subproblems, indeed powerful parallel algorithms can be designed [18]. All these techniques have been thoroughly worked out and tested in the framework of finite difference or finite element discretizations, and they represent efficient tools for the treatment of elliptic and parabolic problems.

Many of the phenomena modelled by PDEs or integral equations exhibit singularities. Prominent examples are elliptic equations on domains with non-smooth boundary, e.g., re-entrant corners (cf. [53, 69, 70, 86]), or problems with a right-hand side f having singularities. In such a situation, for the efficient numerical approximation of the solution, *adaptive schemes* are often indispensable. In this thesis, by an adaptive scheme, we understand an algorithm that iteratively generates a sequence of approximations in $\mathcal{H}$ to the exact solution u. Here, starting from some initial guess, for instance, from zero, these approximations are adapted to the characteristics of u. To get from one iterate to the next one, only information is used that is already contained in the previous approximation, or it is acquired from the right-hand side f or $\mathcal{L}$. At no point in this process, any kind of a priori information about u is used.

This thesis deals with the development of new adaptive (domain decomposition) multilevel methods for elliptic operator equations. These methods will be based on special *redundant wavelet discretizations*.

Design of adaptive wavelet algorithms

In recent years, discretizations of operator equations based on *wavelets* have been brought into focus. Wavelets are commonly designed to form special stable multi-scale bases for $L_2(\mathbb{R}^n)$, where each function is obtained by translation, dilation, and scaling of one or more *mother wavelets*. For example, in one dimension this looks like

$$\psi_{j,k}(x) := 2^{j/2} \psi(2^j x - k), \quad j, k \in \mathbb{Z}, \quad x \in \mathbb{R}.$$

Wavelets with strong analytical properties can be constructed. In particular, the great benefit of wavelets is that they can be designed such that

- they possess an arbitrary *smoothness* in combination with *compact supports*,

- they enable a *characterization of smoothness spaces*, e.g., Sobolev or Besov spaces, by means of an equivalence of the norm in the smoothness space with a weighted sequence norm of the expansion coefficients in the basis,

- they satisfy *cancellation properties*, meaning that the L_2-inner product of a smooth function with a wavelet decays exponentially in modulus when the scale j of the wavelet increases.

Thanks to these properties, wavelets are predestined to resolve well local phenomena, such as singularities, while smooth data can be coded with very few coefficients. Therefore, wavelet schemes have been successfully applied to several tasks. In signal/image analysis and compression, wavelet algorithms are by now well-accepted and compete with other methods [89].

Moreover, in a series of studies, the applicability of wavelets for the adaptive solution of operator equations has been demonstrated [8,27–29,34,35,40,42,64,65,97,102]. The success of these methods essentially relies on two properties. One important advantage is that many operators have a (quasi-)sparse representation in wavelet coordinates, including local differential as well as non-local integral operators [105]. This means that the arising system matrices exhibit an exponential off-diagonal decay. Furthermore, by a simple diagonal preconditioning, wavelets usually induce uniformly bounded condition numbers of these matrices [42,44].

A typical situation is that $\mathcal{H}$ is continuously and densely embedded in $L_2(\Omega)$, so that also $L_2(\Omega)$ (identified with its dual space) can be embedded in $\mathcal{H}'$. Thus, $\mathcal{H} \subset L_2(\Omega) \subset \mathcal{H}'$ forms a *Gelfand triple* (cf. Chapter 1). The wavelet basis is traditionally first constructed as a Riesz basis in $L_2(\Omega)$. Then, from the mentioned norm equivalences, it can be inferred that a properly scaled version of the basis actually forms a Riesz basis for $\mathcal{H}$. We shall briefly demonstrate in the sequel how an operator equation can be discretized with a wavelet basis $\Psi = \{\psi_\lambda\}_{\lambda \in \Lambda} \subset \mathcal{H}$ and how adaptive schemes are then designed. To this end, we fix some common notation. Throughout this thesis, by $A \lesssim B$ we mean that A can be bounded by a multiple of B, independently of parameters which A and B may depend on. Obviously, $A \gtrsim B$ is defined as $B \lesssim A$, and $A \eqsim B$ as $A \lesssim B$ and $B \lesssim A$.

The discretization of $\mathcal{L}$ can be performed with respect to the full wavelet Riesz basis Ψ, resulting in a boundedly invertible bi-infinite *stiffness matrix* $\mathbf{A} : \ell_2(\Lambda) \to \ell_2(\Lambda)$, where $\mathbf{A} = \{\mathcal{L}(\psi_\mu)(\psi_\lambda)\}_{\lambda,\mu \in \Lambda}$. With $\mathbf{f} = \{f(\psi_\lambda)\}_{\lambda \in \Lambda} \in \ell_2(\Lambda)$ being the discrete representation of f in the dual basis $\tilde{\Psi} \subset \mathcal{H}'$, using the ellipticity of $\mathcal{L}$ and the Riesz basis property of Ψ in $\mathcal{H}$, one can show (see Chapter 2)

$$\left\| u - \mathbf{v}^\top \Psi \right\|_{\mathcal{H}} \eqsim \left\| \mathcal{L}(u - \mathbf{v}^\top \Psi) \right\|_{\mathcal{H}'} = \left\| f - \mathcal{L}(\mathbf{v}^\top \Psi) \right\|_{\mathcal{H}'} \eqsim \left\| \mathbf{f} - \mathbf{A}\mathbf{v} \right\|_{\ell_2(\Lambda)}, \quad \mathbf{v} \in \ell_2(\Lambda).$$

The adaptive schemes are then designed as solvers for the associated linear system of equations $\mathbf{A}\mathbf{u} = \mathbf{f}$.

An important milestone was the development of an efficient and reliable a posteriori error estimator for Galerkin schemes (cf. §1.5) in [34]. As a continuation of this research, in [27], an adaptive wavelet Galerkin method has been derived, in which the generated approximations $\{u_{\Lambda_j}\}_{j \geq 0} \subset \mathcal{H}$ are given as finite linear combinations $u_{\Lambda_j} = \sum_{\lambda \in \Lambda_j} \mathbf{u}_\lambda \psi_\lambda$, $\#\Lambda_j < \infty$, $\Lambda_j \subset \Lambda$. The algorithm could be proven to have *asymptotically optimal complexity* for a wide range of elliptic, symmetric, positive definite operators $\mathcal{L}$, i.e., $\mathcal{L} = \mathcal{L}'$, $\inf_{0 \neq v \in \mathcal{H}} \mathcal{L}(v)(v)/\|v\|_{\mathcal{H}}^2 > 0$. By the term "asymptotically optimal complexity" we understand that

(O1) *the relation between the attained accuracy $\|u - u_{\Lambda_j}\|_{\mathcal{H}}$ and the number of degrees of freedom $\#\Lambda_j$ is optimal, in the sense that it holds $\|u - u_{\Lambda_j}\|_{\mathcal{H}} \leq C_1(u)(\#\Lambda_j)^{-s}$, with the largest possible $s > 0$, for some constant $C_1 > 0$ depending on u.*

The optimal *convergence rate* $s > 0$ is given by the decay rate of the *best N-term approximation error* $\|u - u_N\|_{\mathcal{H}} = \inf\{\|u - v\|_{\mathcal{H}} : v = \sum_{\lambda \in \mathcal{J}} \mathbf{v}_\lambda \psi_\lambda, \#\mathcal{J} \leq N, \mathcal{J} \subset \Lambda\}$ of the solution u. Moreover, the method in [27] could be shown to require $\mathcal{O}(\#\Lambda_j)$ floating point operations and storage locations to compute u_{Λ_j}, when suitable numerical techniques for the computation of the matrix $\mathbf{A}$ are applied; see also [66]. This implies that the scheme is of

(O2) *linear complexity.*

Thus, within the wavelet framework, usually a clear comprehension of the convergence properties of adaptive schemes is available. Moreover, the numerical results in [8] have practically confirmed (O1) and (O2), and they have shown that the wavelet scheme from [27] may outperform established adaptive finite element solvers in terms of the number of degrees of freedom spent. In the context of adaptive mesh refinement based on a posteriori error estimators in the finite element framework, a clear understanding of the rate of convergence, or even the convergence alone, has been lacking for a long time. Quite recently, compared to the long history of finite element methods, for special refinement strategies and a relevant class of right-hand sides, convergence with optimal rates has been proven rigorously [12, 92, 107].

Usually, the convergence rate of the best N-term wavelet approximations of the solution u in $\mathcal{H}$ depends on its Besov regularity in a certain scale. Fortunately, for many instances of elliptic problems with singular solutions, it is known that the Besov regularity of u is significantly higher than its Sobolev regularity which governs the convergence rates of uniform refinement strategies. In these cases optimal adaptive algorithms really pay off, because higher convergence rates can be achieved. We refer to Chapter 3 for a detailed presentation of this issue.

Adaptive wavelet algorithms have also been developed for non-symmetric or indefinite problems, saddle point problems, and stationary nonlinear problems; see [28, 29, 35, 40, 64]. In addition, the adaptive solution of parabolic problems has been investigated in [97, 102]. However, we shall be concerned in this work with the case of an elliptic, symmetric and positive definite operator $\mathcal{L}$.

So far, mainly two types of adaptive wavelet solvers for the symmetric case can be distinguished. The first one is the adaptive Galerkin approach in [27] mentioned above. The second, different strategy is to directly apply well-known iterative solvers to the bi-infinite linear system $\mathbf{Au} = \mathbf{f}$ (see [28, 104]), in which the single iterative steps are performed only inexactly, using implementable routines for the finite approximation of the matrix-vector products and of the right-hand side $\mathbf{f}$. This proceeding is successful because the perturbation of the kth step of an ℓ_2-linearly convergent iterative solver with convergence rate $\rho \in (0, 1)$ up to some suitably chosen, geometrically decreasing tolerances δ_k, $k \geq 0$, does not spoil the convergence. Indeed, let $\varepsilon_0 := \|\mathbf{u}\|_{\ell_2}$, $\varepsilon_k := \rho(1 + K)\varepsilon_{k-1}$, where $K > 0$ such that $\rho(1 + K) < 1$, and set $\delta_k := K\rho\varepsilon_{k-1}$. We denote the iterates of the perturbed solver by $\mathbf{u}_k$, $k \geq 0$, set $\mathbf{u}_0 := 0$, and assume $\|\mathbf{u} - \mathbf{u}_{k-1}\|_{\ell_2} \leq \varepsilon_{k-1}$. Let further $\mathbf{u}_k^e$ be the result of an *exact* step of the iteration starting from $\mathbf{u}_{k-1}$. Then, we get

$$\|\mathbf{u} - \mathbf{u}_k\|_{\ell_2} \leq \|\mathbf{u} - \mathbf{u}_k^e\|_{\ell_2} + \|\mathbf{u}_k^e - \mathbf{u}_k\|_{\ell_2} \leq \rho\|\mathbf{u} - \mathbf{u}_{k-1}\|_{\ell_2} + \delta_k$$
$$\leq \rho\varepsilon_{k-1} + K\rho\varepsilon_{k-1} = \rho\varepsilon_{k-1}(1 + K) = \varepsilon_k.$$

Hence, the perturbed iterates also converge linearly. Thanks to its (quasi-)sparsity, an efficient approximation of $\mathbf{A}$ by finite dimensional sparse portions of itself is enabled; we say $\mathbf{A}$ is *compressible* (cf. §5.1). Based on this observation, in [7, 27, 104], a numerical routine

$$\mathbf{APPLY}[\mathbf{A}, \mathbf{v}, \varepsilon] \rightarrow \mathbf{z}_\varepsilon$$

for the inexact application of the system matrix $\mathbf{A}$ to a given finitely supported sequence $\mathbf{v}$ has been constructed, where the output vector $\mathbf{z}_\varepsilon$ approximates $\mathbf{Av}$ up to accuracy ε in the ℓ_2-norm with the (near) minimal amount of coefficients $\# \text{supp} \, \mathbf{z}_\varepsilon$ in linear time. The output vector $\mathbf{z}_\varepsilon$ is always adapted to the vector $\mathbf{v}$ and the accuracy ε, so that we may speak of an adaptive scheme whenever $\mathbf{APPLY}$ is used in a numerical algorithm. Implementable adaptive methods can then be generated by replacing in an iterative solver any application of $\mathbf{A}$ to a vector by a call of $\mathbf{APPLY}$ and a suitable finite approximation of $\mathbf{f}$ in each step. By adding a thresholding step to maintain an optimal balance between the number of nonzero coefficients of $\mathbf{u}_k$ and the attained accuracy, again convergent methods of asymptotically optimal complexity in the sense of (O1) and (O2) can be designed [28, 104].

However, there are still some bottlenecks. Usually, the equation under consideration is posed on some non-trivially shaped bounded domain or on a closed manifold Ω. Now, the construction of a wavelet Riesz basis for the associated function space over Ω, with all the desired features mentioned above, is particularly more complicated than the case $\Omega = \mathbb{R}^n$. It is not possible to resort to the powerful Fourier techniques. The common construction principle uses *non-overlapping* domain decompositions, where each subdomain is a smooth parametric image of the n-dimensional unit cube [19, 32, 48]. Wavelet bases on this cube are lifted to the subdomains, and continuously connected to form wavelet bases on the whole domain or manifold, called *composite wavelets*. Unfortunately, wavelets with supports that intersect interfaces between subdomains generally have no vanishing moments (no vanishing L_2-inner products of the wavelet with a monomial) and their smoothness does not exceed continuity. This prohibits the treatment of higher order equations. Proposals to circumvent these problems have been made in [79, 106], however, resulting in wavelets with larger supports, or requiring a more complicated construction. Another difficulty is that continuous "gluing" of wavelets over the interfaces requires some matching condition on the parametrizations that in practical situations might be difficult to fulfill. An elegant construction that does not require this matching, and yields wavelets that satisfy all requirements concerning smoothness and vanishing moments, was proposed in [49]. Unfortunately, so far with this approach it seems not easy to obtain wavelets with competitive quantitative properties. A recent investigation of this approach has been made in [87].

Existing adaptive wavelet methods suffer from the often bad condition numbers of the used bases or from their complicated construction and handling. Consequently, one is seeking for alternative, easier ways to create more stable wavelet systems on bounded domains. One idea is to give up the linear independence of the elements and to consider redundant collections, namely wavelet *frames*.

Adaptive frame methods

The difficulties with the construction of suitable well-conditioned wavelet Riesz bases on general domains have motivated to consider the alternative weaker concept of *frames* [23, 50, 51, 80]. Frames can be seen as some kind of "overcomplete basis", i.e., any element of a vector space can still be represented, but this representation might no longer be unique. The redundancy of a frame has turned out to play an important role, e.g., in signal processing [89], because it increases the robustness of an expansion against perturbations. In addition, with a richer system of functions one may potentially be able to find sparser representations of an object.

The hope is that this concept facilitates the geometric construction of wavelet frames on general domains combined with reasonable quantitative properties. In [104] a first adaptive method for elliptic operator equations based on a wavelet frame discretization has been proposed. Indeed, suitable wavelet frames can be constructed by decomposing the domain Ω into $m \in \mathbb{N}$ *overlapping* subdomains Ω_i, $\Omega = \bigcup_{i=0}^{m-1} \Omega_i$, each of them being the parametric image of the n-dimensional unit cube, and by lifting reference Riesz bases or even frames on the unit cube to the subdomains. The collection of all elements of the resulting systems on Ω_i forms an *aggregated frame* for the global function space over Ω. In the context of such a setting, in principle, all essential ingredients for the development of asymptotically optimal schemes are available. The main deviation from the classical Riesz basis setting is that the stiffness matrix $\mathbf{A}$, in general, will have a non-trivial kernel, so that multiple solutions $\mathbf{u}$ of the generally underdetermined bi-infinite linear system $\mathbf{Au} = \mathbf{f}$ might exist that represent the continuous solution u. In [104] a damped Richardson iteration for the solution of symmetric problems has been proposed. Performing only inexact steps of the iteration as discussed above, the iterates $\mathbf{u}_k$ will contain components in the kernel of $\mathbf{A}$. This does not spoil the convergence of the scheme with respect to the norm in the function space, because these kernel components are automatically removed by the reconstruction of a continuous approximation u_k from the discrete iterate $\mathbf{u}_k$, i.e., by taking $u_k = \mathbf{u}_k^\top \Psi$ (see §2.5 for details). However, once occurred in an iterate, the kernel components are not damped by subsequent steps, but still cause some computational effort in the call of the inexact matrix-vector product **APPLY**. Hence, they have to be regarded in the complexity analysis. The verification of optimality in the sense of (O1) and (O2) in [104] is therefore based on a technical assumption on the ℓ_2-orthogonal projection **Q** onto the range of **A**, which is hard to rigorously verify in most cases of interest. This is due to the fact that **Q** is directly related to the *canonical dual frame* (cf. Chapter 2) which usually is not explicitly given. Therefore, an alternative adaptive algorithm was presented in [104], in which, regularly, after a fixed number of iterations, an explicitly given and principally implementable projector $\mathbf{P} \neq \mathbf{Q}$ with the same kernel as **A** is applied to the current iterate to control the redundant parts. This technique enables the unconstrained verification of optimality, but though can be expected to have worse quantitative properties due to the additional amount of computational work.

These frame methods have been further investigated in [38], where special emphasis was placed on the conception of a proper notion of frames, the *Gelfand frames*. This concept represents a natural generalization of the classical setting of wavelet Riesz bases that, as mentioned, are usually constructed as a basis in $L_2(\Omega)$, and then lead to characterizations of the solution space by means of norm equivalences. Indeed, Gelfand frames describe the class of frames that simultaneously allow for the characterization of a Hilbert space V, typically $L_2(\Omega)$, as well as a continuously and densely embedded subspace $\mathcal{H} \subset V$, e.g., some Sobolev space.

Main objectives

In the context of the adaptive frame methods for operator equations mentioned above, there are still some principal obstacles that have to be overcome. Besides the theoretical difficulties in proving optimality, the first numerical tests have shown that the quantitative performance of the frame schemes so far is not satisfactory, although the *asymptotically* optimal complexity can be observed (cf. [97, 117]). Therefore, the fundamental aim of this work is to develop new efficient, asymptotically optimal, adaptive frame methods with which these theoretical and practical problems can be solved. On the one hand, a rigorous and thorough convergence and complexity analysis of the new methods plays an important role. On the other hand, emphasis will also be placed on the practical realization and the examination of their numerical properties. The main intention is to give clear evidence that with the concept of aggregated wavelet frames the known problems of wavelet methods based on existing wavelet constructions on bounded domains can be overcome. We also want to show that frame methods can be constructed that compete with standard solvers based on finite element discretizations, in the sense that less degrees of freedom have to be spent to resolve the solution up to a given accuracy.

This thesis mainly focuses on those elliptic operator equations that are induced by a variational formulation of an elliptic PDE over a bounded domain with homogeneous Dirichlet boundary conditions. In this situation, $\mathcal{H}$ usually is some Sobolev space $H_0^t(\Omega)$, $t \in \mathbb{N}_0$ (see Chapter 1).

The not fully convincing performance of the existing frame methods is due to the rather slow convergence of the damped Richardson iteration, that also requires an a priori calculation of the optimal relaxation parameter, which, in turn, is hardly possible in a frame setting (cf. [117]). In particular, a good estimate of the spectral norm of the pseudo-inverse of $\mathbf{A}$ is needed, but this quantity is very hard to access. The first central topic can thus be formulated as follows.

(T1) Using a discretization with respect to an aggregated wavelet frame, we want to develop a new adaptive solver based on the well-known *steepest descent iteration* which is convergent and optimal as in (O1), (O2). We shall investigate, to which extent this approach ameliorates the theoretical and practical deficiencies of the damped Richardson method considered so far. In a series of numerical tests, the practical properties of the new algorithm are analyzed.

The consideration of the steepest descent method is mainly motivated by the fact that the cumbersome a priori computation of the optimal relaxation parameter, as needed for the Richardson scheme, is avoided. The numerical results confirm the optimality of the scheme, and they indicate that the new algorithm may outperform the Richardson method in case here the relaxation parameter is underestimated.

The second, even more promising type of algorithms considered in this thesis are *overlapping Schwarz domain decomposition methods*. The basic idea of these goes back to an early work of H. A. Schwarz [103]. Here, the global problem in Ω is treated by an iterated updating of the current global approximation on the subdomains Ω_i of an overlapping covering. These local updates are obtained by solving elliptic auxiliary problems in Ω_i (see §6.1.1 for details). Two main advantages are introduced by such an approach. Firstly, an additional preconditioning of the global problem is achieved. Mainly, one distinguishes between *multiplicative* and *additive* preconditioners (cf. §6.1 and §6.2). Particularly the additive methods allow for the straightforward development of efficient *parallel* solvers by the simultaneous treatment of local subproblems. Using an aggregated frame as introduced above, on each subdomain, one is equipped with a lifted wavelet basis. The local problems can thus be solved using an adaptive wavelet method with respect to the local wavelet collection on Ω_i, e.g., again the adaptive Galerkin method from [27]. The second essential topic can be formulated as follows.

(T2) We want to develop new adaptive multiplicative and additive Schwarz frame domain decomposition methods. Their convergence and asymptotically optimal complexity shall be theoretically proved and practically confirmed by various numerical tests. We also want to compare the developed adaptive multiplicative Schwarz method with a standard adaptive solver based on a finite element discretization (see [6, 83]) and with the steepest descent method. In addition, the results of numerical tests obtained with a first parallel implementation of the additive Schwarz method are presented and discussed.

We shall consider a modified version of the original multiplicative Schwarz method. This modification consists in the additional removing of certain degrees of freedom in the subdomain Ω_i, before the current global approximation is updated here. By this proceeding, the sparsity of the iterates is significantly enforced compared to the steepest descent scheme, and it will also be possible to drop the theoretical assumption needed for the proof of optimality in [104]. The practical results show the efficiency of this method. Moreover, for the Poisson equation in a two-dimensional L-shaped domain with homogeneous Dirichlet data, we will see that the multiplicative domain decomposition method outperforms the finite element solver in terms of the degrees of freedom spent. The optimality of the multiplicative scheme is also confirmed for a fourth order biharmonic problem in the mentioned L-shaped domain. Note that this problem can not be treated with the *composite wavelets* from [48] because of the lacking smoothness of these functions. The additive algorithm, as a sequential solver, turns out to be almost as efficient as the multiplicative case. Moreover, the numerical results will show that the additive solver may also outperform the multiplicative scheme, when the local problems are solved in parallel.

Besides these fundamental topics, we will also be concerned with the treatment of two further problems arising in the context of adaptive frame methods.

One delicate question concerns the computation of the entries of the system matrix **A**, for which usually quadrature is needed. For instance, taking the Poisson equation, one entry reads as

$$\mathbf{A}_{\lambda,\mu} = \mathcal{L}(\psi_\mu)(\psi_\lambda) = \sum_{i=1}^{n} \int_\Omega \frac{\partial \psi_\lambda}{\partial x_i}(x) \frac{\partial \psi_\mu}{\partial x_i}(x) \mathrm{d}x.$$

In the overlapping region of the subdomains, usually, there exist collections of lifted wavelets whose elements are piecewise smooth with respect to images of dyadic square meshes on $(0,1)^n$ under *different* parametrizations. Hence, in this case, quadrature is a potential problem, in particular with regard to the desired linear complexity of the adaptive solver. Therefore, in Chapter 5 we construct special composite quadrature rules that can be used within the routine **APPLY** for the approximation of the entries $\mathbf{A}_{\lambda,\mu}$, so that the crucial properties of **APPLY** can be guaranteed, and thus the construction of schemes with linear complexity is made possible.

The basic tool for the verification of the frame property of an aggregated frame is a *sufficiently smooth* partition of unity $\{\sigma_i\}_{i=0}^{m-1}$ associated to the overlapping domain covering, i.e., $\operatorname{supp}\sigma_i \subset \overline{\Omega_i}$, $\sum_{i=0}^{m-1} \sigma_i \equiv 1$. In particular, for practically relevant coverings avoiding non-matching grid constellations as mentioned above, often no arbitrarily smooth partitions exist. One example is a two-dimensional L-shaped domain covered by two congruent rectangles. From a practical point of view, this covering is convenient. However, the existence of a suitable partition $\{\sigma_0, \sigma_1\}$ is not obvious (see §2.7). Moreover, a partition of unity is also used to derive results on the approximation properties of the aggregated frame. In fact, to have a benchmark for the optimal convergence rate of an adaptive frame scheme (cf. (O1)), one needs to know the convergence rate of best N-term frame approximations of the solution u, as N tends to infinity. It is known that in the basis case the convergence rate hinges on the Besov regularity of u in a certain scale. We also study that coherence in the case of aggregated frames in this thesis. The basic technique will be to decompose the solution u into local components $\sigma_i u$ over Ω_i and to apply well-known results on best N-term approximation of $\sigma_i u$ with respect to the wavelet basis on Ω_i. Consequently, the regularity of the functions σ_i again plays an important role. For those domain coverings which admit a smooth partition of unity, the well-known results from the theory of wavelet Riesz bases immediately carry over to the case of aggregated frames. As a prototypical situation for which this is not the case, we study the L-shaped domain composed of two overlapping subdomains.

This work provides a unified presentation of the results published in [37, 39, 108, 109], and they are complemented by further theoretical findings, mainly concerning adaptive Schwarz frame methods (§6.2), and extended numerical tests. In particular, this thesis is structured as follows.

Layout

The introductory **Chapter 1** is dedicated to a presentation of the scope of problems covered in this thesis. Moreover, the most relevant results on Sobolev spaces are collected, and we present a short description of the standard Galerkin approach from which we deduce desirable properties of the ansatz functions.

Afterwards, in **Chapter 2**, attention is drawn to the mentioned construction of suitable wavelet frames for Sobolev spaces $H_0^t(\Omega)$, $t \geq 0$, on bounded domains $\Omega \subset \mathbb{R}^n$. Firstly, the fundamental results from the classical frame theory in separable Hilbert spaces are discussed, as far as it is needed for our purposes. Then, common construction principles for wavelet Riesz bases for this non-shift-invariant setting are described, and it is roughly highlighted how these concepts can be applied to the case of $\Omega = (0,1)$, and $\Omega = (0,1)^n$, $n > 1$. In §2.3 aggregated frames are introduced as special instances of *stable space splittings*, which is in the spirit of [94] (cf. Proposition 2.7). Then, an explicit construction of aggregated wavelet (Gelfand) frames for $H_0^t(\Omega)$, $t \geq 0$, is presented. The main result in this context is Proposition 2.11. After a short summary of the fundamental wavelet properties that are preserved by this procedure, we show how elliptic operator equations can be discretized with such a collection of functions. The technical difficulty concerning the computation of $\mathbf{A}$ is revealed in §2.6, from which the advantage of matching grids in the overlapping regions becomes clear. Afterwards, the availability of a suitable partition of unity is studied. Motivated by the discussion in §2.6 we consider a special prototypical domain decomposition of a two-dimensional L-shaped domain into two congruent overlapping rectangles. A sufficiently (but not arbitrarily) smooth partition of unity is constructed (see Lemma 2.4). This result shows that also in this situation a proper aggregated frame for $H_0^t(\Omega)$, $t \in \mathbb{N}_0$, exists.

The study of the convergence rates of best N-term aggregated frame approximations is addressed in **Chapter 3**. Some of the fundamental results of the theory of function spaces, as well as interpolation and approximation spaces, and their interrelation, are depicted in §3.1 and §3.2. Subsequent to that, in §3.3, classical results on linear and nonlinear (N-term) approximation with classical wavelet Riesz bases in $H^t(\Omega)$ are collected. In §3.4, we turn to the case of aggregated frames. Here, we study the decay rate of the best N-term aggregated frame approximation error. First, the existence of a smooth partition of unity is assumed. Then, for the prototypical L-shaped domain with the mentioned decomposition into two congruent rectangles, this issue is also investigated. We shall prove that the well-known results from the basis case carry over, provided that the object being approximated exhibits a slightly higher Besov regularity as customary, and provided that it decays sufficiently fast in the vicinity of the re-entrant corner. The main result in §3.4 is stated in Theorem 3.6. We also show that, for the solutions of the Poisson and the biharmonic problems with homogeneous Dirichlet data, these extended properties can actually be guaranteed, so that Theorem 3.6 is applicable (cf. Corollary 3.5 and 3.6).

In **Chapter 4**, the building blocks needed for the practical realization of an optimal adaptive method are presented and discussed. Then, topic (T1) is addressed. An adaptive steepest descent frame algorithm is developed. The main result in this chapter is represented by Theorem 4.2, stating convergence and asymptotically optimal complexity of this method. In §4.4 the outcome of numerical tests for one- and two-dimensional Poisson problems with singular solutions is presented, including a comparison with the damped Richardson method from [104]. The chapter is closed with some motivation for further improvement acquired from the quite instructive outcome of the practical experiments.

In **Chapter 5**, a sufficient compressibility (§5.1) of the representation of a general linear partial differential operator of order $2t \in \mathbb{N}$ in wavelet frame coordinates is verified. After that, the construction of suitable quadrature rules is outlined in §5.2, and the developed error estimates are confirmed by some numerical results in §5.3.

The most important chapter of this thesis, namely **Chapter 6**, is dedicated to (T2). The development of a multiplicative adaptive domain decomposition method is contained in §6.1. The convergence and optimality of this algorithm will be stated in the main results Proposition 6.1, and Theorem 6.3, respectively. In §6.2 an additive domain decomposition solver is considered. Similar to the strategies used in §6.1, we develop an adaptive algorithm, and we prove its convergence in Proposition 6.7. Afterwards, its optimality is separately discussed for the case of two or more than two subdomains. In the first case, it is proved that the global continuous iterates are composed of local contributions over Ω_i which form convergent sequences to explicitly known limits $u^{(i)}$ which have a sufficiently sparse wavelet expansion in the local basis on Ω_i. This is also the basic property from which optimality is inferred in the multiplicative case. In the second case of more than two patches, the situation seems to be more complicated. Incorporating the projector $\mathbf{P}$, similar to the strategy mentioned on page 7, certain theoretical difficulties can be overcome. We outline how a modified algorithm can be constructed. We shall also present a convenient strategy for the application of the projector $\mathbf{P}$ (cf. §6.2.4).

In a series of numerical tests, the theoretical findings from Chapter 6 are confirmed in **Chapter 7**. Similar to §4.4, the optimal convergence rates and the linear complexity of the domain decomposition schemes are confirmed for non-trivial one- and two-dimensional Poisson and biharmonic model problems with singular solutions. We compare the multiplicative scheme with the steepest descent method, and with an adaptive finite element solver. Moreover, we show and discuss the results of a first parallel implementation of the additive Schwarz scheme and compare it with the sequential multiplicative method.

Finally, this thesis is closed with a summary and a discussion of the achieved theoretical and practical results, and a collection of perspectives for future research projects in the field of adaptive frame methods for operator equations.

Chapter 1

Elliptic Operator Equations

In the following, we present the range of problems this thesis will be concerned with, i.e., the class of elliptic operator equations. Many problems in numerical analysis, typically given in terms of an equation with an unknown solution u, can be formulated as an abstract operator equation

$$\mathcal{L}u = f, \tag{1.0.1}$$

where $\mathcal{L}$ maps between appropriate function spaces. Existence and uniqueness of a solution then depend on the mapping properties of $\mathcal{L}$ and on f. Prominent classes of such problems are differential and integral equations. This thesis particularly deals with the development of adaptive solvers for *elliptic boundary value problems* over a bounded domain $\Omega \subset \mathbb{R}^n$. In this context, usually $\mathcal{L}$ is a mapping from a Hilbert space $\mathcal{H}$ into its normed dual $\mathcal{H}'$, where typically $\mathcal{H}$ is some Sobolev space over Ω. An operator $\mathcal{L}$ is called *elliptic* if

$$\|\mathcal{L}v\|_{\mathcal{H}'} \approx \|v\|_{\mathcal{H}}, \quad \text{for all } v \in \mathcal{H}, \tag{1.0.2}$$

ensuring that (1.0.1) represents a well-posed problem.

We shall briefly describe in this chapter how an elliptic boundary value problem can be formulated in terms of (1.0.1) and (1.0.2). At first, the discussion will be confined to the case of homogeneous Dirichlet boundary data. After an introduction to the theory of Sobolev spaces, the weak formulation of the boundary value problem will be presented, and the treatment of inhomogeneous Dirichlet data is addressed subsequent to that. Finally, we recall the standard Galerkin approach and also consider an instance of a boundary integral equation.

1.1 Elliptic boundary value problems

For an n-tuple of non-negative integers $\alpha = (\alpha_1, \ldots, \alpha_n) \in \mathbb{N}_0^n$, let

$$D^\alpha := \frac{\partial^{|\alpha|}}{\partial x_1^{\alpha_1} \cdots \partial x_n^{\alpha_n}} \tag{1.1.1}$$

denote a partial differential operator of order $|\alpha| := \alpha_1 + \cdots + \alpha_n$. Using this notation, let us further define the general linear partial differential operator of order $2t \in \mathbb{N}$ written in divergence form

$$L := \sum_{|\alpha| \leq t} \sum_{|\beta| \leq t} (-1)^{|\beta|} D^\beta (a_{\alpha,\beta} D^\alpha) \qquad (1.1.2)$$

with sufficiently smooth coefficients $a_{\alpha,\beta} : \Omega \to \mathbb{R}$ in a bounded domain $\Omega \subset \mathbb{R}^n$. We assume L to be *uniformly elliptic* in Ω, meaning that there exists an $\varepsilon > 0$ such that

$$\sum_{|\alpha|=|\beta|=t} a_{\alpha,\beta}(x)\xi^{\alpha+\beta} \geq \varepsilon \|\xi\|^{2t}, \quad \text{for all } x \in \Omega, \ \xi \in \mathbb{R}^n, \qquad (1.1.3)$$

where $\|\xi\|$ stands for the Euclidean norm on $\mathbb{R}^n$. This thesis is aimed at the development of adaptive schemes for the solution of elliptic boundary value problems

$$Lu = g \quad \text{in } \Omega, \qquad (1.1.4)$$

$$\frac{\partial^k u}{\partial n^k} = 0 \quad \text{on } \partial\Omega, \quad k = 0, \ldots, t-1, \qquad (1.1.5)$$

with homogeneous Dirichlet boundary conditions. In (1.1.5) it is required that all normal derivatives $\partial/\partial n$ of u up to order $t-1$ vanish on the boundary $\Gamma := \partial\Omega$. The first order normal derivative of u on Γ is defined as

$$\frac{\partial u}{\partial n}(x) = \sum_{i=1}^{n} n_i(x)\frac{\partial u}{\partial x_i}(x) = \langle n(x), \nabla u(x) \rangle, \quad x \in \Gamma, \qquad (1.1.6)$$

with the outward normal vector field $n(x) = (n_1(x), \ldots, n_n(x))$, $x \in \Gamma$. Surely, a sufficient smoothness of Γ such that (1.1.5) is well-defined has to be assumed. In this thesis we will be concerned with the development of solvers for the case that the coefficients of L are *symmetric*, i.e., $a_{\alpha,\beta} = a_{\beta,\alpha}$, $0 \leq |\alpha|, |\beta| \leq t$.

If all partial derivatives are interpreted in a distributional sense, it is possible to reformulate (1.1.4), (1.1.5) as a well-posed operator equation of the form (1.0.1), where $\mathcal{L}$ is defined on an appropriate Sobolev space containing all possible classical solutions $u \in C^{2t}(\Omega)$. This process is known as the *weak or variational formulation*. We shall summarize the basic ideas of this methodology in the following. To this end, we give a short introduction to the theory of Sobolev spaces first.

1.2 Sobolev spaces

A complete presentation of the theory of Sobolev spaces can be found in, e.g., [2,112]. Let Ω be a domain in $\mathbb{R}^n$. For $0 < p \leq \infty$, as usual, we denote with $L_p(\Omega)$ the space of all equivalence classes of measurable functions $f : \Omega \to \mathbb{C}$ for which

$$\|f\|_{L_p(\Omega)} := \begin{cases} \left(\displaystyle\int_\Omega |f(x)|^p \, \mathrm{d}x \right)^{1/p} , & 0 < p < \infty \\ \operatorname{ess\,sup}_{x \in \Omega} |f(x)| , & p = \infty \end{cases} \qquad (1.2.1)$$

is finite. For $1 \leq p \leq \infty$ the latter expression is a norm and $L_p(\Omega)$ forms a Banach space. For $p < 1$, (1.2.1) represents a quasi-norm. Moreover, $L_2(\Omega)$ is a Hilbert space with the inner product

$$\langle f, g \rangle_{L_2(\Omega)} := \int_\Omega f(x)\overline{g(x)} \, \mathrm{d}x. \qquad (1.2.2)$$

Let us define the concept of weak or distributional derivative of a function $f \in L_1^{\mathrm{loc}}(\Omega)$, the latter being the space of functions which are contained in $L_1(A)$ for any measurable set $A \subset\subset \Omega$. Here $A \subset\subset \Omega$ means that the closure $\overline{A}$ is compact and $\overline{A}$ is a subset of Ω. Let $C_0^\infty(\Omega) := \{ f \in C^\infty(\Omega) : \operatorname{supp} f \subset\subset \Omega \}$ be the space of test functions, the dual space of which is called the *space of distributions* over Ω. If there exists a function $g_\alpha \in L_1^{\mathrm{loc}}(\Omega)$, for which it holds that

$$\int_\Omega f(x) D^\alpha \varphi(x) \, \mathrm{d}x = (-1)^{|\alpha|} \int_\Omega g_\alpha(x)\varphi(x) \, \mathrm{d}x, \quad \text{for all } \varphi \in C_0^\infty(\Omega), \qquad (1.2.3)$$

then g_α is unique up to sets of measure zero. This function is called the *weak or distributional derivative* of f and is denoted with $D^\alpha f$.

Now, for a non-negative integer m and $1 \leq p \leq \infty$, the functional

$$\|f\|_{W_p^m(\Omega)} := \left(\sum_{0 \leq |\alpha| \leq m} \|D^\alpha f\|_{L_p(\Omega)}^p \right)^{1/p} , \quad 1 \leq p < \infty, \qquad (1.2.4)$$

$$\|f\|_{W_\infty^m(\Omega)} := \max_{0 \leq |\alpha| \leq m} \|D^\alpha f\|_{L_\infty(\Omega)} \qquad (1.2.5)$$

represents a norm on each function space on which the right-hand sides are finite. In particular, let

$$W_p^m(\Omega) := \{ f \in L_p(\Omega) : D^\alpha f \in L_p(\Omega), \text{ for } 0 \leq |\alpha| \leq m \}, \qquad (1.2.6)$$

and with $\mathring{W}_p^m(\Omega)$ we denote the closure of $C_0^\infty(\Omega)$ in the space $W_p^m(\Omega)$. With the corresponding norm given in (1.2.4) and (1.2.5), these spaces are called *Sobolev spaces*, measuring m orders of smoothness in $L_p(\Omega)$. The spaces $W_p^m(\Omega)$ are separable, reflexive Banach spaces. For the special case $p = 2$, we use the abbreviation $H^m(\Omega) := W_2^m(\Omega)$

and $H_0^m(\Omega) := \mathring{W}_2^m(\Omega)$, respectively. Moreover, it holds that $H^m(\Omega)$ is a Hilbert space with the inner product

$$\langle f, g \rangle_{H^m(\Omega)} = \sum_{0 \le |\alpha| \le m} \langle D^\alpha f, D^\alpha g \rangle_{L_2(\Omega)}, \quad f, g \in H^m(\Omega). \tag{1.2.7}$$

Clearly, $W_p^0(\Omega) = L_p(\Omega)$, and for $1 \le p < \infty$, it is $\mathring{W}_p^0(\Omega) = L_p(\Omega)$, because $C_0^\infty(\Omega)$ is dense in $L_p(\Omega)$ in that case. In case Ω is bounded, an equivalent norm on $\mathring{W}_p^m(\Omega)$, $1 \le p < \infty$, is given by

$$|f|_{W_p^m(\Omega)} := \left(\sum_{|\alpha|=m} \|D^\alpha f\|_{L_p(\Omega)}^p \right)^{1/p}. \tag{1.2.8}$$

Let $1 \le p < \infty$ and $0 < s$ not an integer with the decomposition $s = [s] + \{s\}$, $[s]$ integer and $0 < \{s\} < 1$. Then, the fractional order Sobolev spaces $W_p^s(\Omega)$ are defined as the set of all functions in $f \in W_p^{[s]}(\Omega)$ for which

$$\|f\|_{W_p^s(\Omega)} := \left(\|f\|_{W_p^{[s]}(\Omega)}^p + \sum_{|\alpha|=[s]} \int_\Omega \int_\Omega \frac{|D^\alpha f(x) - D^\alpha f(y)|^p}{\|x-y\|^{n+\{s\}p}} \, dx dy \right)^{1/p} \tag{1.2.9}$$

is finite. We further define $\mathring{W}_p^s(\Omega)$, for $s \ge 0$, as the closure of $C_0^\infty(\Omega)$ in $W_p^s(\Omega)$, and for $s < 0$ we set

$$W_p^s(\Omega) := [\mathring{W}_{p'}^{-s}(\Omega)]', \quad \frac{1}{p} + \frac{1}{p'} = 1. \tag{1.2.10}$$

Thus, in particular for $p = 2$, we have $H^s(\Omega) = (H_0^{-s}(\Omega))'$, $s < 0$. Note that the spaces $W_p^s(\Omega)$, for $s < 0$, are duals of spaces having $C_0^\infty(\Omega)$ as a dense subset. Thus, they are indeed spaces of distributions over Ω.

Sometimes it is necessary to prescribe homogeneous boundary conditions only on subsets of Γ. For $\check{\Gamma} \subset \Gamma$ and $s > 0$, we define as usual

$$\mathring{W}_{p,\check{\Gamma}}^s(\Omega) := \mathrm{clos}_{W_p^s(\Omega)}\{u \in W_p^s(\Omega) \cap C^\infty(\Omega) : \mathrm{supp}\, u \cap \check{\Gamma} = \emptyset\}. \tag{1.2.11}$$

We also state that a proper definition of Sobolev spaces on surfaces or manifolds can be found in [69, §1.3.3] and [81]. The study of boundary value problems requires also a meaningful definition of the restriction $\gamma f = f|_\Gamma$ of a function f to the boundary, which is well-defined if f is continuous on $\overline{\Omega}$. Let us assume that the boundary Γ is k times Lipschitz continuously differentiable. Assume that $s \le k+1$, $s - \frac{1}{p} = l + \{s\}$, for a $0 < \{s\} < 1$, with an integer $l \ge 0$. Then, the mapping

$$f \mapsto \left\{ \gamma f, \gamma \left(\frac{\partial f}{\partial n} \right), \dots, \gamma \left(\frac{\partial^l f}{\partial n^l} \right) \right\}, \tag{1.2.12}$$

which is well-defined for all k times Lipschitz continuously differentiable functions on $\overline{\Omega}$, has a unique continuous extension to a so-called *trace operator*, mapping from $W_p^s(\Omega)$ to $\prod_{j=0}^{l} W_p^{s-j-1/p}(\Gamma)$; see [69, Theorem 1.5.1.2]. Furthermore, we state that one has the alternative characterization

$$\mathring{W}_p^s(\Omega) = \left\{ f \in W_p^s(\Omega) : \gamma_j f := \gamma\left(\frac{\partial^j f}{\partial n^j}\right) = 0, \text{ for } j = 0, \dots, l \right\}; \qquad (1.2.13)$$

see [69, Corollary 1.5.1.6]. This result can also be shown to carry over to the case of domains with piecewise smooth boundary such as polygonal domains in two dimensions; cf. [69, §1.5.2].

For our purposes, the case $p = 2$ will be the most important one. Note that for $s \geq 0$, $H_0^s(\Omega) \subset L_2(\Omega)$ in the sense of a continuous and dense embedding, hence, also $(L_2(\Omega))'$ can be continuously and densely embedded into $H^{-s}(\Omega)$. If we identify $L_2(\Omega)$ with its dual via the canonical Riesz isomorphism, we get a *Gelfand triple*

$$H_0^s(\Omega) \subset L_2(\Omega) \subset H^{-s}(\Omega), \quad s \geq 0, \qquad (1.2.14)$$

in the sense of continuous and dense embeddings. This scale of Sobolev spaces will play an important role throughout this thesis. The space $H^{-s}(\Omega)$ is equipped with the norm

$$\|f\|_{H^{-s}(\Omega)} := \sup\{|\langle f, g\rangle_{L_2(\Omega)}|/\|g\|_{H^s(\Omega)} : 0 \neq g \in H_0^s(\Omega)\}. \qquad (1.2.15)$$

Note that in this setting $\langle \cdot, \cdot \rangle_{L_2(\Omega)}$ always denotes the unique continuous extension of the L_2 inner product onto $H^{-s}(\Omega) \times H_0^s(\Omega)$ which, moreover, coincides with dual form $\langle \cdot, \cdot \rangle_{H^{-s}(\Omega) \times H_0^s(\Omega)}$. Consequently, a function $f \in H_0^s(\Omega)$ can be identified with the distribution $\langle f, \cdot \rangle_{L_2(\Omega)} \in H^{-s}(\Omega)$. In the following sections it will become clear that the Gelfand triple framework is the key to a distributional formulation of (1.1.4), (1.1.5), and it gives rise to a convenient strategy for discretization.

1.3 Weak formulation

Let us now reconsider our fundamental problem (1.1.4), (1.1.5). Due to (1.2.13) for $p = 2$ and $s = t$, any classical solution u must be contained in $C^{2t}(\Omega) \cap H_0^t(\Omega)$. Assuming only boundedness (and symmetry) of the coefficients $a_{\alpha,\beta}$, it is immediate to see that by

$$a(u, v) := \sum_{|\alpha|, |\beta| \leq t} \int_{\Omega} a_{\alpha,\beta}(x)(D^\alpha u)(x)(D^\beta v)(x) \, \mathrm{d}x \qquad (1.3.1)$$

a (symmetric) bilinear form is defined, which is bounded on $H_0^t(\Omega) \times H_0^t(\Omega)$, in the sense that for a constant $C_S > 0$,

$$|a(u, v)| \leq C_S \cdot \|u\|_{H^t(\Omega)} \|v\|_{H^t(\Omega)}, \quad \text{for all } u, v \in H_0^t(\Omega). \qquad (1.3.2)$$

Now, defining for $g \in L_2(\Omega)$ the linear functional

$$f(v) := \int_\Omega g(x)v(x) \, dx, \tag{1.3.3}$$

the *weak formulation* of the problem (1.1.4), (1.1.5) reads as follows:

$$\text{Find a } u \in H_0^t(\Omega) \text{ such that } a(u,v) = f(v), \text{ for all } v \in H_0^t(\Omega). \tag{1.3.4}$$

Exploiting the density of $C_0^\infty(\Omega)$ in $H_0^t(\Omega)$ and using integration by parts, it can be easily shown that the latter problem and (1.1.4), (1.1.5) share the same classical solutions, though not all solutions to (1.3.4) are classical ones, and they are called *weak solutions*.

By setting $\langle \mathcal{L}u, v \rangle_{H^{-t}(\Omega) \times H_0^t(\Omega)} := a(u,v)$, (1.3.4) can be written as an operator equation

$$\mathcal{L}u = f, \tag{1.3.5}$$

where $\mathcal{L} : H_0^t(\Omega) \to H^{-t}(\Omega)$ is a bounded linear operator, and $f \in H^{-t}(\Omega)$. The bounded bilinear form $a(\cdot, \cdot)$ is called $H_0^t(\Omega)$-*elliptic* if there exists a constant $C_E > 0$ such that

$$a(v,v) \geq C_E \|v\|_{H^t(\Omega)}^2, \quad \text{for all } v \in H_0^t(\Omega). \tag{1.3.6}$$

Boundedness and ellipticity of $a(\cdot, \cdot)$ can be expressed in terms of $\mathcal{L}$, i.e.,

$$C_E \|v\|_{H^t(\Omega)} \leq \|\mathcal{L}v\|_{H^{-t}(\Omega)} \leq C_S \|v\|_{H^t(\Omega)}, \quad \text{for all } v \in H_0^t(\Omega).$$

Consequently, $\mathcal{L}$ is a symmetric, positive definite, boundedly invertible operator with $\|\mathcal{L}\|_{H^{-t}(\Omega) \leftarrow H_0^t(\Omega)} \leq C_S$ and $\|\mathcal{L}^{-1}\|_{H_0^t(\Omega) \leftarrow H^{-t}(\Omega)} \leq C_E^{-1}$. Hence, the elliptic operator equation (1.3.5), or equivalently (1.3.4), has a unique solution $u \in H_0^t(\Omega)$. The latter coherence is known as the *Lax-Milgram theorem*. Note that in this context, in (1.3.4) or (1.3.5), the right-hand side can also be an arbitrary functional $f \in H^{-t}(\Omega)$. It does not have to be induced by an element from $L_2(\Omega)$ as in (1.3.3). A sufficient condition for the bilinear form $a(\cdot, \cdot)$ to be $H_0^t(\Omega)$-elliptic is given in the following theorem; see, for instance, [74].

Theorem 1.1. *Let the coefficients $a_{\alpha,\beta}$ be constant for $|\alpha| = |\beta| = t$, $a_{\alpha,\beta} = 0$, for $0 < |\alpha| + |\beta| \leq 2t - 1$, and $a_{00} \geq 0$. Furthermore, let the operator L be uniformly elliptic in the sense of (1.1.3). Then, $a(\cdot, \cdot)$ is $H_0^t(\Omega)$-elliptic.*

Example 1.1. *We give two simple but prominent examples for which Theorem 1.1 can be applied.*

- **Poisson equation**
  The Poisson equation *with homogeneous Dirichlet boundary conditions reads as*

$$
\begin{aligned}
-\Delta u &= g \quad in \quad \Omega, \\
u &= 0 \quad on \quad \partial\Omega.
\end{aligned}
\tag{1.3.7}
$$

The Laplace operator $\Delta = \sum_{i=1}^{n} \frac{\partial^2}{\partial x_i^2}$ *is a uniformly elliptic constant coefficient operator. The bilinear form corresponding to this boundary value problem is given by*

$$
a(u,v) = \int_{\Omega} \nabla u(x) \cdot \nabla v(x) \, \mathrm{d}x.
\tag{1.3.8}
$$

- **Biharmonic equation**
  Let us also mention the biharmonic equation

$$
\begin{aligned}
\Delta^2 u &= g \quad in \quad \Omega, \\
u = \frac{\partial u}{\partial n} &= 0 \quad on \quad \partial\Omega,
\end{aligned}
$$

which represents a fourth order problem, $2t = 4$. In this case we obtain that

$$
a(u,v) = \int_{\Omega} \Delta u(x)\Delta v(x) \, \mathrm{d}x
\tag{1.3.9}
$$

defines an $H_0^2(\Omega)$-elliptic bilinear form.

A weaker concept which can be sufficient to guarantee the existence of a weak solution is the notion of *coercivity*. The bounded bilinear form $a(\cdot,\cdot)$ is called $H_0^t(\Omega)$-coercive if there exist constants $C_1 > 0$ and $C_2 \in \mathbb{R}$ such that

$$
a(v,v) \geq C_1 \|v\|_{H^t(\Omega)}^2 - C_2 \|v\|_{L_2(\Omega)}^2, \quad \text{for all } v \in H_0^t(\Omega).
\tag{1.3.10}
$$

Indeed, using the compactness of the embedding $H_0^t(\Omega) \subset L_2(\Omega)$; cf. [2, Chapter 6], and the classical *Riesz-Schauder theory*, it can be proved that (1.3.10) implies that the spectrum $\sigma(\mathcal{L})$ is a set of measure zero, so that in case $0 \notin \sigma(\mathcal{L})$ existence of a weak solution is guaranteed. Under the assumption (1.1.3) and $a_{\alpha,\beta} \in L_\infty(\Omega)$, $|\alpha|, |\beta| \leq t$, and if the coefficients $a_{\alpha,\beta}$ are uniformly continuous in $\overline{\Omega}$ for $|\alpha| = |\beta| = t$, indeed (1.3.10) holds true, and in that case it is known as *Gârding's inequality* [74]. However, throughout this thesis, we will always assume the bilinear form to be $H_0^t(\Omega)$-elliptic.

1.4 Inhomogeneous boundary conditions

So far, in (1.1.5) only homogeneous Dirichlet boundary conditions have been considered. In this section, it will be pointed out how the boundary value problem

$$
Lu = g \quad in \ \Omega, \quad \frac{\partial^k u}{\partial n^k} = \varphi_k \quad on \ \Gamma, \quad k = 0,\ldots,t-1,
\tag{1.4.1}
$$

can be handled. The weak formulation of this problem amounts to solving the task

Find a $u \in H^t(\Omega)$ with $\dfrac{\partial^k u}{\partial n^k} = \varphi_k$ on Γ, $k = 0, \ldots, t-1$, such that   (1.4.2)

$a(u,v) = f(v)$, for all $v \in H_0^t(\Omega)$.   (1.4.3)

Here, the equalities $\frac{\partial^k u}{\partial n^k} = \varphi_k$ on Γ, $k = 0, \ldots, t-1$, clearly hold as equalities in the trace spaces $H^{t-k-1/2}(\Gamma)$. Obviously, the existence of a function $u_0 \in H^t(\Omega)$ such that

$$\frac{\partial^k u_0}{\partial n^k} = \varphi_k \text{ on } \Gamma, \ k = 0, \ldots, t-1,$$ (1.4.4)

is necessary. An equivalent formulation of (1.4.2), (1.4.3) reads as follows.

Let $u_0 \in H^t(\Omega)$ such that (1.4.4) holds.

Find a $w \in H_0^t(\Omega)$ such that   (1.4.5)

$a(w,v) = f(v) - a(u_0,v)$, for all $v \in H_0^t(\Omega)$.   (1.4.6)

Indeed, if u_0 and w are the solutions to the latter problem, then $u := u_0 + w$ solves (1.4.2), (1.4.3). Conversely, if u solves (1.4.2), (1.4.3), by setting $u_0 = u$ and $w = 0$, a solution to (1.4.5), (1.4.6) can be given. Using these observations, it is immediate to see the following result stating existence and uniqueness of a weak solution to (1.4.1).

Proposition 1.1. *If the problem (1.3.4) has a unique solution for any $f \in H^{-t}(\Omega)$, then a unique solution of (1.4.2), (1.4.3) exists if and only if condition (1.4.4) holds for a suitable $u_0 \in H^t(\Omega)$.*

Hence, the inhomogeneous Dirichlet problem (1.4.1) can be transferred into a homogeneous one. This issue will play a pivotal role in the development of adaptive domain decomposition methods in Chapter 6.

1.5 The Galerkin approach

The basic idea of the well-known *Galerkin approach* consists in the projection of the problem (1.3.4) onto a finite dimensional subspace $V_N \subset H_0^t(\Omega)$, $\dim V_N = N$, i.e.,

Find a $u^N \in V_N$ such that $a(u^N, v) = f(v)$, for all $v \in V_N$.   (1.5.1)

The *Galerkin solution* u^N to (1.5.1) can be written as $u^N = (\mathbf{u}^N)^\top \Phi := \sum_{i=1}^N (\mathbf{u}^N)_i \phi_i$, where $\mathbf{u}^N \in \mathbb{R}^N$ and $\Phi = (\phi_1, \ldots, \phi_N)^\top$ represents a vector of functions, the coefficients of which form a basis of V_N, i.e., $V_N = \mathrm{span}\{\phi_1, \ldots, \phi_N\}$. By plugging this basis expansion of u^N and the corresponding basis expansion of a $v \in V_N$ into (1.5.1), one can easily verify that an equivalent discrete formulation of (1.5.1) reads as

$\mathbf{A}\mathbf{u}^N = \mathbf{f}$,   (1.5.2)

where $\mathbf{A} = \{a(\phi_j, \phi_i)\}_{i,j=1}^N$ is called *stiffness matrix* and $\mathbf{f} = \{f(\phi_i)\}_{i=1}^N$ *load vector*. With the coefficient operator $\mathbf{P} : \mathbb{R}^N \to V_N$, $\mathbf{v} = (v_1, \ldots, v_N)^\top \mapsto \sum_{i=1}^N v_i \phi_i$ we get $a(u, v) = \langle \mathbf{Au}, \mathbf{v} \rangle$ for $u = \mathbf{Pu}$, $v = \mathbf{Pv}$, $\langle \cdot, \cdot \rangle$ denoting the Euclidean scalar product. Naturally, (1.5.1) has a unique solution if and only if $\mathbf{A}$ is nonsingular. Now, due to the ellipticity of $a(\cdot, \cdot)$, for all $0 \neq \mathbf{v} \in \mathbb{R}^N$, one has $\langle \mathbf{Av}, \mathbf{v} \rangle = a(\mathbf{Pv}, \mathbf{Pv}) \geq C_E \|\mathbf{Pv}\|_{H^t(\Omega)} > 0$. Thus, $\mathbf{A}$ is a symmetric, positive definite matrix. Then, *Cea's Lemma* (cf. [74]) allows for an estimate of the *error of discretization* $\|u - u^N\|_{H^t(\Omega)}$ according to

$$\|u - u^N\|_{H^t(\Omega)} \leq \left(1 + \frac{C_S}{C_E} \right) \inf_{w \in V_N} \|u - w\|_{H^t(\Omega)}, \tag{1.5.3}$$

with the constants C_S and C_E from (1.3.2) and (1.3.6), respectively. Obviously, up to a constant factor, u^N is a best approximation to u from V_N. As a consequence, if for a sequence of subspaces $V_j := V_{N_j}$, $j \geq 1$,

$$\lim_{j \to \infty} \text{dist}_{H^t(\Omega)}(u, V_j) = 0, \tag{1.5.4}$$

with $\text{dist}_{H^t(\Omega)}(u, V_j) := \inf_{w \in V_j} \|u - w\|_{H^t(\Omega)}$, convergence of the Galerkin solutions to u follows, and the speed of convergence depends on the approximation properties of the *ansatz spaces* V_j. In case the spaces V_j form a dense hierarchy in the sense that

$$V_1 \subset V_2 \subset \cdots \subset V_j \subset \cdots \subset H_0^t(\Omega), \quad \text{clos}_{H^t(\Omega)} \bigcup_{j=1}^\infty V_j = H_0^t(\Omega), \tag{1.5.5}$$

(1.5.4) can be shown to hold. Naturally, in this context one is interested in the efficiency in which u is approximated by such a process. It depends on

(i) the **approximation power** of the ansatz spaces V_j,

(ii) the **sparsity** and **condition number** of the stiffness matrix, as well as on the existence of efficient tools for the computation of the entries of $\mathbf{A}$ and $\mathbf{f}$.

Therefore, the most important issue is now the appropriate choice of the spaces V_j and their bases. One possibility are *finite elements*, being piecewise polynomial, compactly supported functions with respect to some partition associated to the underlying domain Ω, with which sparse stiffness matrices can be obtained and complicated domain geometries can be handled [14, 17, 25, 74]. One fundamental problem is that the (spectral) condition numbers of the arising stiffness matrices grow with the number of degrees of freedom N_j, so that *preconditioning strategies*, such as *multigrid, multilevel* and *domain decomposition methods* are mandatory; cf. [15, 16, 72, 93, 119–123], as well as [22, 68, 96, 110, 118].

As we shall see in the proceeding chapter, in the context of wavelets, which form a *multilevel basis*, a simple diagonal scaling of the elements yields uniform stability of the wavelets with respect to the H^t-norm, and thus gives rise to uniformly bounded condition numbers of the Galerkin systems; see also [42, 44]. Moreover, they lead to sparse representations of differential as well as certain integral operators [105].

1.6 Boundary integral equations

Another class of problems fitting into the framework (1.0.1), (1.0.2) are boundary integral equations, arising as reformulations of boundary value problems for partial differential equations. For instance, the exterior boundary value problem

$$\Delta u = 0 \quad \text{in } \mathbb{R}^3 \setminus \Omega, \quad u = f \quad \text{on } \Gamma, \tag{1.6.1}$$

can be treated by first solving the boundary integral equation

$$u(x) = f(x) = -\frac{1}{4\pi} \int_\Gamma \frac{q(y)}{\|x - y\|} d\Gamma(y), \quad \text{for all } x \in \Gamma \tag{1.6.2}$$

and by setting

$$u(x) = -\frac{1}{4\pi} \int_\Gamma \frac{q(y)}{\|x - y\|} d\Gamma(y), \quad \text{for } x \in \mathbb{R}^3 \setminus \Omega$$

afterwards; see the survey [5] for details. The weak formulation of the involved integral operator, which is known as the *single layer potential operator*, reads as

$$\mathcal{V} : H^{-1/2}(\Gamma) \to H^{1/2}(\Gamma), \quad v \mapsto \left(x \mapsto \frac{1}{4\pi} \int_\Gamma \frac{v(y)}{\|x - y\|} d\Gamma(y) \right), \tag{1.6.3}$$

and it is well-defined and elliptic; see [73, §8.1] for details.

More general, one may also consider integral operators with a global Schwartz kernel of the form

$$(\mathcal{A}v)(x) = \int_\Omega \mathcal{K}(x, y) v(y) dy, \tag{1.6.4}$$

where $\mathcal{K}(x, y)$ is smooth off the diagonal $x = y$, and where it is required that for a suitable parameter ρ, whenever $n + \rho + |\mu| + |\nu| > 0$, one has

$$|\partial_x^\mu \partial_y^\nu \mathcal{K}(x, y)| \lesssim \text{dist}(x, y)^{-(n + \rho + |\mu| + |\nu|)}. \tag{1.6.5}$$

Estimates like (1.6.5) hold for a wide range of operators such as classical pseudo-differential operators and Calderón-Zygmund operators [46].

However, this thesis focuses on the case of positive order operators, $t \geq 0$. The ideas developed in the following chapters can, in principle, also be carried over to the case of integral equations. A first study in this direction has been provided in [1]. Adaptive methods for boundary integral equations have also been studied in [78, 85, 101, 116] and in a series of recent works by Harbrecht, Schneider and collaborators.

Chapter 2

Wavelet Frames on Bounded Domains

The intention of this chapter is to illustrate how wavelet frames on bounded domains can be constructed which are well suited for the discretization of elliptic operator equations. In particular, in view of §1.3 and §1.5, the crucial issues are a *sparse stiffness matrix*, the entries of which should be easy to compute (or it should be possible to approximate them at a rather low cost), the *stability of the collection of wavelets* in the solution space $H_0^t(\Omega)$, and, moreover, from the practical point of view we also aim at a *simple construction principle*. It will be depicted that the notion of *aggregated wavelet frames* permits all these desired features.

2.1 Basic frame theory

In order to give the discussions in the following chapters a well-founded background, we summarize here the fundamental coherences of the classical frame theory in Hilbert spaces on a rather abstract level.

2.1.1 Frames for Hilbert spaces

Let H be a separable Hilbert space with inner product $\langle \cdot, \cdot \rangle$ and the induced norm $\| \cdot \|$. Let furthermore $\mathcal{I}$ be a countable index set.

Definition 2.1. A collection of functions $\mathcal{F} = \{f_i\}_{i \in \mathcal{I}} \subset H$ is called a *frame sequence* if there exist constants $0 < A \leq B < \infty$ such that the equivalence

$$A\|f\|^2 \leq \sum_{i \in \mathcal{I}} |\langle f, f_i \rangle|^2 \leq B\|f\|^2 \tag{2.1.1}$$

holds for all $f \in \text{clos}_H(\text{span}(\mathcal{F}))$. It is called a *(Hilbert) frame* for H in case (2.1.1) holds for all $f \in H$.

The constants A and B are called *frame bounds*. Obviously, they are not unique. The *optimal lower frame bound* is the supremum over all lower frame bounds, the *optimal upper frame bound* is the infimum over all upper frame bounds. Frames for which one can choose $A = B$ are called *tight*.

The second bound in (2.1.1) means that $\mathcal{F}$ is a *Bessel system*, i.e., the operator

$$F : H \to \ell_2(\mathcal{I}), \quad f \mapsto \{\langle f, f_i \rangle\}_{i \in \mathcal{I}} \tag{2.1.2}$$

is well-defined and bounded with $\|F\| \leq \sqrt{B}$, where here and in the remainder of this thesis, for $0 < p \leq \infty$, with $\ell_p(\mathcal{I})$ we denote the space of all sequences $\mathbf{c} = \{c_i\}_{i \in \mathcal{I}}$ for which

$$\|\mathbf{c}\|_{\ell_p(\mathcal{I})} := \begin{cases} \left(\sum\limits_{i \in \mathcal{I}} |c_i|^p \right)^{1/p}, & 0 < p < \infty \\ \sup\limits_{i \in \mathcal{I}} |c_i|, & p = \infty \end{cases} \tag{2.1.3}$$

is finite. It is easy to check that the adjoint of F is given by

$$F^* : \ell_2(\mathcal{I}) \to H, \quad \{c_i\}_{i \in \mathcal{I}} \mapsto \sum_{i \in \mathcal{I}} c_i f_i, \tag{2.1.4}$$

hence $\|F^*\| = \|F\| = \sqrt{B}$. F and F^* are often called *analysis* and *synthesis operator*, respectively. Their composition

$$S := F^* F : H \to H, \quad f \mapsto \sum_{i \in \mathcal{I}} \langle f, f_i \rangle f_i \tag{2.1.5}$$

is called *frame operator*, and it plays a fundamental role in frame theory as the following theorem shows; cf. [23].

Theorem 2.1.   (i) *S is a self-adjoint positive definite operator satisfying $A\,I \leq S \leq B\,I$. Moreover, S is invertible and $B^{-1}I \leq S^{-1} \leq A^{-1}I$.*

(ii) *The collection of functions $\tilde{\mathcal{F}} = \{\tilde{f}_i\}_{i \in \mathcal{I}} := \{S^{-1}f_i\}_{i \in \mathcal{I}}$ is again a frame with frame bounds B^{-1}, A^{-1}. It is called the canonical dual frame.*

Using the fact that S^{-1} is self-adjoint, one can immediately infer the following statement.

Theorem 2.2. *Any $f \in H$ can be written as*

$$f = S^{-1}Sf = \sum_{i \in \mathcal{I}} \langle f, f_i \rangle \tilde{f}_i = SS^{-1}f = \sum_{i \in \mathcal{I}} \langle f, \tilde{f}_i \rangle f_i. \tag{2.1.6}$$

Moreover, the analysis operator $\tilde{F}$ and synthesis operator $\tilde{F}^*$ corresponding to the canonical dual frame can be written as $\tilde{F} = FS^{-1}$ and $\tilde{F}^* = S^{-1}F^*$, see [51, §3.2], with which (2.1.6) reads as

$$f = \tilde{F}^* F f = F^* \tilde{F} f. \tag{2.1.7}$$

Consequently, the simple *frame condition* (2.1.1), which can be interpreted as a generalized Parseval identity, is sufficient to guarantee that any element $f \in H$ can be written as a superposition of primal or canonical dual frame elements. Thus, one may decompose a function f into a discrete set of components and perfectly reconstruct it from the respective coefficients. From (2.1.7) it follows that F^* is onto, and the lower estimate in (2.1.1) gives the injectivity of F, whereas the injectivity of F^* is not guaranteed in general. Thus, frames allow redundancies. Besides the canonical one, various *non-canonical dual frames* might exist for which the reconstruction formulas (2.1.7) hold; see [51, §3.2] for a very simple example. Note that for a tight frame the canonical dual coincides with the primal one up to a constant, i.e., $S^{-1}f_i = \frac{1}{A}f_i$, $i \in \mathcal{I}$; cf. [23, §1.1].

Naturally, the different representations $f = \sum_{i \in \mathcal{I}} c_i f_i$ can have completely different properties. In particular, because of the identity

$$\sum_{i \in \mathcal{I}} |c_i|^2 = \sum_{i \in \mathcal{I}} |\langle f, S^{-1} f_i \rangle|^2 + \sum_{i \in \mathcal{I}} |c_i - \langle f, S^{-1} f_i \rangle|^2 \qquad (2.1.8)$$

(see [23, Lemma 5.4.2] for a proof), the expansion coefficients induced by the canonical dual frame are those with the smallest ℓ_2-norm, but if one is interested in, say, *sparsity* of the sequence $\mathbf{c} = \{c_i\}_{i \in \mathcal{I}}$, in the sense that as many c_i vanish as possible, $\mathbf{c} = \{\langle f, S^{-1} f_i \rangle\}_{i \in \mathcal{I}}$ might not be the optimal choice.

The next statement shows that any set of coefficients $\mathbf{c} \in \ell_2(\mathcal{I})$ can be decomposed into a redundant part and a non-redundant part, where the decomposition depends on the chosen dual frame.

Proposition 2.1. *Let $\tilde{\mathcal{F}}$ be any dual frame for $\mathcal{F}$. Then, the operator $\mathbf{Q} := \tilde{F}F^*$: $\ell_2(\mathcal{I}) \to \ell_2(\mathcal{I})$ is a bounded projector with $\mathrm{ran}(\mathbf{Q}) = \mathrm{ran}(\tilde{F})$ and $\ker(\mathbf{Q}) = \ker(F^*)$, hence $\ell_2(\mathcal{I}) = \mathrm{ran}(\tilde{F}) \oplus \ker(F^*)$ is a direct sum decomposition of $\ell_2(\mathcal{I})$.*

For the special case that $\tilde{\mathcal{F}} = S^{-1}\mathcal{F}$, we have $\mathrm{ran}(\tilde{F}) = \mathrm{ran}(F)$, and $\mathbf{Q}$ is the orthogonal projection onto $\mathrm{ran}(F)$, i.e., $\mathrm{ran}(\mathbf{Q}) = \mathrm{ran}(F)$ and $\ker(\mathbf{Q}) = \ker(F^*)$.

Proof. Since $(\tilde{F}F^*)(\tilde{F}F^*) = \tilde{F}(F^*\tilde{F})F^* = \tilde{F}F^*$, $\mathbf{Q}$ is a projector. F^* is onto, hence $\mathrm{ran}(\mathbf{Q}) = \mathrm{ran}(\tilde{F})$, and $\ker(\mathbf{Q}) = \ker(F^*)$ holds because $\tilde{F}$ is injective.

In case $\tilde{F} = FS^{-1}$, it is $\mathrm{ran}(\tilde{F}) = \mathrm{ran}(F)$, because S^{-1} is clearly onto. Noting that $\ell_2(\mathcal{I}) = \mathrm{ran}(F) \oplus^\perp \ker(F^*)$ is an orthogonal decomposition of $\ell_2(\mathcal{I})$ completes the proof. $\qquad\square$

We close this section by giving two equivalent characterizations of frames. For a proof of the first one we refer the reader to [23, Theorem 5.5.1].

Theorem 2.3. *The collection of functions $\mathcal{F} = \{f_i\}_{i \in \mathcal{I}} \subset H$ is a frame for H if and only if the operator*

$$F^* : \ell_2(\mathcal{I}) \to H, \quad \{c_i\}_{i \in \mathcal{I}} \mapsto \sum_{i \in \mathcal{I}} c_i f_i \qquad (2.1.9)$$

is well-defined and onto.

The second equivalent characterization reads as follows.

Proposition 2.2. *$\mathcal{F} \subset H$ is a frame for H if and only if*

(i) $\operatorname{clos}_H(\operatorname{span}(\mathcal{F})) = H$ *and*

(ii) $B^{-1}\|f\|^2 \leq \displaystyle\inf_{\{c \in \ell_2(\mathcal{I}),\ F^*c=f\}} \|c\|_{\ell_2(\mathcal{I})}^2 \leq A^{-1}\|f\|^2, \quad$ *for all $f \in H$.*

Proof. Suppose first that $\mathcal{F}$ is a frame. If $f \in \operatorname{span}(\mathcal{F})^{\perp}$, the lower frame bound in (2.1.1) gives $f = 0$, thus (i) follows. Furthermore, because of the relation (2.1.8), it is

$$\inf_{\{c \in \ell_2(\mathcal{I}),\ F^*c=f\}} \|c\|_{\ell_2(\mathcal{I})}^2 = \sum_{i \in \mathcal{I}} |\langle f, S^{-1}f_i \rangle|^2.$$

Hence, (ii) follows from Theorem 2.1 (ii).

For the converse statement, by Theorem 2.3, it is sufficient to show that $F^* : \ell_2(\mathcal{I}) \to H$ is well-defined and onto. From the lower estimate in (ii) one can infer that indeed F^* is well-defined and bounded. To show this, we consider without loss of generality the case $\mathcal{I} = \mathbb{N}$. The sequence of operators $\{(F^*)^{(n)}\}_{n \in \mathbb{N}}$, $(F^*)^{(n)} : \ell_2(\mathbb{N}) \to H$, $(F^*)^{(n)}c := \sum_{k \leq n} c_k f_k$ is uniformly bounded, because

$$B^{-1}\|(F^*)^{(n)}c\|^2 \leq \inf_{\{d \in \ell_2(\mathbb{N}),\ F^*d=(F^*)^{(n)}c\}} \|d\|_{\ell_2(\mathbb{N})}^2 \leq \sum_{k \leq n} |c_k|^2 \leq \|c\|_{\ell_2(\mathbb{N})}^2.$$

An analogous argument shows that for a fixed $c \in \ell_2(\mathbb{N})$, $\{(F^*)^{(n)}c\}_{n \in \mathbb{N}}$ is a Cauchy sequence and thus convergent in H. By the Banach-Steinhaus theorem, the operator F^* is then well-defined and bounded. Moreover, using the upper estimate in (ii), we will show below that $\operatorname{ran}(F^*)$ is closed. Because of $\operatorname{span}(\mathcal{F}) \subset \operatorname{ran}(F^*) \subset H$ and (i), F^* must then be onto, which completes the proof. Let now $\{g_n\}_{n \in \mathbb{N}} \subset \operatorname{ran}(F^*)$ be a convergent sequence with the limit $g \in H$. We have to find a sequence $c \in \ell_2(\mathbb{N})$ such that $g = F^*c \in \operatorname{ran}(F^*)$. Without loss of generality, let $g_1 = 0$ and $\|g_{n+1}-g_n\|^2 \leq 2^{-n}$ for all $n \geq \bar{n}$, for some $\bar{n} \in \mathbb{N}$. Then, using the upper bound in (ii) and $g_{n+1} - g_n \in \operatorname{ran}(F^*)$, one finds a sequence $\{d^{(n)}\}_{n \in \mathbb{N}} \subset \ell_2(\mathbb{N})$ with $F^*d^{(n)} = g_{n+1} - g_n$ such that

$$\|d^{(n)}\|_{\ell_2(\mathbb{N})}^2 \leq \inf_{\{e \in \ell_2(\mathbb{N}),\ F^*e=g_{n+1}-g_n\}} \|e\|_{\ell_2(\mathbb{N})}^2 + 2^{-n}$$

$$\leq A^{-1}\|g_{n+1} - g_n\|^2 + 2^{-n} \leq (A^{-1}+1)2^{-n}, \quad n \geq \bar{n}.$$

Hence, the series $c := \sum_{n \in \mathbb{N}} d^{(n)}$ is convergent in $\ell_2(\mathbb{N})$, and

$$F^*c = F^* \sum_{n \in \mathbb{N}} d^{(n)} = \sum_{n \in \mathbb{N}} F^*d^{(n)} = \sum_{n \in \mathbb{N}} g_{n+1} - g_n = g.$$

$\square$

Prominent types of frames are *Gabor frames* (see [52], [23, Chapter 8]) or wavelet frames (see [9, 98], [23, Chapter 11–14]) for $L_2(\mathbb{R}^n)$.

2.1.2 Riesz bases

The most prominent subclass of frames are *Riesz bases*.

Definition 2.2. A collection of functions $\mathcal{F} = \{f_i\}_{i \in \mathcal{I}}$ is called *Riesz basis* for H if

(i) $\mathrm{clos}_H(\mathrm{span}(\mathcal{F})) = H$ and

(ii) there exist constants $0 < C \leq D < \infty$ such that

$$C\|\mathbf{c}\|_{\ell_2(\mathcal{I})}^2 \leq \left\| \sum_{i \in \mathcal{I}} c_i f_i \right\|^2 \leq D\|\mathbf{c}\|_{\ell_2(\mathcal{I})}^2, \quad \mathbf{c} = \{c_i\}_{i \in \mathcal{I}} \in \ell_2(\mathcal{I}). \qquad (2.1.10)$$

An extensive discussion of the relation between frames and Riesz bases can be found in [23, §5.4, §6]. We state first that indeed Riesz bases are special frames.

Proposition 2.3. *Any Riesz basis $\mathcal{F}$ for H is also frame for H. The Riesz bounds C and D coincide with the frame bounds, i.e., $A = C$, $B = D$. The dual frame is unique, it can thus be obtained by $\{S^{-1}f_i\}_{i \in \mathcal{I}}$, and it also forms a Riesz basis for H.*

Proof. [23, Theorem 5.4.1]. □

Remark 2.1. The characterization of frames given by Proposition 2.2 is somewhat closer to Definition 2.2 than Definition 2.1, because condition (ii) in Proposition 2.2 immediately follows from (2.1.10) by noting that any $f \in H$ has its unique expansion coefficients for the case of a Riesz basis.

As usual, in the following we say that a collection of functions $\{f_i\}_{i \in \mathcal{I}} \subset H$ has a biorthogonal sequence if there exists a sequence $\{g_i\}_{i \in \mathcal{I}} \subset H'$ such that

$$g_k(f_j) = \delta_{k,j} := \begin{cases} 1, & k = j, \\ 0, & k \neq j. \end{cases}$$

We will also make use of the following essential coherences; see [23, Theorem 6.1.1].

Theorem 2.4. *Let $\{f_i\}_{i \in \mathcal{I}}$ be a frame for H. Then, the following statements are equivalent:*

(i) $\{f_i\}_{i \in \mathcal{I}}$ *is a Riesz basis.*

(ii) $\{f_i\}_{i \in \mathcal{I}}$ *has a biorthogonal sequence.*

(iii) $\ker(F^*) = \{0\}$.

(iv) $\{f_i\}_{i \in \mathcal{I}}$ *is a Schauder basis, i.e., for each $f \in H$ there exist unique scalar coefficients $\{c_i(f)\}_{i \in \mathcal{I}}$ such that $f = \sum_{i \in \mathcal{I}} c_i(f) f_i$.*

In this context we also recall a well-known fact concerning existence and uniqueness of a biorthogonal sequence of a Schauder basis; cf. [23, Theorem 3.3.2].

Theorem 2.5. *Suppose that $\{f_i\}_{i \in \mathcal{I}}$ is a Schauder basis for H. Then, there exists a unique sequence $\{g_i\}_{i \in \mathcal{I}}$ in H' such that $f = \sum_{i \in \mathcal{I}} \langle f, g_i \rangle_{H \times H'} f_i$ for all $f \in H$. Moreover, $\{g_i\}_{i \in \mathcal{I}}$ itself is also a Schauder basis for H'. In addition, $\{f_i\}_{i \in \mathcal{I}}$ and $\{g_i\}_{i \in \mathcal{I}}$ are biorthogonal.*

2.1.3 Banach frames

The concept of frames can also be studied in a more general setting of Banach spaces; see, for instance, [61, 62, 71]. This point of view will also be helpful for our purposes.

Let $\mathcal{B}$ be a separable and reflexive Banach space, $\mathcal{B}'$ its normed dual space, and $\langle \cdot, \cdot \rangle_{\mathcal{B} \times \mathcal{B}'}$ the associated dual form.

Definition 2.3. A *Banach frame* for $\mathcal{B}$ is a countable set $\mathcal{F} = \{f_i\}_{i \in \mathcal{I}} \subset \mathcal{B}'$ with an associated sequence space b such that the following properties hold.

 (i) The coefficient operator $F : \mathcal{B} \to b$ defined by $Ff := \{\langle f, f_i \rangle_{\mathcal{B} \times \mathcal{B}'}\}_{i \in \mathcal{I}}$ is bounded.

 (ii) Norm equivalence:

$$\|f\|_{\mathcal{B}} \approx \| \{\langle f, f_i \rangle_{\mathcal{B} \times \mathcal{B}'}\}_{i \in \mathcal{I}} \|_b, \quad f \in \mathcal{B}. \tag{2.1.11}$$

 (iii) There exists a bounded reconstruction operator $R : b \to \mathcal{B}$ such that

$$R\left(\{\langle f, f_i \rangle_{\mathcal{B} \times \mathcal{B}'}\}_{i \in \mathcal{I}}\right) = f. \tag{2.1.12}$$

With this more general notion at hand, we reconsider the concept of Hilbert frames.

2.1.4 Hilbert frames revisited

Let now again $\mathcal{F} = \{f_i\}_{i \in \mathcal{I}} \subset H$ be a Hilbert frame for the Hilbert space H. For some considerations in this thesis it will be convenient to consider the dual frame as a collection of coefficient functionals in H', forming a Hilbert frame for H'. To this end, we give an adapted equivalent description of Hilbert frames as in [104].

The classical frame condition (2.1.1) for $\mathcal{F} \subset H$ can be equivalently expressed as

$$A\|f\|_{H'}^2 \le \sum_{i \in \mathcal{I}} |f(f_i)|^2 \le B\|f\|_{H'}^2, \quad f \in H', \tag{2.1.13}$$

using the canonical Riesz isomorphism between H and H'. From this can be inferred that the analysis operator

$$F : H' \to \ell_2(\mathcal{I}), \quad f \mapsto \{f(f_i)\}_{i \in \mathcal{I}}, \tag{2.1.14}$$

and its dual synthesis operator

$$F' : \ell_2(\mathcal{I}) \to H, \quad \{c_i\}_{i \in \mathcal{I}} \mapsto \sum_{i \in \mathcal{I}} c_i f_i, \tag{2.1.15}$$

are bounded. Then, the frame operator reads as $S := F'F : H' \to H$, it is also boundedly invertible, and we may define

$$\tilde{F} := FS^{-1} : H \to \ell_2(\mathcal{I}) \quad \text{and} \quad \tilde{F}' := S^{-1}F' : \ell_2(\mathcal{I}) \to H'. \tag{2.1.16}$$

The canonical dual frame is then given by the collection $S^{-1}\mathcal{F}$, and it is a frame for H'. Moreover, one has the relations

$$\tilde{F}'F = \mathrm{id}_{H'} \quad \text{and} \quad F'\tilde{F} = \mathrm{id}_H. \tag{2.1.17}$$

Clearly, as in Proposition 2.1, it still holds that $\mathbf{Q} = \tilde{F}F' : \ell_2(\mathcal{I}) \to \ell_2(\mathcal{I})$ is the orthogonal projector onto $\mathrm{ran}(F)$ and $\mathrm{ran}(F) \oplus^{\perp} \ker(F') = \ell_2(\mathcal{I})$.

This description of a Hilbert frame is consistent with the concept of Banach frames. $\mathcal{F}$ is a Banach frame for H' with respect to $b = \ell_2(\mathcal{I})$ in the sense of Definition 2.3. Indeed, to be a Banach frame for H', $\mathcal{F}$ has to be in $H'' = H$, which is the case. The boundedness of the operator $F : H' \to \ell_2(\mathcal{I})$, the equivalence (2.1.13), as well as (2.1.17) verify (i), (ii), and (iii) in Definition 2.3, respectively.

2.2 Classical wavelet bases

In order to discretize the elliptic operator $\mathcal{L}$ from (1.3.5) which is defined on $H_0^t(\Omega)$, a frame or a basis for (subspaces of) $H_0^t(\Omega)$ is needed. Therefore, in this section, we will first of all present the fundamental tools and principles commonly used for the construction of pairs of biorthogonal wavelet Riesz bases $(\Psi, \tilde{\Psi})$ for $L_2(\Omega)$, where $\Omega \subset \mathbb{R}^n$ is a bounded domain or a closed manifold; see, for instance, [45, 47, 48, 95]. We shall see that these bases also lead to characterizations of certain scales of Sobolev and Besov spaces in a natural way.

After the collection of the abstract machinery, it will be summarized how the results can be applied to the special case of the n-dimensional unit cube $[0,1]^n$ and $L_2([0,1]^n)$. These wavelet bases will form the building block in the construction of wavelet frames on bounded domains, as we will show later in §2.3.

2.2.1 Common construction principles

Part of the theory developed in the following can be considered in the general context of a separable Hilbert space H with scalar product $\langle \cdot, \cdot \rangle$ and norm $\| \cdot \| = \langle \cdot, \cdot \rangle^{1/2}$. We will enter the concrete setting of $H = L_2(\Omega)$ at places where it is convenient to do so. The principal tool for the construction of wavelet bases which is used in wavelet theory is the concept of *multiresolution analysis* (MRA) [88, 91]. In the present non-shift-invariant setting we take a similar notion as point of departure.

Definition 2.4. A sequence $\mathcal{V} = \{V_j\}_{j \geq j_0}$, $j_0 \in \mathbb{N}_0$, of closed subspaces of H is called *multiresolution sequence* (MS) for H if

(i) $V_j \subset V_{j+1}$, $\quad j \geq j_0$,

(ii) $\mathrm{clos}_H \bigcup_{j \geq j_0} V_j = H$.

29

The spaces V_j are usually generated by a system of functions $\Phi_j = \{\phi_{j,k} : k \in \Delta_j\}$, where Δ_j is a finite or infinite (but countable) set of indices, i.e.,

$$V_j := S(\Phi_j) := \mathrm{clos}_H \, \mathrm{span}(\Phi_j).$$

The concept of multiresolution analysis plays an important role in the classical wavelet theory on $\mathbb{R}^n$, where the spaces V_j are typically generated by integer translation and dilation of a function ϕ, i.e.,

$$V_j = \mathrm{clos}_{L_2(\mathbb{R}^n)} \, \mathrm{span}\{\phi_{j,k} := 2^{jn/2}\phi(2^j \cdot -k) : k \in \mathbb{Z}^n\}. \qquad (2.2.1)$$

It is also possible to consider the case of spaces V_j being spanned by the translates and dilates of more than one function. This represents the starting point in *multiwavelet theory*; see, e.g., [84] and the references therein. In this context, frequent use is made of powerful Fourier techniques, to which, in our setting of a bounded domain, we cannot resort to. However, the techniques developed in [21] and [41] represent a rather general approach for the construction of biorthogonal wavelet Riesz bases which are also applicable in the non-shift-invariant setting. In the following, we will describe a recipe for a wavelet construction which is based on the main ideas developed there.

As we will see, it is convenient to consider the case where $\{\Phi_j\}_{j\geq j_0}$ forms a sequence of *uniformly stable bases*, meaning that

$$\|\mathbf{c}\|_{\ell_2(\Delta_j)} \approx \|\mathbf{c}^\top \Phi_j\|, \quad \text{for all } \mathbf{c} \in \ell_2(\Delta_j), \ j \geq j_0, \qquad (2.2.2)$$

holds uniformly in j. Note that here we use the notation $\mathbf{c}^\top \Phi_j := \sum_{k \in \Delta_j} c_k \phi_{j,k}$, where $\mathbf{c} = \{c_k\}_{k \in \Delta_j}$. Thus, the set of functions Φ_j is interpreted as a column vector. The systems Φ_j are typically called *generator bases* or *single scale bases* and its elements $\phi_{j,k}$ *scaling functions*.

For normed linear spaces X and Y, let us denote with $L(X, Y)$ the space of bounded linear operators from X to Y. Using the nestedness and the uniform stability (2.2.2), one can easily prove the existence of *refinement matrices*

$$\mathbf{M}_{j,0} \in L(\ell_2(\Delta_j), \ell_2(\Delta_{j+1})), \ \|\mathbf{M}_{j,0}\| = \mathcal{O}(1), \qquad (2.2.3)$$

where $\|\mathbf{M}_{j,0}\| := \sup_{\mathbf{v} \in \ell_2(\Delta_j), \|\mathbf{v}\|_{\ell_2(\Delta_j)}=1} \|\mathbf{M}_{j,0}\mathbf{v}\|_{\ell_2(\Delta_{j+1})}$ such that

$$\Phi_j = \mathbf{M}_{j,0}^\top \Phi_{j+1}, \quad j \geq j_0. \qquad (2.2.4)$$

Let us furthermore consider a second multiresolution sequence $\tilde{\mathcal{V}} := \{\tilde{V}_j\}_{j\geq j_0}$ with

$$\tilde{V}_j := S(\tilde{\Phi}_j) := \mathrm{clos}_H \, \mathrm{span}(\tilde{\Phi}_j), \quad \tilde{\Phi}_j = \{\tilde{\phi}_{j,k} : k \in \Delta_j\},$$

which is *biorthogonal*, meaning that

$$\langle \Phi_j, \tilde{\Phi}_j \rangle = \mathbf{I}, \quad j \geq j_0. \qquad (2.2.5)$$

Note that here and in the remainder we use the abbreviation $\langle \Phi, \Theta \rangle := \{\langle \phi, \theta \rangle\}_{\phi \in \Phi, \theta \in \Theta}$ for arbitrary vectors of functions $\Phi, \Theta \in H$. Clearly, $\mathbf{I}$ represents the identity matrix. In case the sequence of systems $\{\tilde{\Phi}_j\} := \{\tilde{\Phi}_j\}_{j \geq j_0}$ is also uniformly stable, the collections $\tilde{\Phi}_j$ are also refinable

$$\tilde{\Phi}_j = \tilde{\mathbf{M}}_{j,0}^\top \tilde{\Phi}_{j+1}, \quad j \geq j_0, \tag{2.2.6}$$

where $\tilde{\mathbf{M}}_{j,0}$ are uniformly bounded mappings from $\ell_2(\Delta_j)$ to $\ell_2(\Delta_{j+1})$.

From now on, we assume the uniform stability of the collections Φ_j and $\tilde{\Phi}_j$. A sufficient criterion for this in the case $H = L_2(\Omega)$ would be (2.2.5) in combination with *uniform boundedness*

$$\|\phi_{j,k}\|_H, \|\tilde{\phi}_{j,k}\|_H \lesssim 1, \quad k \in \Delta_j, \tag{2.2.7}$$

and the *uniform locality* of the primal and dual generators, i.e.,

$$\#\{k' \in \Delta_j : \square_{j,k'} \cap \square_{j,k} \neq \emptyset\}, \ \#\{k' \in \Delta_j : \tilde{\square}_{j,k'} \cap \tilde{\square}_{j,k} \neq \emptyset\} \lesssim 1, \tag{2.2.8}$$

$$\operatorname{diam} \square_{j,k}, \operatorname{diam} \tilde{\square}_{j,k} \lesssim 2^{-j}, \ \text{for all } k \in \Delta_j, \tag{2.2.9}$$

where $\square_{j,k} := \operatorname{supp} \phi_{j,k}$ and $\tilde{\square}_{j,k} := \operatorname{supp} \tilde{\phi}_{j,k}$; cf. [45, Lemma 2.1]. Naturally, the locality property plays an important role for numerical computations as well. It can also be expressed in terms of certain sparsity properties of the refinement matrices $\mathbf{M}_{j,0}, \tilde{\mathbf{M}}_{j,0}$. For instance, in the shift-invariant setting, the *refinement relation* (2.2.4) for $j = j_0 = 0$ reads as

$$\phi(x) = \sum_{k \in \mathbb{Z}^n} a_k \phi(2x - k), \quad x \in \mathbb{R}^n. \tag{2.2.10}$$

A function satisfying (2.2.10) is called a *refinable function*. It is clearly desirable to choose ϕ as a compactly supported function with a *mask* $\mathbf{a} = \{a_k\}_{k \in \mathbb{Z}^n}$ containing only finitely many non-zero coefficients. Fortunately, this is indeed possible, and we shall see an example in §2.2.4 below. Hence, in this case $\mathbf{M}_{j,0}$ is a bi-infinite matrix with a band-like structure. It is furthermore desirable to ensure similar locality properties for the primal and dual wavelets. The following proceeding will take care of this target.

By setting

$$Q_j v := \langle v, \tilde{\Phi}_j \rangle^\top \Phi_j, \quad Q_j^* := \langle v, \Phi_j \rangle^\top \tilde{\Phi}_j, \quad j \geq j_0, \tag{2.2.11}$$

we define projectors Q_j onto V_j and their adjoints Q_j^*, projecting onto $\tilde{V}_j$. Because of the uniform stability of the systems $\{\Phi_j\}$ and $\{\tilde{\Phi}_j\}$, surely Q_j and Q_j^* are uniformly bounded. By setting $Q_{j_0-1} := 0$, one can write any $f \in H$ according to

$$f = \sum_{j=j_0}^{\infty} (Q_j - Q_{j-1}) f,$$

where the sum converges strongly in H. The difference $(Q_j - Q_{j-1})f$ can be viewed as the *detail* needed to update the information of f which is contained in V_{j-1} to a higher scale V_j. The refinability of $\{\Phi_j\}$ and $\{\tilde{\Phi}_j\}$ implies

$$Q_l Q_j f = Q_l f, \quad \text{for all } l \leq j, \tag{2.2.12}$$

ensuring the independence of the details to be added. In order to be able to prove stability of the details $(Q_j - Q_{j-1})f$, i.e., $\|f\|_H^2 \approx \sum_{j=j_0}^{\infty} \|(Q_j - Q_{j-1})f\|_H^2$, specific approximation and regularity properties of the spaces V_j have to be required.

Theorem 2.6. *Let us fix $H = L_2(\Omega)$. Let the sequences $\{\Phi_j\}$ and $\{\tilde{\Phi}_j\}$ as well as the projections Q_j be defined as above. Assume that for $\mathcal{S} = \{S_j\}_{j \geq j_0} \in \{\mathcal{V}, \tilde{\mathcal{V}}\}$ the Jackson-type estimate*

$$\inf_{v_j \in S_j} \|v - v_j\|_{L_2(\Omega)} \lesssim 2^{-sj} \|v\|_{H^s(\Omega)}, \quad v \in H^s(\Omega), \; s \leq \sigma_{\mathcal{S}}, \tag{2.2.13}$$

and the Bernstein-type estimate

$$\|v_j\|_{H^s(\Omega)} \lesssim 2^{sj} \|v_j\|_{L_2(\Omega)}, \quad v_j \in S_j, \; s \leq \mu_{\mathcal{S}}, \tag{2.2.14}$$

hold. Then, with $\gamma := \min\{\sigma_{\mathcal{V}}, \mu_{\mathcal{V}}\}$ and $\tilde{\gamma} := \min\{\sigma_{\tilde{\mathcal{V}}}, \mu_{\tilde{\mathcal{V}}}\}$, we have the norm equivalence

$$\|v\|_{H^s(\Omega)}^2 \approx \sum_{j=j_0}^{\infty} 2^{2sj} \|(Q_j - Q_{j-1})v\|_{L_2(\Omega)}^2, \quad s \in (-\tilde{\gamma}, \gamma). \tag{2.2.15}$$

In (2.2.15), although it is a slight abuse of the notation introduced in §1.2, for $s < 0$ the space $H^s(\Omega)$ represents $(H^{-s}(\Omega))'$. Estimates of the form (2.2.13) and (2.2.14) are also called *direct* and *inverse estimates*, respectively.

Now, the outline for the construction of a pair of biorthogonal wavelet Riesz bases is the following. One has to find stable bases $\Psi_j = \{\psi_{j,k} : k \in \nabla_j\}$ and $\tilde{\Psi}_j = \{\tilde{\psi}_{j,k} : k \in \nabla_j\}$ of $W_{j+1} := (Q_{j+1} - Q_j)H$ and $\tilde{W}_{j+1} = (Q_{j+1}^* - Q_j^*)H$, respectively, such that the biorthogonality relation

$$\langle \Psi_j, \tilde{\Psi}_j \rangle = \mathbf{I} \tag{2.2.16}$$

is satisfied. Note that because of the relation (2.2.12) it is $W_{j+1} = (Q_{j+1} - Q_j)V_{j+1}$ and $\tilde{W}_{j+1} = (Q_{j+1}^* - Q_j^*)\tilde{V}_{j+1}$. Moreover, the necessary relation $W_{j+1} \perp \tilde{V}_j$ follows from (2.2.5) and (2.2.6); see, for instance, [45]. The guideline is then, in a first step, to construct stable bases Ψ_j of *some* complement $S(\Psi_j)$ of V_j in V_{j+1} such that the wavelets are compactly supported and local in the sense of (2.2.8), (2.2.9). Then, the idea is to project the initial complement onto $(Q_{j+1} - Q_j)V_{j+1}$, preserving the local supports and the stability. An adequate criterion for the stability of the complement bases Ψ_j is the notion of *stable completion*.

Definition 2.5. Let $\{\Phi_j\}$ be uniformly stable and assume (2.2.4). Let furthermore $\Psi_j \subset V_{j+1}$ with

$$\Psi_j = \mathbf{M}_{j,1}^\top \Phi_{j+1}, \qquad (2.2.17)$$

where $\mathbf{M}_{j,1} \in L(\ell_2(\nabla_j), \ell_2(\Delta_{j+1}))$. Then, the matrix $\mathbf{M}_{j,1}$ is called *stable completion* of $\mathbf{M}_{j,0}$ if $\mathbf{M}_j := (\mathbf{M}_{j,0}, \mathbf{M}_{j,1}) \in L(\ell_2(\Delta_j \cup \nabla_j), \ell_2(\Delta_{j+1}))$ is invertible and satisfies $\|\mathbf{M}_j\|, \|\mathbf{M}_j^{-1}\| = \mathcal{O}(1)$, $j \geq j_0$.

Proposition 2.4 (cf. [21]). $\{\Phi_j \cup \Psi_j\}$ *is uniformly stable if and only if there exists a stable completion* $\mathbf{M}_{j,1}$ *of* $\mathbf{M}_{j,0}$.

The uniform stability of $\{\Phi_j \cup \Psi_j\}$ does not yet imply the Riesz basis property of the union of the complement bases Ψ_j for all $j \geq j_0$. The difficulty is now to extract a suitable stable completion such that the corresponding bases Ψ_j span W_{j+1} and to simultaneously specify systems $\tilde{\Psi}_j$, spanning $\tilde{W}_{j+1}$ such that also (2.2.16) holds. In [21] it has been shown that, given some *initial stable* completion, all others can be written as the image of a linear transformation of the initial one. In the following proposition from [21], a special stable completion is identified, from which a biorthogonal pair of wavelet Riesz bases can be derived.

Proposition 2.5 (cf. [21]). *Let* $\{\Phi_j\}$, $\{\tilde{\Phi}_j\}$, *as well as* $\mathbf{M}_{j,0}$, $\tilde{\mathbf{M}}_{j,0}$, *be defined as above. Suppose that* $\breve{\mathbf{M}}_{j,1}$ *is some initial stable completion of* $\mathbf{M}_{j,0}$ *and that* $\breve{\mathbf{G}}_j = \breve{\mathbf{M}}_j^{-1} = \left(\mathbf{M}_{j,0}, \breve{\mathbf{M}}_{j,1} \right)^{-1}$. *Then,*

$$\mathbf{M}_{j,1} := (\mathbf{I} - \mathbf{M}_{j,0}\tilde{\mathbf{M}}_{j,0}^\top)\breve{\mathbf{M}}_{j,1} \qquad (2.2.18)$$

is also a stable completion of $\mathbf{M}_{j,0}$, *and* $\mathbf{G}_j = \mathbf{M}_j^{-1}$ *has the form* $\mathbf{G}_j = \begin{pmatrix} \tilde{\mathbf{M}}_{j,0}^\top \\ \breve{\mathbf{G}}_{j,1} \end{pmatrix}$. *Moreover, the collections*

$$\Psi_j := \mathbf{M}_{j,1}^\top \Phi_{j+1}, \quad \tilde{\Psi}_j := \breve{\mathbf{G}}_{j,1}\tilde{\Phi}_{j+1} \qquad (2.2.19)$$

form biorthogonal systems,

$$\langle \Psi_j, \tilde{\Psi}_j \rangle = \mathbf{I}, \quad \langle \Psi_j, \tilde{\Phi}_j \rangle = \langle \Phi_j, \tilde{\Psi}_j \rangle = \mathbf{0}, \qquad (2.2.20)$$

so that $(Q_{j+1} - Q_j)S(\Phi_{j+1}) = S(\Psi_j)$, $(Q_{j+1}^* - Q_j^*)S(\tilde{\Phi}_{j+1}) = S(\tilde{\Psi}_j)$.

Consequently, (2.2.5) and (2.2.20) imply that the systems of functions

$$\Psi := \Phi_{j_0} \cup \bigcup_{j \geq j_0} \Psi_j, \quad \tilde{\Psi} := \tilde{\Phi}_{j_0} \cup \bigcup_{j \geq j_0} \tilde{\Psi}_j \qquad (2.2.21)$$

form biorthogonal collections, i.e.,

$$\langle \Psi, \tilde{\Psi} \rangle = \mathbf{I}. \qquad (2.2.22)$$

For the latter we have used the notation

$$\Psi := \left\{ \psi_\lambda := \psi_{j,k} : \lambda = (j,k) \in \nabla := \bigcup_{j \geq j_0-1} \{j\} \times \nabla_j \right\},$$

where $\nabla_{j_0-1} := \Delta_{j_0}$, $\psi_{j_0-1,k} := \phi_{j_0,k}$, and with $\tilde{\psi}_{j_0-1,k} := \tilde{\phi}_{j_0,k}$, $\tilde{\Psi}$ is analogously defined. Moreover, in the following, for an index $\lambda \in \nabla$, with $|\lambda|$, as usual, we denote the scale of the corresponding wavelet ψ_λ, i.e., $|\lambda| = |(j,k)| := j$. It is important to note that in case $\check{\mathbf{M}}_j$, $\check{\mathbf{G}}_j$ and $\tilde{\mathbf{M}}_{j,0}$ are sparse matrices, then this also holds for $\mathbf{M}_j$ and $\mathbf{G}_j$, thus the biorthogonal wavelets Ψ and $\tilde{\Psi}$ will have compact support. It becomes obvious that the construction of the initial stable completion such that the mentioned sparsity constraints are fulfilled is the demanding part of the construction and has to be reconsidered for a special situation at hand. Once this step has been accomplished, the remaining part amounts to an application of Proposition 2.5.

2.2.2 Norm equivalences

Finally, from Theorem 2.6 and Proposition 2.5 one can deduce a classical result, stating that the norm of a function $v \in H^s(\Omega)$ is equivalent to a weighted ℓ_2-norm of the expansion coefficients in the basis.

Corollary 2.1. *Under the same assumptions as in Theorem 2.6 it holds that*

$$\|v\|_{H^s(\Omega)} \approx \left(\sum_{j=j_0-1}^{\infty} \sum_{k \in \nabla_j} 2^{2sj} |\langle v, \tilde{\psi}_{j,k} \rangle_{L_2(\Omega)}|^2 \right)^{1/2}, \quad s \in (-\tilde{\gamma}, \gamma). \tag{2.2.23}$$

Now, for $s = 0$, the Riesz basis property of Ψ and $\tilde{\Psi}$ for $L_2(\Omega)$ follows easily. Indeed, (2.2.23) tells us that $\tilde{\Psi}$ is frame for $L_2(\Omega)$. Moreover, $\tilde{\Psi}$ has a biorthogonal sequence, namely Ψ. These two properties are equivalent to the Riesz basis property of $\tilde{\Psi}$ by Theorem 2.4. Interchanging the role of the primal and dual multiresolution sequence in Theorem 2.6 shows that the primal system Ψ is also a frame for $L_2(\Omega)$ and thus also a Riesz basis. More general, we get the following result for which we introduce the diagonal scaling matrix $\mathbf{D} := \operatorname{diag}(2^{s|\lambda|})_{\lambda \in \nabla}$.

Proposition 2.6. *Under the same assumptions as in Corollary 2.1 it holds that for $s \in [0, \gamma)$ the weighted collection of wavelets $\Psi_{\mathbf{D}} := \mathbf{D}^{-1}\Psi = \{2^{-s|\lambda|}\psi_\lambda\}_{\lambda \in \nabla}$ forms a Riesz basis for $H^s(\Omega)$.*

Proof. The case $s = 0$ has already been discussed above. Let now $s \in (0, \gamma)$. By Theorem 2.3 it is sufficient to prove that the operator $F^* : \ell_2(\nabla) \to H^s(\Omega)$, $\{c_\lambda\}_{\lambda \in \Lambda} \mapsto \sum_{\lambda \in \nabla} c_\lambda 2^{-s|\lambda|} \psi_\lambda$ is well-defined and onto to show that $\Psi_{\mathbf{D}}$ is a frame for $H^s(\Omega)$. Using biorthogonality and (2.2.23), it is easy to prove that indeed the series converges in

$H^s(\Omega)$. We may thus skip the details here. Furthermore, since Ψ is a Riesz basis for $L_2(\Omega)$, we can write any $v \in H^s(\Omega)$ as

$$v = \sum_{\lambda \in \nabla} \langle v, \tilde{\psi}_\lambda \rangle_{L_2(\Omega)} \psi_\lambda = \sum_{\lambda \in \nabla} 2^{s|\lambda|} \langle v, \tilde{\psi}_\lambda \rangle_{L_2(\Omega)} 2^{-s|\lambda|} \psi_\lambda \qquad (2.2.24)$$

with convergence of the series in $L_2(\Omega)$. Because of (2.2.23), the sequence $\{2^{s|\lambda|} \langle v, \tilde{\psi}_\lambda \rangle_{L_2(\Omega)}\}_{\lambda \in \nabla}$ is contained in $\ell_2(\nabla)$, hence the series also converges in $H^s(\Omega)$, thus F^* is onto. If for a $\mathbf{c} := \{c_\lambda\}_{\lambda \in \nabla} \in \ell_2(\nabla)$, $\sum_{\lambda \in \nabla} c_\lambda 2^{-s|\lambda|} \psi_\lambda = 0$, since Ψ is a Riesz basis in $L_2(\Omega)$ and surely $\mathbf{D}^{-1}\mathbf{c} \in \ell_2(\nabla)$, then follows $\mathbf{D}^{-1}\mathbf{c} = 0$, thus $\mathbf{c} = 0$. This shows that F^* is injective and the claim follows by Theorem 2.4. $\qquad \square$

Direct and inverse estimates

It is important to note that even for the verification of the Riesz basis property in $L_2(\Omega)$, the availability of the Jackson and Bernstein estimates (2.2.13), (2.2.14) for a non-empty range of s has to be guaranteed. For a complete discussion of this matter in the shift-invariant case (2.2.1) with compactly supported primal and dual tensor product type generators ϕ and $\tilde{\phi}$, we refer to [26, §3.3 and §3.4].

The basic ingredients for the verification of the direct estimates are results on *local polynomial approximation* (Whitney estimates) and the *polynomial reproduction* of the spaces V_j. For instance, (2.2.13) with $\sigma_{\mathcal{V}} = d$, $d \in \mathbb{N}$, can be verified using (2.2.8), (2.2.9), if the functions $\phi_{j,k}$ reproduce polynomials up to order d, i.e.,

$$\Pi_{d-1} \subset V_j, \qquad (2.2.25)$$

where Π_{d-1} denotes the space of all polynomials up to degree $d-1$ on Ω [45]. We emphasize that (2.2.25) really makes sense, because we are dealing with spaces over a bounded domain. In case (2.2.25) holds, we also say that V_j (or Φ_j) is *exact of order d*. For a discussion of the inverse estimate in the shift-invariant case, we refer to [26, Theorem 3.4.1].

Fortunately, for many cases of practical interest, with some $0 < \gamma < d$, one can show that for $0 < s \le t \le d$, $s < \gamma$, one has

$$\inf_{v_j \in V_j} \|v - v_j\|_{H^s(\Omega)} \lesssim 2^{-j(t-s)} \|v\|_{H^t(\Omega)}, \quad v \in H^t(\Omega), \qquad (2.2.26)$$

with scaling functions in $H^s(\Omega)$, and for $0 < s \le t < \gamma$ it is

$$\|v_j\|_{H^t(\Omega)} \lesssim 2^{j(t-s)} \|v_j\|_{H^s(\Omega)}, \quad v_j \in V_j, \qquad (2.2.27)$$

with scaling functions in $H^t(\Omega)$, can be shown to hold.

2.2.3 Cancellation properties

The biorthogonality relations (2.2.20) in combination with the polynomial exactness of the dual multiresolution sequence of order $\tilde{d} \in \mathbb{N}$ imply

$$\langle P, \psi_{j,k} \rangle_{L_2(\Omega)} = 0, \quad \text{for all } P \in \Pi_{\tilde{d}-1}, \quad j > j_0 - 1. \tag{2.2.28}$$

In view of the cancellation of the monomial basis of $\Pi_{\tilde{d}-1}$, in this case we say that $\psi_{j,k}$ possesses $\tilde{d}$ *vanishing moments*. Moreover, one can prove

$$|\langle f, \psi_{j,k} \rangle_{L_2(\Omega)}| \lesssim 2^{-j(\tilde{d}+\frac{n}{2})} \|f\|_{W_\infty^{\tilde{d}}(\operatorname{supp} \psi_{j,k})}, \quad j > j_0 - 1. \tag{2.2.29}$$

This coherence is commonly called *cancellation property* of order $\tilde{d}$. It means that L_2-inner products of smooth functions with wavelets decay exponentially in modulus when the scale j increases. This can be deduced from (2.2.28), Hölder's inequality, a classical Whitney estimate, the locality $\operatorname{diam}(\operatorname{supp} \psi_{j,k}) \lesssim 2^{-j}$, hence $\operatorname{vol}(\operatorname{supp} \psi_{j,k}) \lesssim 2^{-jn}$, and $\|\psi_{j,k}\|_{L_\infty(\Omega)} \lesssim 2^{j\frac{n}{2}}$. Here, the needed Whitney estimate states that for any cube I_h of side-length $h > 0$, it is

$$\inf_{g \in \Pi_m} \|f - g\|_{L_p(I_h)} \lesssim h^{m+1} \|f\|_{W_p^{m+1}(I_h)}, \quad 1 \le p \le \infty. \tag{2.2.30}$$

Properties of the kind (2.2.29) are one of the keys to prove that certain elliptic operators such as specific classes of differential and integral operators have a (quasi-) sparse representation with respect to a wavelet basis, allowing for efficient compression strategies in numerical algorithms; cf. [66, 67, 105, 108]. We will review the mentioned chain of arguments used to deduce (2.2.29) in §5.1.

2.2.4 Spline wavelet bases on the interval

As a first application of the results presented above, it will be now depicted how biorthogonal wavelet bases for $L_2([0,1])$ can be constructed. We are particularly interested in bases where the primal functions are actually B-splines. Throughout this thesis, we will work with the wavelets developed in [45, 47] and [95] which are quite similar in spirit. It shall be briefly described in the sequel how theses bases are designed.

As an initial step, a pair of biorthogonal multiresolution sequences $\{V_j\}$, $\{\tilde{V}_j\}$ in $L_2([0,1])$ has to be constructed. The idea in [45] is to adapt an existing pair of biorthogonal classical MRA on the real line $\mathbb{R}$ to the interval. Such a pair, generated by spline functions, has been developed in [30], and for convenience we recall some of its basic properties here.

Let for a sequence of knots $t_i \le \cdots \le t_{i+d}$, $[t_i, \ldots, t_{i+d}]f$ denote the d-th order divided difference of a function $f \in C^d(\mathbb{R})$ at $t_i, \ldots, t_{i+d}$, and define $x_+ := \max\{x, 0\}$. Then, the *cardinal B-spline* of order d is defined as the function

$$_d\varphi(x) := d[0, \ldots, d](\cdot - x - \lfloor \tfrac{d}{2} \rfloor)_+^{d-1}, \quad x \in \mathbb{R}. \tag{2.2.31}$$

It has several beneficial properties. $_d\varphi$ is symmetric around $\frac{\mu(d)}{2}$, $\mu(d) = d \mod 2$, it has compact support, $\operatorname{supp}_d \varphi = [l_1, l_2] := [-\lfloor \frac{d}{2} \rfloor, \lceil \frac{d}{2} \rceil]$, and it is a refinable function with finitely supported mask $\mathbf{a} = \{a_k\}_{k\in\mathbb{Z}}$.

In [30] it has been shown that for any d and $\tilde{d} \geq d$, $\tilde{d} \in \mathbb{N}$ with $d + \tilde{d}$ even, there exists a *dual* function $_{d,\tilde{d}}\varphi$ such that

$$\langle _d\varphi(\cdot), _{d,\tilde{d}}\varphi(\cdot - k)\rangle_{L_2([0,1])} = \delta_{0,k}, \quad k \in \mathbb{Z}. \tag{2.2.32}$$

We say that $(_d\varphi, _{d,\tilde{d}}\varphi)$ forms a *dual pair*. Moreover, also $_{d,\tilde{d}}\varphi$ is compactly supported and refinable with a finitely supported mask $\tilde{\mathbf{a}} = \{\tilde{a}_k\}_{k\in\mathbb{Z}}$. The integer translates of the primal function $_d\varphi$ are exact of order d, whereas $_{d,\tilde{d}}\varphi$ reproduces polynomials up to order $\tilde{d}$. In addition, it is

$$_d\varphi \in H^s([0,1]), \quad \text{for all } s < \gamma := d - \frac{1}{2}. \tag{2.2.33}$$

The L_2-Sobolev regularity

$$\tilde{\gamma} := \sup\{s : {}_{d,\tilde{d}}\varphi \in H^s([0,1])\} \tag{2.2.34}$$

of $_{d,\tilde{d}}\varphi$ increases proportionally with $\tilde{d}$. The spaces $V_j := \operatorname{clos}_{L_2([0,1])} \operatorname{span}\{\varphi_{j,k} := 2^{j/2}{}_d\varphi(2^j \cdot -k) : k \in \mathbb{Z}\}$ and $\tilde{V}_j := \operatorname{clos}_{L_2([0,1])} \operatorname{span}\{\tilde{\varphi}_{j,k} := 2^{j/2}{}_{d,\tilde{d}}\varphi(2^j \cdot -k) : k \in \mathbb{Z}\}$ both form an MRA in the classical sense which are, moreover, biorthogonal.

In order to obtain multiresolution analyses $V_{j,[0,1]}$, $\tilde{V}_{j,[0,1]}$ for $L_2([0,1])$, one keeps all functions $\varphi_{j,k}, \tilde{\varphi}_{j,k}$, which are supported in $[0,1]$. Then, in a second step, in order to ensure polynomial reproduction of the order d and $\tilde{d}$, respectively, on the primal and dual side at both ends of the interval on each scale j a fixed number of *boundary functions* has to be constructed and added, retaining refinability and stability. Finally, the dual generators are adapted in such a way that also the biorthogonality (2.2.5) holds; see [45] for details.

The proceeding in [95] is slightly different. While in [45] the primal boundary functions are generated by linear combinations of some functions $\varphi_{j,k}$ restricted to the interval, in [95] the primal spline MRA is directly made up of the *Schoenberg spline basis* (cf. [95, §3.2], [54]). It has been shown in [95] that the bases constructed there exhibit good Riesz constants and give rise to well-conditioned stiffness matrices stemming from one-dimensional Poisson problems. Anyway, for both cases the wavelets are constructed using the method of stable completion as demonstrated in §2.2.1. The norm equivalence (2.2.23) holds with γ and $\tilde{\gamma}$ defined in (2.2.33) and (2.2.34), respectively, so that the Riesz basis property can be deduced as outlined above.

For related literature in the context of wavelets on the interval we also refer to [4, 13, 24, 31]. There also exist a couple of constructions based on multi-wavelets. In this direction we mention the papers [3, 43, 59]; see also [100] for a comparison (and extension) of the latter.

2.2.5 Wavelet bases on the unit cube

Biorthogonal wavelet Riesz bases for $L_2([0,1]^n)$, $n > 1$, can be obtained by forming tensor products of the univariate scaling functions and wavelets. For $e \in E := \{0,1\}$ we set

$$\nabla_{j,e} := \begin{cases} \Delta_j, & e = 0 \\ \nabla_j, & e = 1 \end{cases},$$

and $\Lambda_{j,\mathbf{e}} := \nabla_{j,e_1} \times \nabla_{j,e_2} \times \cdots \times \nabla_{j,e_n}$, for $\mathbf{e} \in E^n$. Furthermore, we define the set of *wavelet indices*

$$\begin{aligned} \Lambda := &\{\lambda = (j_0, \mathbf{0}, \mathbf{k}) : \mathbf{k} \in \Lambda_{j_0,\mathbf{0}},\ \mathbf{0} = (0,\ldots,0) \in E^n\} \cup \\ &\{\lambda = (j, \mathbf{e}, \mathbf{k}) : j \geq j_0,\ \mathbf{e} \in E^n \setminus \{\mathbf{0}\},\ \mathbf{k} \in \Lambda_{j,\mathbf{e}}\}. \end{aligned} \tag{2.2.35}$$

For a $\lambda = (j, \mathbf{e}, \mathbf{k}) \in \Lambda$ let us also define $e(\lambda) := \mathbf{e}$ and the scale mapping $|\lambda| := j$. For a tuple (j, e, k), $j \geq j_0$. $e \in E, k \in \nabla_{j,e}$, we set

$$\psi_{j,e,k} := \begin{cases} \phi_{j,k}, & e = 0 \\ \psi_{j,k}, & e = 1. \end{cases}$$

Moreover, for an index $\lambda \in \Lambda$, let

$$\psi_\lambda^\square(x) := \psi_{j,e_1,k_1}(x_1)\psi_{j,e_2,k_2}(x_2)\cdots\psi_{j,e_n,k_n}(x_n), \quad x \in [0,1]^n. \tag{2.2.36}$$

Then, the collection

$$\Psi^\square := \{\psi_\lambda^\square : \lambda \in \Lambda\} \tag{2.2.37}$$

forms a Riesz basis for $L_2([0,1]^n)$. The resulting tensor product basis inherits the important features such as cancellation properties and norm equivalences from the univariate basis. The dual, biorthogonal basis is obtained by the analogous tensor product approach for the dual univariate scaling functions and wavelets.

Applying this procedure, biorthogonal spline wavelet bases of order d can be easily constructed using, for example, the bases from [45, 47] or [95] as point of departure.

Incorporating boundary conditions

For the numerical treatment of certain problems, such as those described in §1.1, §1.3, homogeneous boundary conditions (on parts of $\partial\Omega$) have to be incorporated in the construction of the primal wavelets. Moreover, to ensure that especially the boundary wavelets will have vanishing moments, it is desirable to construct the dual generators such that they still reproduce polynomials up to a certain degree, i.e., also near the boundary; recall the discussion around (2.2.28). Therefore, these functions must not satisfy homogeneous boundary conditions themselves. In other words, the primal and dual generators as well as the wavelets should satisfy *complementary boundary conditions.* For $\Omega = [0,1]^n$, this issue has been worked out in [47]. Besides the boundary constraints, these bases have similar properties as the ones from [45, 95]. They have been also used as the building block for the construction of wavelet bases on manifolds in [49].

Finally, it is important to note that in the mentioned constructions on the interval the boundary wavelets can be designed in such a way that they satisfy boundary conditions of order $r \in \{0, \ldots, d-2\}$, meaning that

$$\psi_{j,k}^{(m)}(0) = \psi_{j,k}^{(m)}(1) = 0, \quad m = 0, \ldots, r. \tag{2.2.38}$$

In case homogeneous boundary conditions of order r are prescribed on the primal side, near the boundaries, the primal generators are only able to reproduce polynomials that also vanish at the same order r. Note that the order of the boundary conditions at the left end, and at the right end, may also be chosen differently.

2.3 Construction of wavelet frames

This section addresses the issue of constructing frames for $H_0^t(\Omega)$, $t \geq 0$, over a bounded domain $\Omega \subset \mathbb{R}^n$. In such a setting, straightforward multiresolution based constructions such as [60, 76, 77] are not applicable. The main idea is then to decompose the domain Ω into overlapping subdomains Ω_i, on which a frame construction is relatively easy, and to prove that the collection of the elements of all frames over Ω_i make up a frame for the whole space over Ω. These ideas have been developed in [104], and they have been further investigated in [38]. In this section, a more abstract view on this topic is chosen which centers around the fact that frames allow for stable splittings of an object into components, a property by which frames can actually be characterized [94].

The case of Ω being a manifold which is relevant, e.g., for the case of integral equations (see §1.6) can in principle be treated in a similar fashion. Nevertheless, for the sake of simplicity, we restrict the discussion to bounded domains.

2.3.1 Aggregated frames and stable space splittings

For a start we stay in the abstract Hilbert space setting from above, but we denote the canonical norm in H with $\| \cdot \|_H$.

A general Hilbert space setting

The idea is to decompose H into simpler spaces for which a Riesz basis or a frame is available and to aggregate these *local* collections into a *global* one for which the frame property can be shown. This approach will be successful if and only if the local spaces allow for a stable decomposition of elements $f \in H$ into contributions from the local spaces, as we shall now particularize.

Let us consider Hilbert spaces H_i, $i = 0, \ldots, m - 1$, $m \in \mathbb{N}$, not necessarily subspaces of H, each of them being equipped with a norm $\| \cdot \|_{H_i}$ such that there exist bounded linear mappings $R_i : H_i \to H$. With $\bar{H}$ we denote the *Hilbert sum* of the spaces H_i, being the Cartesian product $H_0 \times \cdots \times H_{m-1}$ equipped with the norm $\|(f_0, \ldots, f_{m-1})\|_{\bar{H}} := \left(\sum_{i=0}^{m-1} \|f_i\|_{H_i}^2 \right)^{1/2}$. Let us also define the linear operator $R : \bar{H} \to H$, $(f_0, \ldots, f_{m-1}) \mapsto \sum_{i=0}^{m-1} R_i f_i$. Clearly, R is well-defined and bounded.

Definition 2.6. The system $\{H_i; \| \cdot \|_{H_i}, R_i\}_{i=0}^{m-1}$ forms a *stable space splitting* of $\{H; \| \cdot \|_H\}$ if $R : \bar{H} \to H$ is onto and

$$\|f\|_H^2 \eqsim \inf_{\{f = \sum_{i=0}^{m-1} R_i f_i, \ f_i \in H_i\}} \sum_{i=0}^{m-1} \|f_i\|_{H_i}^2, \quad f \in H. \tag{2.3.1}$$

Proposition 2.7. *Let $\mathcal{F}^{(i)} = \{g_j^{(i)}\}_{j \in \mathcal{I}^{(i)}}$ be frames for H_i, $i = 0, \ldots, m - 1$. Then, the aggregated system*

$$\mathcal{F} := \bigcup_{i=0}^{m-1} R_i \mathcal{F}^{(i)} = \left\{ R_i g_j^{(i)} =: f_{(i,j)} : (i,j) \in \mathcal{I} := \bigcup_{i=0}^{m-1} \{i\} \times \mathcal{I}^{(i)} \right\} \tag{2.3.2}$$

is a frame for H if and only if the system $\{H_i; \| \cdot \|_{H_i}, R_i\}_{i=0}^{m-1}$ forms a stable space splitting of $\{H; \| \cdot \|_H\}$.

Proof. Clearly, any $\mathbf{c} = \{c_{(i,j)}\}_{(i,j) \in \mathcal{I}} \in \ell_2(\mathcal{I})$ can be uniquely resorted as an m-tuple $\mathbf{c} = (\mathbf{c}^{(0)}, \ldots, \mathbf{c}^{(m-1)})$ with $\mathbf{c}^{(i)} = \{c_j^{(i)}\}_{j \in \mathcal{I}^{(i)}} \in \ell_2(\mathcal{I}^{(i)})$ such that $c_j^{(i)} = c_{(i,j)}$. First of all, we show that (2.3.1) and Proposition 2.2 (ii) for $\mathcal{F}$ are actually equivalent. Using Proposition 2.2 (ii) for the frames $\mathcal{F}^{(i)}$ we get

$$\inf_{\{f = \sum_{i=0}^{m-1} R_i f_i, \ f_i \in H_i\}} \sum_{i=0}^{m-1} \|f_i\|_{H_i}^2$$

$$\eqsim \inf_{\{f = \sum_{i=0}^{m-1} R_i f_i, \ f_i \in H_i\}} \sum_{i=0}^{m-1} \inf_{\{\mathbf{c}^{(i)} \in \ell_2(\mathcal{I}^{(i)}), \ (\mathbf{c}^{(i)})^\top \mathcal{F}^{(i)} = f_i\}} \|\mathbf{c}^{(i)}\|_{\ell_2(\mathcal{I}^{(i)})}^2$$

$$= \inf_{\{f = \sum_{i=0}^{m-1} R_i f_i, \ f_i = (\mathbf{c}^{(i)})^\top \mathcal{F}^{(i)}, \ \mathbf{c} = (\mathbf{c}^{(0)}, \ldots, \mathbf{c}^{(m-1)}) \in \ell_2(\mathcal{I})\}} \sum_{i=0}^{m-1} \|\mathbf{c}^{(i)}\|_{\ell_2(\mathcal{I}^{(i)})}^2$$

$$= \inf_{\{f = \sum_{i=0}^{m-1} R_i ((\mathbf{c}^{(i)})^\top \mathcal{F}^{(i)}), \ \mathbf{c} = (\mathbf{c}^{(0)}, \ldots, \mathbf{c}^{(m-1)}) \in \ell_2(\mathcal{I})\}} \sum_{i=0}^{m-1} \|\mathbf{c}^{(i)}\|_{\ell_2(\mathcal{I}^{(i)})}^2$$

$$= \inf_{\{f = \sum_{i=0}^{m-1} (\mathbf{c}^{(i)})^\top R_i \mathcal{F}^{(i)}, \ \mathbf{c} = (\mathbf{c}^{(0)}, \ldots, \mathbf{c}^{(m-1)}) \in \ell_2(\mathcal{I})\}} \|\mathbf{c}\|_{\ell_2(\mathcal{I})}^2$$

$$= \inf_{\{\mathbf{c}^\top \mathcal{F} = f, \ \mathbf{c} \in \ell_2(\mathcal{I})\}} \|\mathbf{c}\|_{\ell_2(\mathcal{I})}^2,$$

for all $f \in H$, in which also the boundedness of the R_i has been used in the fourth step. Now, if (2.3.1) holds and R is onto, we can write any $f \in H$ according to

$$
f = \sum_{i=0}^{m-1} R_i f_i = \sum_{i=0}^{m-1} R_i \sum_{j \in \mathcal{I}^{(i)}} \langle f_i, \tilde{g}_j^{(i)} \rangle_{H_i} g_j^{(i)} = \sum_{i=0}^{m-1} \sum_{j \in \mathcal{I}^{(i)}} \langle f_i, \tilde{g}_j^{(i)} \rangle_{H_i} R_i g_j^{(i)},
$$

where we have utilized boundedness of the operators R_i in the last step. Note that all inner sums in the last term converge in H. Hence, $\mathcal{F}$ is also dense, and thus a frame for H.

Conversely, if $\mathcal{F}$ is a frame, F^* must be onto. Thus, for arbitrary $f \in H$, there exists a $\mathbf{c} = (\mathbf{c}^{(0)}, \dots, \mathbf{c}^{(m-1)}) \in \ell_2(\mathcal{I})$ such that

$$
f = F^* \mathbf{c} = \sum_{i=0}^{m-1} (\mathbf{c}^{(i)})^\top R_i \mathcal{F}^{(i)}
$$

$$
= \sum_{i=0}^{m-1} R_i (\mathbf{c}^{(i)})^\top \mathcal{F}^{(i)} = R\left((\mathbf{c}^{(0)})^\top \mathcal{F}^{(0)}, \dots, (\mathbf{c}^{(m-1)})^\top \mathcal{F}^{(m-1)} \right),
$$

showing that R is onto. □

Application to L_2-Sobolev spaces

Let us now analyze how the technique given by Proposition 2.7 can be applied to the case of the Sobolev spaces $H_0^t(\Omega)$, $t \geq 0$. Naturally, the outline is to identify suitable spaces H_i. The idea is to decompose the underlying domain Ω into finitely many *overlapping* subdomains (we also say patches) Ω_i, i.e.,

$$
\Omega = \bigcup_{i=0}^{m-1} \Omega_i, \quad m \in \mathbb{N}, \tag{2.3.3}
$$

and to choose $H_i := H_0^t(\Omega_i)$. The basic tool, which is used to verify that these spaces make up a stable space splitting, are special partitions of unity.

Definition 2.7. A collection of functions $\{\sigma_i\}_{i=0}^{m-1}$, $\sigma_i : \Omega \to \mathbb{R}$ is called a *partition of unity subordinate to the covering* $\{\Omega_i\} := \{\Omega_i\}_{i=0}^{m-1}$, $\Omega = \bigcup_{i=0}^{m-1} \Omega_i$, if for a given $t \geq 0$

(i) $\operatorname{supp} \sigma_i \subset \overline{\Omega_i}$, $\sum_{i=0}^{m-1} \sigma_i \equiv 1$,

(ii) $\sigma_i v \in H_0^t(\Omega_i)$, $v \in H_0^t(\Omega)$,

(iii) $\|\sigma_i v\|_{H^t(\Omega_i)} \lesssim \|v\|_{H^t(\Omega)}$, $v \in H_0^t(\Omega)$.

Naturally, the spaces $H_0^t(\Omega_i)$ are continuously embedded into $H_0^t(\Omega)$ by means of the zero extension E_i, thus we set $R_i := E_i$.

Proposition 2.8. *Let $\Psi^{(i)}$, $i = 0, \dots, m-1$, be a Riesz basis or a frame for $H_0^t(\Omega_i)$. Furthermore, assume that a partition of unity subordinate to the covering $\{\Omega_i\}$ from (2.3.3) exists. Then, the collection*

$$\Psi := \bigcup_{i=0}^{m-1} E_i \Psi^{(i)} \tag{2.3.4}$$

is a frame for $H_0^t(\Omega)$.

Proof. We have to prove that the spaces $H_0^t(\Omega_i)$ give rise to a stable space splitting as in Definition 2.6, because in that case the assertion follows from Proposition 2.7. We show that R is onto first. Naturally, any $v \in H_0^t(\Omega)$ can be decomposed according to

$$v = \sum_{i=0}^{m-1} \sigma_i v = \sum_{i=0}^{m-1} E_i (\sigma_i v)|_{\Omega_i}, \quad (\sigma_i v)|_{\Omega_i} \in H_0^t(\Omega_i). \tag{2.3.5}$$

Thus, $R : \bar{H} \to H_0^t(\Omega)$ is onto, where $\bar{H}$ denotes the Hilbert sum of the spaces $H_0^t(\Omega_i)$, $i = 0, \dots, m-1$.

Moreover, we have

$$\sum_{i=0}^{m-1} \|\sigma_i v\|_{H^t(\Omega_i)}^2 \lesssim \|v\|_{H^t(\Omega)}^2, \quad v \in H_0^t(\Omega), \tag{2.3.6}$$

by Definition 2.7 (iii). This implies

$$\inf_{\{v_i \in H_0^t(\Omega_i), \, v = \sum_{i=0}^{m-1} E_i v_i\}} \sum_{i=0}^{m-1} \|v_i\|_{H^t(\Omega_i)}^2 \lesssim \|v\|_{H^t(\Omega)}^2, \, v \in H_0^t(\Omega).$$

The converse estimate in (2.3.1) is an immediate consequence of the triangle inequality, the boundedness of the E_i, and $\sum_{i=0}^{m-1} \|\cdot\|_{H^t(\Omega_i)} \approx \left(\sum_{i=0}^{m-1} \|\cdot\|_{H^t(\Omega_i)}^2 \right)^{1/2}$, so that the proof is completed. $\qquad\square$

Obviously, in this particular situation, if the local collections $\Psi^{(i)}$ are chosen to be Riesz bases, the redundancy of the collection Ψ is induced by the fact that in the overlapping regions of the patches Ω_i the elements of Ψ will be no longer linearly independent.

The frame property of the aggregated system Ψ is solely induced by properties of the spaces H_i, which, in the present situation also depend on the covering $\{\Omega_i\}$. The key role is played by the partition of unity, the availability of which is not always obvious for general domain decompositions. Therefore, we will separately address this question in §2.7. For the rest of §2.3 we will tacitly assume that suitable partitions are at our disposal.

It is also important to note that, apart from requiring that $\Psi^{(i)}$ is a frame for $H_0^t(\Omega_i)$, no further assumptions have to be imposed on the elements themselves, meaning that the construction is fairly generic.

In order to construct the *reference bases or frames* $\Psi^{(i)}$ in practice, one may proceed as follows. For $0 \leq i \leq m-1$, let

$$\kappa_i : (0,1)^n \to \Omega_i \qquad (2.3.7)$$

be smooth parametrizations of Ω_i with respect to the open unit cube $\square := (0,1)^n$. Additionally, we assume κ_i to be a C^k-diffeomorphism with $k \geq t$, $i = 0, \ldots, m-1$, satisfying

$$|\det D\kappa_i(x)| \eqsim 1, \quad x \in \square.$$

If for $i = 0, \ldots, m-1$, $\Psi_i^\square$ are (possibly different) Riesz bases or frames for $H_0^t(\square)$, then, by setting

$$\Psi^{(i)} := \Psi_i^\square \circ \kappa_i^{-1}, \qquad (2.3.8)$$

we obtain a Riesz basis or a frame for $H_0^t(\Omega_i)$, respectively; cf. Figure 2.1.

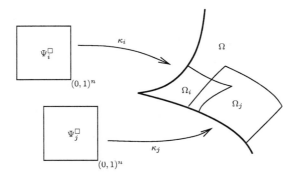

Figure 2.1: Construction of an aggregated frame based on an overlapping domain decomposition.

2.3.2 Aggregated wavelet frames

Let us now turn to the wavelet case. The systems $\Psi_i^\square$ can be constructed as outlined in §2.2.4 and §2.2.5. In particular, we can construct a wavelet Riesz basis $\Psi_{i,L_2}^\square = \{\psi_\mu^\square\}_{\mu \in \Lambda_i^\square}$ for $L_2(\square)$. For an $i \in \{0, \ldots, m-1\}$, $\Lambda_i^\square$ denotes the set of wavelet indices corresponding to $\Psi_{i,L_2}^\square$, see (2.2.35), encoding level, type, and spatial location of the wavelets. For some considerations in the sequel, a proper normalization of the frame elements will be beneficial. In particular, we set

$$\psi_{i,\mu}(x) := \frac{\psi_\mu^\square(\kappa_i^{-1}(x))}{\left| \det D\kappa_i(\kappa_i^{-1}(x)) \right|^{1/2}}, \qquad x \in \Omega_i, \quad \lambda := (i,\mu) \in \Lambda, \tag{2.3.9}$$

where Λ represents the global index set as in Proposition 2.7, i.e.,

$$\Lambda := \bigcup_{i=0}^{m-1} \{i\} \times \Lambda_i^\square. \tag{2.3.10}$$

Furthermore, we lift the dual bases of $\Psi_{i,L_2}^\square$ in $L_2(\square)$, which we denote with $\tilde{\Psi}_{i,L_2}^\square = \{\tilde{\psi}_\mu^\square\}_{\mu \in \Lambda_i^\square}$, to Ω_i by setting

$$\tilde{\psi}_{i,\mu}(x) := \frac{\tilde{\psi}_\mu^\square(\kappa_i^{-1}(x))}{\left| \det D\kappa_i(\kappa_i^{-1}(x)) \right|^{1/2}}, \qquad x \in \Omega_i, \quad \lambda = (i,\mu) \in \Lambda. \tag{2.3.11}$$

It is easy to check that by this procedure a pair of biorthogonal Riesz bases $(\Psi^{(i)}, \tilde{\Psi}^{(i)})$,

$$\Psi^{(i)} := \{\psi_{i,\mu} : \mu \in \Lambda_i^\square\} \text{ and } \tilde{\Psi}^{(i)} := \{\tilde{\psi}_{i,\mu} : \mu \in \Lambda_i^\square\}, \tag{2.3.12}$$

for $L_2(\Omega_i)$ is obtained. Moreover, it holds $\|\psi_{i,\mu}\|_{L_2(\Omega_i)} = \|\psi_\mu^\square\|_{L_2(\square)}$, $\|\tilde{\psi}_{i,\mu}\|_{L_2(\Omega_i)} = \|\tilde{\psi}_\mu^\square\|_{L_2(\square)}$. We define

$$\Psi_{L_2} := \{\psi_\lambda : \lambda \in \Lambda\}, \quad \psi_{(i,\mu)} := E_i \psi_{i,\mu}, \text{ for } (i,\mu) \in \Lambda, \tag{2.3.13}$$

which, by Proposition 2.8, is a frame for $L_2(\Omega)$.

In the sequel, with $|\lambda| := |\mu|$ for $\lambda = (i,\mu) \in \Lambda$, again the level of a wavelet index will be addressed. Furthermore, with

$$p(\lambda) := i \tag{2.3.14}$$

we shall denote the number of the patch appearing in $\lambda = (i,\mu)$, and with

$$e(\lambda) := e(\mu) = \mathbf{e}, \text{ for } \mu = (j,\mathbf{e},\mathbf{k}) \in \Lambda_i^\square, \tag{2.3.15}$$

the type of the wavelet. Moreover, Riesz bases for $H_0^t(\square)$ can be obtained just by scaling $\Psi_{i,L_2}^\square$ according to Proposition 2.6, i.e., by setting $\Psi_i^\square := \mathbf{D}_i^{-1} \Psi_{i,L_2}^\square$, $\mathbf{D}_i = \mathrm{diag}(2^{t|\mu|})_{\mu \in \Lambda_i^\square}$.

An application of Proposition 2.8 yields that the scaled collection

$$\Psi := \bigcup_{i=0}^{m-1} E_i \left(\frac{\Psi_i^\square \circ \kappa_i^{-1}}{\left| \det D\kappa_i \circ \kappa_i^{-1} \right|^{1/2}} \right) = \{2^{-t|\lambda|} \psi_\lambda : \lambda \in \Lambda\} = \mathbf{D}^{-1} \Psi_{L_2}, \tag{2.3.16}$$

where $\mathbf{D} = \mathrm{diag}(\mathbf{D}_0, \ldots, \mathbf{D}_{m-1})$, forms a frame for $H_0^t(\Omega)$.

The frame property for the aggregated system $\Psi_{L_2} = \bigcup_{i=0}^{m-1} E_i \Psi^{(i)}$ in $L_2(\Omega)$ can also be shown to hold without the aid of a partition of unity.

Lemma 2.1. *The aggregated system* Ψ_{L_2} *is a frame for* $L_2(\Omega)$.

Proof. For $v \in L_2(\Omega)$, a transformation of coordinates implies

$$\|v\|_{L_2(\Omega_i)}^2 = \left\| v \circ \kappa_i(\cdot) | \det D\kappa_i(\cdot) |^{1/2} \right\|_{L_2(\square)}^2.$$

Inserting the frame condition for $\Psi_{i,L_2}^{\square}$ in $L_2(\square)$, we get the frame condition for $\Psi^{(i)}$ in $L_2(\Omega_i)$

$$\|v\|_{L_2(\Omega_i)}^2 \approx \sum_{\mu \in \Lambda_i^{\square}} \left| \left\langle v \circ \kappa_i(\cdot) | \det D\kappa_i(\cdot) |^{1/2}, \psi_\mu^{\square} \right\rangle_{L_2(\square)} \right|^2 = \sum_{\mu \in \Lambda_i^{\square}} \left| \langle v, \psi_{i,\mu} \rangle_{L_2(\Omega_i)} \right|^2.$$

$$(2.3.17)$$

Using the inequalities $\|v\|_{L_2(\Omega)}^2 \leq \sum_{i=0}^{m-1} \|v\|_{L_2(\Omega_i)}^2 \leq m\|v\|_{L_2(\Omega)}^2$, the frame condition for Ψ_{L_2} follows by summing up (2.3.17) over i. $\qquad\square$

2.3.3 Gelfand frames

Motivation

In the classical wavelet setting, (2.2.24) has shown that the expansion coefficients of any $v \in H_0^t(\Omega)$ with respect to the scaled basis $\Psi_{\mathbf{D}}$ for $H_0^t(\Omega)$ (cf. Proposition 2.6) can be obtained by computing the weighted L_2-inner products of v with respect to the L_2-dual wavelets of the initial basis Ψ for $L_2(\Omega)$. Naturally, the question arises whether a similar statement holds for aggregated wavelet frames. It turns out that this is the case if an appropriate dual frame for Ψ_{L_2} from §2.3.2 in $L_2(\Omega)$ can be shown to exist, as it will now be explained.

To this end, for a diagonal matrix $D = \mathrm{diag}(d_i)_{i \in \mathcal{I}}$, with $d_i > 0$, let us define the weighted ℓ_2-spaces

$$\ell_{2,D}(\mathcal{I}) := \{ \mathbf{c} = \{c_i\}_{i \in \mathcal{I}} \in \ell_2(\mathcal{I}) : \|D\mathbf{c}\|_{\ell_2(\mathcal{I})}^2 = \sum_{i \in \mathcal{I}} d_i^2 |c_i|^2 < \infty \}. \qquad (2.3.18)$$

In the case of wavelets, for a $t > 0$, we denote the spaces corresponding to $\mathbf{D} = \mathrm{diag}(2^{t|\lambda|})_{\lambda \in \Lambda}$ by $\ell_{2,2^t}(\Lambda) := \ell_{2,\mathbf{D}}(\Lambda)$. Note that the dual space $(\ell_{2,D}(\Lambda))'$ is always given by $\ell_{2,D^{-1}}(\Lambda)$, and generally, $D : \ell_{2,D}(\mathcal{I}) \to \ell_2(\mathcal{I})$ as well as $D^* = D : \ell_2(\mathcal{I}) \to \ell_{2,D^{-1}}(\mathcal{I})$ are isomorphisms.

Lemma 2.2. *Let* $\Psi_{L_2} = \{\psi_\lambda\}_{\lambda \in \Lambda}$ *be a frame for* $L_2(\Omega)$ *as constructed in* §2.3.2. *Let further* $\tilde{\Psi}_{L_2} = \{\tilde{\psi}_\lambda\}_{\lambda \in \Lambda}$ *be a dual frame in* $L_2(\Omega)$. *And let* F^* *be the synthesis operator with respect to* Ψ_{L_2} *and* $\tilde{F}$ *the analysis operator with respect to* $\tilde{\Psi}_{L_2}$. *Then, if*

$$F^* : \ell_{2,2^t}(\Lambda) \to H_0^t(\Omega), \quad F^*\mathbf{c} = \mathbf{c}^\top \Psi_{L_2}, \quad \tilde{F} : H_0^t(\Omega) \to \ell_{2,2^t}(\Lambda), \quad \tilde{F}v = \langle v, \tilde{\Psi}_{L_2} \rangle_{L_2(\Omega)}$$

are well-defined and bounded operators, the relation

$$v = F^* \tilde{F} v = \sum_{\lambda \in \Lambda} \langle v, \tilde{\psi}_\lambda \rangle_{L_2(\Omega)} \psi_\lambda = \sum_{\lambda \in \Lambda} 2^{t|\lambda|} \langle v, \tilde{\psi}_\lambda \rangle_{L_2(\Omega)} 2^{-t|\lambda|} \psi_\lambda \qquad (2.3.19)$$

holds for all $v \in H_0^t(\Omega)$ with convergence of the series in $H_0^t(\Omega)$. Moreover, one has the norm equivalence

$$\|v\|_{H^t(\Omega)} \approx \left(\sum_{\lambda \in \Lambda} 2^{2t|\lambda|} |\langle v, \tilde{\psi}_\lambda \rangle_{L_2(\Omega)}|^2 \right)^{1/2}, \quad v \in H_0^t(\Omega). \qquad (2.3.20)$$

Proof. Let $v \in H_0^t(\Omega)$. Since Ψ_{L_2} is a frame for $L_2(\Omega)$, we have $v = F^* \tilde{F} v = \sum_{\lambda \in \Lambda} \langle v, \tilde{\psi}_\lambda \rangle_{L_2(\Omega)} \psi_\lambda$ with convergence of the series in $L_2(\Omega)$. By the assumed boundedness of $\tilde{F}$ and F^*, the series must also converge in $H_0^t(\Omega)$, completing the proof of (2.3.19). Furthermore, by

$$\|v\|_{H^t(\Omega)} = \|F^* \tilde{F} v\|_{H^t(\Omega)} \lesssim \|\tilde{F} v\|_{\ell_{2,2^t}(\Lambda)} \lesssim \|v\|_{H^t(\Omega)}$$

(2.3.20) is proved. □

Definition and properties

This observation leads to the following general definition; cf. [38].

Definition 2.8. Let $H \subset V = V' \subset H'$ be a Gelfand triple of Hilbert spaces. Then, a frame $\mathcal{F} = \{f_i\}_{i \in \mathcal{I}}$ for V with the dual frame $\tilde{\mathcal{F}} = \{\tilde{f}_i\}_{i \in \mathcal{I}}$ is called *Gelfand frame* for the Gelfand triple (H, V, H') if $\mathcal{F} \subset H$, $\tilde{\mathcal{F}} \subset H'$, and there exists a Gelfand triple of sequence spaces $(b, \ell_2(\mathcal{I}), b')$ such that

$$F^* : b \to H, \ F^* \mathbf{c} = \mathbf{c}^\top \mathcal{F} \quad \text{and} \quad \tilde{F} : H \to b, \ \tilde{F} f = \{\langle f, \tilde{f}_i \rangle_{H \times H'}\}_{i \in \mathcal{I}} \qquad (2.3.21)$$

are bounded operators. We further require that an isomorphism $D : b \to \ell_2(\Lambda)$ exists such that its adjoint is an isomorphism $D^* : \ell_2(\Lambda) \to b'$.

Assumption 2.1. In what follows we shall always assume that D is a diagonal matrix as in (2.3.18) with $D = D^*$.

By duality, also the operators

$$\tilde{F}^* : b' \to H', \ \tilde{F}^* \mathbf{c} = \mathbf{c}^\top \tilde{\mathcal{F}} \quad \text{and} \quad F : H' \to b', \ F f = \{\langle f, f_i \rangle_{H' \times H}\}_{i \in \mathcal{I}} \qquad (2.3.22)$$

are bounded. The relation of the involved spaces and operators in the latter definition is clarified by the diagram given in Figure 2.2.

Thus, in Lemma 2.2 we have assumed Ψ_{L_2} to be a Gelfand frame for the triple $(H_0^t(\Omega), L_2(\Omega), H^{-t}(\Omega))$ with respect to the Gelfand triple of sequence spaces

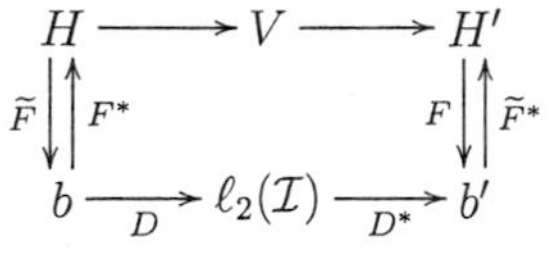

$$
\begin{array}{ccc}
H & \longrightarrow V & \longrightarrow H' \\
\tilde{F} \Big\| \Big\downarrow F^* & & F \Big\| \Big\downarrow \tilde{F}^* \\
b & \xrightarrow{\ \ D\ \ } \ell_2(\mathcal{I}) \xrightarrow{\ \ D^*\ \ } & b'
\end{array}
$$

Figure 2.2: Spaces and operators associated with a Gelfand frame.

$(\ell_{2,2^t}(\Lambda), \ell_2(\Lambda), \ell_{2,2^{-t}}(\Lambda))$, which obviously represents the right assumption to obtain (2.2.23) and (2.2.24) also for aggregated wavelet frames.

Let us collect in the sequel some important properties of a Gelfand frame. Firstly, for a general Gelfand frame, the reconstruction formulas

$$
f = \sum_{i \in \mathcal{I}} \langle f, \tilde{f}_i \rangle_{H \times H'} f_i, \quad f \in H \tag{2.3.23}
$$

$$
g = \sum_{i \in \mathcal{I}} \langle g, f_i \rangle_{H' \times H} \tilde{f}_i, \quad g \in H' \tag{2.3.24}
$$

can be shown to hold analogous to the proof Lemma 2.2 with convergence of the series in H and H', respectively. That means, Gelfand frames allow for expansions of the form (2.1.6) in the space V as well as in the densely embedded Hilbert space H. What is more, the scaled version $D^{-1}\mathcal{F}$ of the frame $\mathcal{F}$ forms a frame for the smaller space H.

Proposition 2.9. *Let $\mathcal{F} = \{f_i\}_{i \in \mathcal{I}}$ be a Gelfand frame as in Definition 2.8, D as in (2.3.18), and $b = \ell_{2,D}(\mathcal{I})$, $b' = \ell_{2,D^{-1}}(\mathcal{I})$. Then, the following statements hold.*

(i) *The collection $\mathcal{G} := D^{-1}\mathcal{F} = \{d_i^{-1} f_i\}_{i \in \mathcal{I}} \subset H$ is a Hilbert frame for H, and $\tilde{\mathcal{G}} := D\tilde{\mathcal{F}} = \{d_i \tilde{f}_i\}_{i \in \mathcal{I}} \subset H'$ is a Hilbert frame for H'.*

(ii) *The operator $\mathbf{P} = D\tilde{F} F^* D^{-1} : \ell_2(\mathcal{I}) \to \ell_2(\mathcal{I})$ is a bounded projector with $\operatorname{ran}(\mathbf{P}) = \operatorname{ran}(D\tilde{F})$ and $\ker(\mathbf{P}) = \ker(F^* D^{-1})$.*

Proof. For a proof of (i) it is sufficient to show that the operators $F_{\mathcal{G}}^* : \ell_2(\mathcal{I}) \to H$, $\mathbf{c} \mapsto \mathbf{c}^\top (D^{-1}\mathcal{F})$ and $\tilde{F}_{\mathcal{G}}^* : \ell_2(\mathcal{I}) \to H'$, $\mathbf{c} \mapsto \mathbf{c}^\top (D\tilde{\mathcal{F}})$ are well-defined and onto (cf. Theorem 2.3). If $\mathbf{c} = \{c_i\}_{i \in \mathcal{I}} \in \ell_2(\mathcal{I})$, then clearly it is $\{d_i^{-1} c_i\}_{i \in \mathcal{I}} \in \ell_{2,D}(\mathcal{I})$ and $F_{\mathcal{G}}^* \mathbf{c} = \sum_{i \in \mathcal{I}} c_i (d_i^{-1} f_i) = F^*\{d_i^{-1} c_i\}$. Hence, by (2.3.21), $F_{\mathcal{G}}^*$ is well-defined and even bounded. The reconstruction formula (2.3.23) yields

$$
f = \sum_{i \in \mathcal{I}} \langle f, d_i \tilde{f}_i \rangle_{H \times H'} d_i^{-1} f_i, \quad \text{for all } f \in H. \tag{2.3.25}
$$

Thus, $F_{\mathcal{G}}^*$ is onto. $\tilde{F}_{\mathcal{G}}^*$ can be treated analogously. Concerning (ii), note that by (2.3.23) $F^* : b \to H$ is onto and $\tilde{F} : H \to b$ injective, hence $\operatorname{ran}(\mathbf{P}) = \operatorname{ran}(D\tilde{F})$ and $\ker(\mathbf{P}) = \ker(F^* D^{-1})$. Again by (2.3.23), $\mathbf{P}$ is a projector. $\qquad \square$

Remark 2.2. The reconstruction formula (2.3.25) shows that the collection $D\tilde{\mathcal{F}} \subset H'$ forms a dual collection of coefficient functionals for $D^{-1}\mathcal{F} \subset H$.

Proposition 2.10. *The Hilbert frames $\mathcal{G} = D^{-1}\mathcal{F} \subset H$, and $\tilde{\mathcal{G}} = D\tilde{\mathcal{F}} \subset H'$ are Banach frames in the sense of §2.1.3 for H' and H, respectively. The associated sequence space is $\ell_2(\mathcal{I})$ in both cases.*

Proof. We show the result for $D^{-1}\mathcal{F} \subset H$. The rest of the claim follows by analogous dual arguments. $D^{-1}\mathcal{F} \subset H'' = H$, and $F_{\mathcal{G}} : H' \to \ell_2(\mathcal{I})$, $F_{\mathcal{G}}f = \{\langle f, d_i^{-1}f_i\rangle_{H' \times H}\}_{i \in \mathcal{I}}$ is bounded because $F_{\mathcal{G}} = (D^*)^{-1}F$ with F from (2.3.22), and $D = D^* = \mathrm{diag}(d_i)_{i \in \mathcal{I}}$. Moreover, using (2.3.24), we get

$$\|f\|_{H'} = \|\tilde{F}^*D^*(D^*)^{-1}Ff\|_{H'} \lesssim \|(D^*)^{-1}Ff\|_{\ell_2(\mathcal{I})} = \|F_{\mathcal{G}}f\|_{\ell_2(\mathcal{I})} \lesssim \|f\|_{H'}, \quad (2.3.26)$$

for all $f \in H'$. That means (2.1.11) is verified. Finally, the existence of a bounded reconstruction operator from $\ell_2(\mathcal{I})$ to H' is ensured by $\tilde{F}^*D^*F_{\mathcal{G}}f = f$, $f \in H'$. $\qquad \square$

Remark 2.3. We emphasize that (2.3.26) shows the equivalence

$$\|f\|_{H'}^2 \approx \sum_{i \in \mathcal{I}} |f(d_i^{-1}f_i)|^2, \quad f \in H', \qquad (2.3.27)$$

which is (2.1.13) for $\mathcal{G} = D^{-1}\mathcal{F}$.

Back to the wavelet case

Now, for the verification of the assumptions in Lemma 2.2, a suitable dual frame has to be detected. Although the systems $\tilde{\Psi}^{(i)}$ (recall (2.3.12)) form a dual basis for $\Psi^{(i)}$ in $L_2(\Omega_i)$, the collection $\bigcup_{i=0}^{m-1} E_i\tilde{\Psi}^{(i)}$ obviously fails to be a dual frame for Ψ_{L_2}. Nevertheless, taking $\sigma_i\tilde{\Psi}^{(i)}$ instead, indeed leads to a dual frame with which Ψ_{L_2} forms a Gelfand frame. In particular, we prove the following statement.

Proposition 2.11. *Let Ψ_{L_2} be defined by (2.3.13), and assume that $\{\sigma_i\}_{i=0}^{m-1}$ is a partition of unity subordinate to the covering $\{\Omega_i\}$. Then, with $\tilde{\psi}_{i,\mu}$ from (2.3.11), the system*

$$\tilde{\Psi}_{L_2} := \{\tilde{\psi}_\lambda : \lambda \in \Lambda\}, \quad \tilde{\psi}_{(i,\mu)} := E_i(\sigma_i\tilde{\psi}_{i,\mu}), \quad (i,\mu) \in \Lambda, \qquad (2.3.28)$$

is a non-canonical global dual frame for Ψ_{L_2} in $L_2(\Omega)$. More specific, Ψ_{L_2} together with $\tilde{\Psi}_{L_2}$ forms a wavelet Gelfand frame for $(H_0^t(\Omega), L_2(\Omega), H^{-t}(\Omega))$ with respect to the Gelfand triple of sequence spaces $(\ell_{2,2^t}(\Lambda), \ell_2(\Lambda), \ell_{2,2^{-t}}(\Lambda))$.

Proof. By the partition of unity property, it is $\|v\|_{L_2(\Omega)}^2 \approx \sum_{i=0}^{m-1} \|\sigma_i v\|_{L_2(\Omega)}^2$. Since the system $\tilde{\Psi}^{(i)} = \{\tilde{\psi}_{i,\mu}\}_{\mu \in \Lambda_i^\square}$ from (2.3.12) forms a Riesz basis and thus a frame for $L_2(\Omega_i)$, this implies the frame property of $\tilde{\Psi}_{L_2}$ for $L_2(\Omega)$,

$$\|v\|_{L_2(\Omega)}^2 \approx \sum_{i=0}^{m-1} \sum_{\mu \in \Lambda_i^\square} |\langle v, \sigma_i\tilde{\psi}_{i,\mu}\rangle_{L_2(\Omega_i)}|^2, \quad v \in L_2(\Omega).$$

The validity of the duality relation (2.1.7) is also induced by the partition property of the σ_i. Indeed, we have that any $v \in L_2(\Omega)$ can be written as

$$
v = \sum_{i=0}^{m-1} \sigma_i v = \sum_{i=0}^{m-1} \sum_{\mu \in \Lambda_i^\square} \langle \sigma_i v, \tilde{\psi}_{i,\mu} \rangle_{L_2(\Omega_i)} \psi_{i,\mu} = \sum_{i=0}^{m-1} \sum_{\mu \in \Lambda_i^\square} \langle v, \sigma_i \tilde{\psi}_{i,\mu} \rangle_{L_2(\Omega_i)} \psi_{i,\mu}
$$

$$
= \sum_{i=0}^{m-1} \sum_{\mu \in \Lambda_i^\square} \langle v, \tilde{\psi}_{(i,\mu)} \rangle_{L_2(\Omega)} \psi_{(i,\mu)} = F^* F_{\tilde{\Psi}_{L_2}} v, \tag{2.3.29}
$$

hence $F^* F_{\tilde{\Psi}_{L_2}} = \mathrm{id}_{L_2(\Omega)}$. Moreover, $\mathrm{id}_{L_2(\Omega)} = \left(F^* F_{\tilde{\Psi}_{L_2}} \right)^* = F_{\tilde{\Psi}_{L_2}}^* F$, thus we have shown that $\tilde{\Psi}_{L_2}$ is a non-canonical dual frame. Concerning the Gelfand frame property, we will show that the operators $F^* : \ell_{2,2^t}(\Lambda) \to H_0^t(\Omega)$, $F^* \mathbf{c} = \mathbf{c}^\top \Psi_{L_2}$, and $\tilde{F} := F_{\tilde{\Psi}_{L_2}} : H_0^t(\Omega) \to \ell_{2,2^t}(\Lambda)$, $\tilde{F}v = \left(\langle v, \tilde{\psi}_\lambda \rangle_{H_0^t(\Omega) \times H^{-t}(\Omega)} \right)_{\lambda \in \Lambda}$ are bounded. By assumption, the local Riesz basis property of $(2^{-t|\mu|} \psi_{i,\mu})_{\mu \in \Lambda_i^\square}$ in $H_0^t(\Omega_i)$ implies that the operators $(F^{(i)})^* : \ell_{2,2^t}(\Lambda_i^\square) \to H_0^t(\Omega_i)$, $(F^{(i)})^* \mathbf{c} = \mathbf{c}^\top \Psi^{(i)}$, are bounded. Since the operators E_i are continuous, for any $\mathbf{c} = (\mathbf{c}^{(0)}, \quad , \mathbf{c}^{(m-1)}) \in \ell_{2,2^t}(\Lambda)$, $\mathbf{c}^{(i)} = \{c_\mu^{(i)}\}_{\mu \in \Lambda_i^\square}$, $i = 0, \ldots, m-1$, one has the representation

$$
F^* \mathbf{c} = \sum_{i=0}^{m-1} \sum_{\mu \in \Lambda_i^\square} c_\mu^{(i)} \psi_{(i,\mu)} = \sum_{i=0}^{m-1} E_i (F^{(i)})^* \mathbf{c}^{(i)},
$$

from which, by applying the continuity of E_i and $(F^{(i)})^*$, follows

$$
\|F^* \mathbf{c}\|_{H^t(\Omega)} \leq \sum_{i=0}^{m-1} \left\| (F^{(i)})^* \mathbf{c}^{(i)} \right\|_{H^t(\Omega_i)} \lesssim \sum_{i=0}^{m-1} \|\mathbf{c}^{(i)}\|_{\ell_{2,2^t}(\Lambda_i^\square)} \approx \|\mathbf{c}\|_{\ell_{2,2^t}(\Lambda)}.
$$

As regards the continuity of $\tilde{F} : H_0^t(\Omega) \to \ell_{2,2^t}(\Lambda)$, we know that again by the Riesz basis property of $(2^{-t|\mu|} \psi_{i,\mu})_{\mu \in \Lambda_i^\square}$ in $H_0^t(\Omega_i)$, the operators $\tilde{F}^{(i)} : H_0^t(\Omega_i) \to \ell_{2,2^t}(\Lambda_i^\square)$, $\tilde{F}^{(i)} v = \left(\langle v, \tilde{\psi}_{i,\mu} \rangle_{H_0^t(\Omega_i) \times H^{-t}(\Omega_i)} \right)_{\mu \in \Lambda_i^\square}$, are bounded. Together with Definition 2.7 it follows

$$
\|\tilde{F}v\|_{\ell_{2,2^t}(\Lambda)}^2 = \sum_{i=0}^{m-1} \|\tilde{F}^{(i)}(\sigma_i v)\|_{\ell_{2,2^t}(\Lambda_i^\square)}^2 \lesssim \sum_{i=0}^{m-1} \|\sigma_i v\|_{H^t(\Omega_i)}^2 \lesssim \|v\|_{H^t(\Omega)}^2, \quad v \in H_0^t(\Omega).
$$

$\square$

One of the most important features of the frame construction presented in this section is its simplicity which makes the implementation of these frames rather easy. Once suitable interval bases or frames are available and the parametrizations are implemented, the rest of the construction is straightforward.

2.4 Summary of fundamental wavelet properties

Now that we have available a general and rather easy to realize mechanism for the construction of wavelet frames Ψ for $H_0^t(\Omega)$ or, more special, Gelfand frames for $(H_0^t(\Omega), L_2(\Omega), H^{-t}(\Omega))$ out of wavelets on the interval, we want to collect the basic properties these wavelets expose. The findings in the following chapters will be based on these characteristics.

Let $\Psi := \mathbf{D}^{-1}\Psi_{L_2} =: \{\psi_\lambda\}_{\lambda \in \Lambda}$ be the aggregated wavelet frame for $H_0^t(\Omega)$, $t \in \mathbb{N}_0$, from (2.3.16), based on the interval spline wavelet bases of order $d > 0$, constructed, for instance, in [45, 47], or [95]. Throughout this thesis, we will be mostly concerned with the special setup

$$d \geq t+1 \quad \text{and} \quad \frac{d-t}{n} \geq \frac{1}{2}. \tag{2.4.1}$$

The (already scaled) wavelets ψ_λ are *local* in the sense that

$$\begin{cases} \operatorname{diam}(\operatorname{supp}\psi_\lambda) \lesssim 2^{-|\lambda|} \quad \text{and} \\ \sup_{x \in \Omega, l \geq j_0} \#\{|\lambda| = l : B(x; 2^{-l}) \cap \operatorname{supp}\psi_\lambda \neq \emptyset\} < \infty, \end{cases} \tag{2.4.2}$$

where for an $r > 0$, $B(x; r)$ denotes a ball with radius r around the point $x \in \Omega$.

Let us denote by $\operatorname{sing supp}\psi_\lambda$ the singular support of the wavelet. The wavelets are *piecewise smooth*, meaning that $\operatorname{supp}\psi_\lambda \backslash \operatorname{sing supp}\psi_\lambda$ is the disjoint union of K domains $\Xi_{\lambda,1}, \ldots, \Xi_{\lambda,K}$, with $\bigcup_{i=1}^K \overline{\Xi_{\lambda,i}} = \operatorname{supp}\psi_\lambda$, where $\psi_\lambda|_{\Xi_{\lambda,i}}$ is smooth with, for any $\gamma \in \mathbb{N}_0^n$,

$$\sup_{x \in \Xi_{\lambda,i}} |D^\gamma \psi_\lambda(x)| \lesssim 2^{|\lambda|(\frac{n}{2}+|\gamma|-t)}, \tag{2.4.3}$$

provided that sufficiently many derivatives of κ_i stay bounded.

Furthermore, each set $\kappa_{p(\lambda)}^{-1}(\Xi_{\lambda,i})$ is an n-cube aligned with the Cartesian coordinates, and

$$\left((\psi_\lambda \circ \kappa_{p(\lambda)}) |\det D\kappa_{p(\lambda)}|^{1/2} \right) \big|_{\kappa_{p(\lambda)}^{-1}(\Xi_{\lambda,i})} \in Q_{d-1}, \tag{2.4.4}$$

with Q_{d-1} being the n-fold tensor product of the space of univariate polynomials of degree $d-1$. The wavelets have the maximal global smoothness

$$\psi_\lambda \in C^{d-2}(\Omega), \tag{2.4.5}$$

when $d \geq 2$. The latter property necessitates that on interior boundaries $\partial \Omega_i \cap \Omega$, homogeneous boundary conditions of the maximal order $d-2$ are incorporated in the wavelet construction; cf. (2.2.38). By (2.4.3), this shows that for $k \in [0, d-1]$,

$$\|\psi_\lambda\|_{W_\infty^k(\Omega)} \lesssim 2^{|\lambda|(\frac{n}{2}+k-t)}. \tag{2.4.6}$$

For $\tilde{d} > 0$ from §2.2.4, we have

$$(\psi_\lambda \circ \kappa_{p(\lambda)})|\det D\kappa_{p(\lambda)}|^{1/2} \perp \Pi_{\tilde{d}-1}, \tag{2.4.7}$$

with the exception of the λ with $e(\lambda) = 0$.

After fixing our basic tool, namely (aggregated wavelet) Gelfand frames, we turn to the issue of shifting the elliptic operator equations to a discrete setting, which will later serve as the starting point for their numerical treatment.

2.5 Discretization of operator equations

Let again H be a separable Hilbert space and $\mathcal{L} : H \to H'$ a boundedly invertible operator. We aim at a discretization of an operator equation

$$\mathcal{L}u = f, \tag{2.5.1}$$

for a right-hand side $f \in H'$. Moreover, assume that (H, V, H') is a Gelfand triple of Hilbert spaces and that $\mathcal{F} = \{f_i\}_{i \in \mathcal{I}}$ is a Gelfand frame for (H, V, H') as in Definition 2.8.

Under these assumptions the following lemma holds analogously to the classical case of wavelet bases; cf. [42, §6.2].

Lemma 2.3. *Let F and F^* be the analysis and synthesis operators of $\mathcal{F}$ from (2.3.21) and (2.3.22). Then, the operator*

$$\mathbf{A} := (D^*)^{-1}F\mathcal{L}F^*D^{-1} \tag{2.5.2}$$

is bounded from $\ell_2(\mathcal{I})$ to $\ell_2(\mathcal{I})$. Moreover,

$$\mathbf{A}|_{\mathrm{ran}(\mathbf{A})} : \mathrm{ran}(\mathbf{A}) \to \mathrm{ran}(\mathbf{A}) \tag{2.5.3}$$

is boundedly invertible, and $\mathrm{ran}(\mathbf{A}) = \mathrm{ran}((D^*)^{-1}F)$ as well as $\ker(\mathbf{A}) = \ker(F^*D^{-1})$.

Proof. Clearly, $\mathbf{A}$ is bounded as a composition of bounded operators. Of course D^{-1} and $\mathcal{L}$ are onto, which by (2.3.23) also holds for F^*. Hence, $\mathrm{ran}(\mathbf{A}) = \mathrm{ran}((D^*)^{-1}F)$. Similarly, by (2.3.24) F must be injective. Consequently, $\ker(\mathbf{A}) = \ker(F^*D^{-1})$.

It is easy to see that the range of $(D^*)^{-1}F$ is closed, thus, $\ell_2(\mathcal{I}) = \mathrm{ran}((D^*)^{-1}F) \oplus^\perp \ker(F^*D^{-1}) = \mathrm{ran}(\mathbf{A}) \oplus^\perp \ker(\mathbf{A})$ is an orthogonal decomposition of $\ell_2(\mathcal{I})$. An application of the bounded inverse theorem shows that $\mathbf{A}|_{\mathrm{ran}(\mathbf{A})} : \mathrm{ran}(\mathbf{A}) \to \mathrm{ran}(\mathbf{A})$ is boundedly invertible. $\qquad\square$

Remark 2.4. In case $\mathcal{L}$ is symmetric and positive definite, meaning that $\mathcal{L} = \mathcal{L}'$, $\inf_{0 \neq v \in H}(\mathcal{L}v)(v)/\|v\|_H^2 > 0$, then we have

$$\langle \mathbf{A}\mathbf{v}, \mathbf{v}\rangle_{\ell_2(\mathcal{I})} = \langle (D^*)^{-1}F\mathcal{L}F^*D^{-1}\mathbf{v}, \mathbf{v}\rangle_{\ell_2(\mathcal{I})} = \langle \mathbf{v}, (D^*)^{-1}F\mathcal{L}F^*D^{-1}\mathbf{v}\rangle_{\ell_2(\mathcal{I})}$$
$$= \langle \mathcal{L}F^*D^{-1}\mathbf{v}, F^*D^{-1}\mathbf{v}\rangle_{H' \times H},$$

from which we infer that $\mathbf{A} = \mathbf{A}^*$, and that $\mathbf{A}$ is positive semi-definite. In addition, defining for $i \in \mathcal{I}$, $e_i := \{\delta_{k,i}\}_{k \in \mathcal{I}}$, $\mathbf{A}$ can be considered as a bi-infinite matrix with the entries $\mathbf{A}_{i,j} := e_i^\top \mathbf{A} e_j = (\mathcal{L}(d_j^{-1} f_j))(d_i^{-1} f_i)$, $i, j \in \mathcal{I}$.

Proposition 2.12. *Let $u \in H$ be the unique solution of (2.5.1), and let $\mathbf{Q}$ denote the orthogonal projector onto* $\mathrm{ran}(\mathbf{A})$. *Then, the sequence* $\mathbf{u} = D\tilde{F}u \in \ell_2(\mathcal{I})$ *is a solution of the discrete equation*

$$\mathbf{A}\mathbf{u} = \mathbf{f}, \quad \mathbf{f} = (D^*)^{-1} F f. \tag{2.5.4}$$

$\mathbf{Q}\mathbf{u}$ *represents the unique solution of (2.5.4) in* $\mathrm{ran}(\mathbf{A})$. *The solution set is thus given by* $\mathbf{Q}\mathbf{u} + \ker(\mathbf{A}) = \mathbf{Q}\mathbf{u} + \ker(F^* D^{-1})$. *Moreover, from any solution* $\mathbf{v} \in \ell_2(\mathcal{I})$ *of (2.5.4), the solution $u \in H$ can be reconstructed, i.e.,* $u = F^* D^{-1}\mathbf{v}$.

Proof. Using $F^*\tilde{F} = \mathrm{id}_H$, we get $\mathbf{A}D\tilde{F}u = (D^*)^{-1}F\mathcal{L}u = (D^*)^{-1}Ff = \mathbf{f}$. Thus, $\mathbf{u} = D\tilde{F}u$ is a solution of (2.5.4). Let $\mathbf{v} \in \ell_2(\mathcal{I})$ solve (2.5.4), so that $(D^*)^{-1}F\mathcal{L}F^*D^{-1}\mathbf{v} = (D^*)^{-1}Ff$. The injectivity of $(D^*)^{-1}F$ shows that $F^*D^{-1}\mathbf{v}$ solves (2.5.1). The remaining statements are clear. $\qquad\square$

Example 2.1. *Let us consider the elliptic operator $\mathcal{L} : H_0^t(\Omega) \to H^{-t}(\Omega)$ from (1.3.5), with $\langle \mathcal{L}u, v \rangle_{H^{-t}(\Omega) \times H_0^t(\Omega)} = a(u, v)$ defined in (1.3.1). For the aggregated wavelet Gelfand frame Ψ_{L_2} from Proposition 2.11 we obtain*

$$\mathbf{A} = \{2^{-t(|\lambda|+|\mu|)} a(\psi_\lambda, \psi_\mu)\}_{\mu, \lambda \in \Lambda} \quad and \quad \mathbf{f} = \{2^{-t|\lambda|} \langle f, \psi_\lambda \rangle_{L_2(\Omega)}\}_{\lambda \in \Lambda}. \tag{2.5.5}$$

In the following, some comments on the orthogonal projector $\mathbf{Q}$ and its matrix representation shall be added. Recall from Proposition 2.9 (i) that $\mathcal{G} = D^{-1}\mathcal{F}$ is a frame for H with the dual collection $\tilde{\mathcal{G}} = D\tilde{\mathcal{F}}$, being a frame for H'. Recall also the equivalence (2.3.27) stated in Remark 2.3, and consider the operators

$$F_{\mathcal{G}} := (D^*)^{-1}F : H' \to \ell_2(\mathcal{I}) \quad and \quad F_{\mathcal{G}}' := F^*D^{-1} : \ell_2(\mathcal{I}) \to H. \tag{2.5.6}$$

It is easy to check that indeed $F_{\mathcal{G}}'$ is the dual operator of $F_{\mathcal{G}}$ (identifying $\ell_2(\Lambda)$ with its dual). Using Assumption 2.1, we infer that $F_{\mathcal{G}}$ and $F_{\mathcal{G}}'$ represent the analysis and synthesis operators of the Hilbert frame $\mathcal{G}$ for H in the sense of §2.1.4, respectively. As mentioned in §2.1.4, the orthogonal projection $\mathbf{Q} : \ell_2(\mathcal{I}) \to \ell_2(\mathcal{I})$ onto $\mathrm{ran}(F_{\mathcal{G}}) = \mathrm{ran}((D^*)^{-1}F) = \mathrm{ran}(\mathbf{A})$ is given by $\mathbf{Q} = F_{\mathcal{G}}(F_{\mathcal{G}}'F_{\mathcal{G}})^{-1}F_{\mathcal{G}}' = \tilde{F}_{\mathcal{G}}F_{\mathcal{G}}'$. This means that the matrix representation of $\mathbf{Q}$ reads as

$$\mathbf{Q} = \{\langle d_i^{-1} f_i, (F_{\mathcal{G}}'F_{\mathcal{G}})^{-1}(d_j^{-1} f_j)\rangle_{H \times H'}\}_{j, i \in \mathcal{I}}. \tag{2.5.7}$$

Unfortunately, in general the canonical dual elements $(F_{\mathcal{G}}'F_{\mathcal{G}})^{-1}(d_j^{-1} f_j) \in H'$, needed to compute $\mathbf{Q}$, are not explicitly given, so that the verification of properties of $\mathbf{Q}$ other than boundedness on $\ell_2(\mathcal{I})$ is often rather difficult. We will review this matter in Chapter 4.

Remark 2.5. Proposition 2.9 (ii) shows that if one takes any V-dual frame $\tilde{\mathcal{F}} = \{\tilde{f}_i\}_{i \in \mathcal{I}} \subset V$ of a Gelfand frame $\mathcal{F} = \{f_i\}_{i \in \mathcal{I}} \subset V$, then by $\mathbf{P} = \{\langle d_i^{-1} f_i, d_j \tilde{f}_j \rangle\}_{H \times H'}\}_{j,i \in \mathcal{I}}$ a projector is obtained that has the same kernel as $\mathbf{Q}$. Let $\tilde{\mathcal{F}}_{\mathrm{can}}$ be the *canonical V-dual* of $\mathcal{F}$ in $V = V'$. There is no evidence that $D\tilde{\mathcal{F}}_{\mathrm{can}} \subset H'$ actually coincides with the canonical H-dual frame $\{(F_{\mathcal{G}}' F_{\mathcal{G}})^{-1} (d_i^{-1} f_i)\}_{i \in \mathcal{I}} \subset H'$ of $D^{-1} \mathcal{F} \subset H$.

Remark 2.6. In [104] the discretization of $\mathcal{L}$ is directly performed with respect to a frame for H for which (2.1.13) is assumed, but which does not have to evolve from a frame for a space V in which H is densely embedded. In this sense, the approach in [104] is even more general. However, when dealing with wavelets, as Lemma 2.2 has shown, the notion of Gelfand frames seems to be the natural generalization of the classical wavelet basis setting to the case of frames. Now, Proposition 2.9 (i) and (2.3.27) in Remark 2.3 show that we also have created a frame for H, namely $\mathcal{G}$, for which (2.1.13) holds. In this sense, the proceeding in [104] and the Gelfand frame approach fit together.

Summarizing, we can say that we have now available an easy to implement construction of (wavelet) frames on bounded domains, allowing for a transformation of an elliptic operator equation into a consistent operator equation on ℓ_2, thanks to the stability of the elements in the solution space H. The verification of the sparsity of $\mathbf{A}$ in (2.5.5), postulated as an important feature of the construction at the beginning of this chapter, shall be postponed to Chapter 5.

As mentioned above, one has to make sure that a partition of unity as in Definition 2.7 is available, which is next on our agenda. It will turn out that this is not immediately clear for all domain coverings. Before we discuss this matter, it shall be pointed out in §2.6 below that, with regard to an efficient computation of the entries of $\mathbf{A}$, the decomposition of Ω has to be chosen with some care.

2.6 How to select the domain decomposition

Unfortunately, in practice, the above construction of frames entails the following technical difficulty. Thinking of Example 2.1 again, we learn that usually quadrature is needed to approximate the entries of the stiffness matrix. Now, for the case of wavelets in the overlapping region of the patches, there exist collections of lifted wavelets whose elements are piecewise smooth with respect to images of dyadic square meshes on $(0,1)^n$ (cf. (2.4.4)) with respect to *different* parametrizations $\kappa_i \neq \kappa_j$; see Figure 2.3. Hence, in this case, even for constant coefficients in (1.3.1), quadrature is a potential problem. Fortunately, it has turned out that also this difficulty can be overcome. Indeed, in Chapter 5 the basic results from [108] will be collected, where special quadrature rules have been developed, which can be used for the approximation of the stiffness matrix $\mathbf{A}$ in an adaptive algorithm for the solution of (2.5.4). Using these rules, the overall computational cost of the methods, which usually can be shown to depend linearly on the degrees of freedom, will not be spoiled

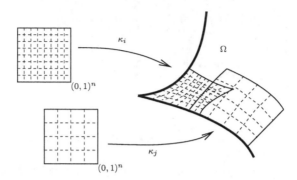

Figure 2.3: Uncorrelated grids in the overlapping region of two subdomains of a domain covering used in the construction of an aggregated wavelet frame.

by the quadrature. However, so far the quantitative performance might not always be satisfactory. Consequently, it is desirable to avoid domain coverings leading to uncorrelated partitions in the overlapping regions whenever this is possible. By this obligation, in the following, we will sometimes be forced to consider special domain coverings $\{\Omega_i\}$, although a different one would be much more comfortable in the theoretical analysis of the properties of an aggregated frame or in the analysis of adaptive frame schemes. We encounter a first instance of such a situation in §2.7. Here we will learn that suitable partitions of unity may be easy to construct for one specific covering of a domain Ω, while for another one, leading, however, to matching grids, this is far less obvious.

2.7 Partitions of unity

We now turn to the issue of constructing partitions of unity as in Definition 2.7. Recall that they form the key tool in the verification of the frame property stated in Proposition 2.8. Let us first of all consider such open coverings $\bigcup_{i=0}^{m-1} \Omega_i = \Omega$ for which the following geometric assumption is satisfied.

Assumption 2.2. The mappings $\kappa_i : (0,1)^n \to \Omega_i$ are smooth parametrizations such that the image of a face of $(0,1)^n$ under κ_i has either empty intersection with $\partial\Omega$ or it is contained in $\partial\Omega$.

For these cases, a straightforward construction of a partition of unity $\{\sigma_i\}_{i=0}^{m-1}$ meeting Definition 2.7, such that even $\sigma_i \in C^\infty(\overline{\Omega})$ holds, has been given in [104, §4.4]. Hence, for these situations we are done.

As a special instance of a covering not meeting Assumption 2.2, consider the two-dimensional L-shaped domain $\Omega := (-1,1)^2 \setminus [0,1)^2$ covered by two rectangles according to $\Omega = \Omega_0 \cup \Omega_1$, $\Omega_0 = (-1,0) \times (-1,1)$, $\Omega_1 = (-1,1) \times (-1,0)$. In view of the discussion in §2.6, this covering is a convenient choice, inducing matching grids. Nevertheless, if we consider any partition $\{\sigma_0, \sigma_1\}$ subordinate to the covering $\{\Omega_0, \Omega_1\}$, σ_0 (and σ_1) will have a discontinuity at the origin, because $\sigma_0|_{(-1,0)\times(0,1)} \equiv 1$ and $\sigma_0|_{(0,1)\times(-1,0)} \equiv 0$. However, it will be proved in the sequel that this discontinuity, occurring at a point on the boundary, where, in addition, homogeneous boundary conditions are prescribed, does not prohibit the properties required in Definition 2.7.

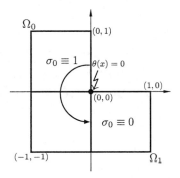

Figure 2.4: The L-shaped domain composed of two overlapping rectangles. A function $\sigma_0 : \Omega \to \mathbb{R}$, satisfying $\sigma_0|_{(-1,0)\times(0,1)} \equiv 1$ and $\sigma_0|_{(0,1)\times(-1,0)} \equiv 0$, is bound to have a discontinuity at the origin.

Indeed, a partition satisfying these conditions can be constructed as follows. Let $\phi : [0, \frac{3\pi}{2}] \to \mathbb{R}$ be a smooth function with $\phi(\theta) = 1$ for $\theta \leq \frac{\pi}{2}$ and $\phi(\theta) = 0$ for $\theta \geq \pi$. Then, with $(r(x), \theta(x))$ denoting polar coordinates of $x \in \Omega$ with respect to the re-entrant corner, the functions $\sigma_0 := \phi \circ \theta$ and $\sigma_1 := 1 - \sigma_0$ form a partition of unity subordinate to the patches Ω_i; see also Figure 2.4.

Lemma 2.4. *For any $u \in H_0^s(\Omega)$, $s \in \mathbb{N}$, it is $\sigma_i u \in H_0^s(\Omega_i)$ and*

$$\|\sigma_i u\|_{H^s(\Omega_i)} \lesssim \|u\|_{H^s(\Omega)}, \quad i = 0, 1. \tag{2.7.1}$$

For the proof of Lemma 2.4, we recall the following theorem from [69].

Theorem 2.7 (see [69, Theorem 1.4.4.4]). *Let Ω be a bounded open subset of $\mathbb{R}^n$ with a Lipschitz boundary Γ, and let $\rho(x)$ denote the distance from a point x to Γ. Then, for all $u \in \mathring{W}_p^s(\Omega)$, $1 \leq p < \infty$ such that $s - \frac{1}{p}$ is not an integer, the following property holds*

$$\rho^{-s+|\alpha|} D^\alpha u \in L_p(\Omega), \quad \text{for all } |\alpha| \leq s. \tag{2.7.2}$$

The proof of this theorem essentially relies on an application of Hardy's inequality. Moreover, what will be most important for later purposes is that it follows from the proof that

$$\|\rho^{-s+|\alpha|} D^\alpha u\|_{L_p(\Omega)} \leq C\|u\|_{W_p^s(\Omega)}, \text{ for all } |\alpha| \leq s, \tag{2.7.3}$$

with a constant $C > 0$ independent of $p \geq 1$.

Proof of Lemma 2.4. It suffices to treat the case $i = 0$, since the claim for $i = 1$ follows analogously. In a first step, we will show that (2.7.1) holds uniformly in $u \in H_0^s(\Omega)$. Afterwards, we verify that also $\sigma_i u \in H_0^s(\Omega_i)$. To this end, for any $\alpha \in \mathbb{N}_0^n$ with $|\alpha| \leq s$, we want to use the multivariate Leibniz rule

$$D^\alpha(\sigma_0 u) = D^\alpha((\phi \circ \theta)u) = \sum_{0 \leq \beta \leq \alpha} \binom{\alpha}{\beta} D^\beta(\phi \circ \theta) D^{\alpha - \beta} u \qquad (2.7.4)$$

which holds as an equality in $L_2(\Omega)$ if the right-hand side is contained in $L_2(\Omega)$. Now, using Theorem 2.7 for $p = 2$, it follows that $\rho^{-s+|\alpha-\beta|} D^{\alpha-\beta} u \in L_2(\Omega)$ with

$$\|\rho^{-s+|\alpha-\beta|} D^{\alpha-\beta} u\|_{L_2(\Omega)} \lesssim \|u\|_{H^s(\Omega)} \qquad (2.7.5)$$

for all $0 \leq \beta \leq \alpha$. It hence remains to prove that the corresponding weak derivatives of the factors $\phi \circ \theta$ in (2.7.4) are compensated by the additional powers of $\rho \leq r$. In particular, we will show that

$$\|D^\beta(\phi \circ \theta) r^{|\beta|}\|_{L_\infty(\Omega_0)} < \infty \qquad (2.7.6)$$

holds for any $\beta \geq 0$. In fact, (2.7.6) and $s - |\alpha - \beta| \geq |\beta|$, for all $0 \leq \beta \leq \alpha$, imply that

$$\|D^\beta(\phi \circ \theta) D^{\alpha-\beta} u\|_{L_2(\Omega_0)} \leq \|D^\beta(\phi \circ \theta) r^{s-|\alpha-\beta|}\|_{L_\infty(\Omega_0)} \|\rho^{-s+|\alpha-\beta|} D^{\alpha-\beta} u\|_{L_2(\Omega)}$$
$$\lesssim \|u\|_{H^s(\Omega)}.$$

In order to verify (2.7.6), we now show that

$$D^\beta(\phi \circ \theta) = (h_\beta \circ \theta) r^{-|\beta|}, \quad \beta \in \mathbb{N}_0^2, \qquad (2.7.7)$$

where $h_\beta \in C^\infty[0, \frac{3}{2}\pi]$, $h_\beta(\theta) = 0$, for $\theta \geq \pi$. In the case of $\beta = 0$, (2.7.7) is ensured by the properties of ϕ. For $|\beta| = 1$, we get

$$\frac{\partial}{\partial x}(\phi \circ \theta)(x, y) = (\phi^{(1)} \circ \theta)(x, y) \frac{\partial}{\partial x} \theta(x, y) = (\phi^{(1)} \circ \theta)(x, y) \frac{\partial}{\partial x} \arctan\left(\frac{y}{x}\right)$$
$$= (\phi^{(1)} \circ \theta)(x, y) \frac{(-\sin \circ \theta)(x, y)}{r(x, y)} = \frac{((-\phi^{(1)} \cdot \sin) \circ \theta)(x, y)}{r(x, y)}.$$

For the partial derivative with respect to y, $-\sin$ is replaced by $\cos$. Hence, (2.7.7) is satisfied with $h_{(1,0)} = -\phi^{(1)} \cdot \sin$ and $h_{(0,1)} = \phi^{(1)} \cdot \cos$. Now let (2.7.7) be true for all $\alpha \in \mathbb{N}_0^2$, $|\alpha| \leq m$, where $m > 0$ and $\beta = (\beta_1, \beta_2) \in \mathbb{N}_0^2$ with $|\beta| = m + 1$. For the case $\beta_1 > 0$ we set $\gamma := (\beta_1 - 1, \beta_2)$ and get

$$D^\beta(\phi \circ \theta)(x, y) = \frac{\partial}{\partial x} D^\gamma(\phi \circ \theta)(x, y) = \frac{\partial}{\partial x}\left((h_\gamma \circ \theta)(x, y) r(x, y)^{-m}\right)$$
$$= (h_\gamma^{(1)} \circ \theta)(x, y) \frac{\partial}{\partial x} \theta(x, y) r(x, y)^{-m}$$
$$- (h_\gamma \circ \theta)(x, y) m r(x, y)^{-(m+1)} \frac{\partial}{\partial x} r(x, y).$$

Together with $\frac{\partial}{\partial x} r(x, y) = (\cos \circ \theta)(x, y)$ one obtains

$$D^\beta(\phi \circ \theta) = \left[\left(-h_\gamma^{(1)} \cdot \sin - m(h_\gamma \cdot \cos)\right) \circ \theta\right] r^{-(m+1)} =: (h_\beta \circ \theta) r^{-(m+1)}.$$

Then, (2.7.7) follows by the induction hypothesis. The case $\beta_2 > 0$ works analogously. Thus, we have verified (2.7.6). This shows that $\sigma_0 u \in H^s(\Omega_0)$ and that (2.7.1) holds uniformly in $u \in H_0^s(\Omega)$. In order to see that $\sigma_0 u \in H_0^s(\Omega_0)$, let us denote with

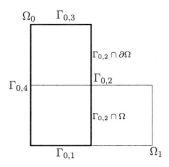

Figure 2.5: Decomposition of the boundary of Ω_0 into four segments.

$\Gamma_{0,1}, \ldots, \Gamma_{0,4}$ the open segments of $\partial \Omega_0$ and with n_l the outward normal with respect to $\Gamma_{0,l}$, $l = 1, \ldots, 4$; cf. Figure 2.5. Moreover, with $\gamma_l v$ we denote the trace of a $v \in H^s(\Omega_0)$ at $\Gamma_{0,l}$. Using this notation it is

$$H_0^s(\Omega_0) = \left\{ v \in H^s(\Omega_0) : \gamma_l \frac{\partial^j v}{\partial n_l^j} = 0, \ 1 \le l \le 4, \ 0 \le j \le s-1 \right\};$$

see [69, Theorem 1.5.2.1, Remark 1.5.2.11]. Since $u \in H_0^s(\Omega)$, we can choose test functions $u_k \in C_0^\infty(\Omega)$, $k \in \mathbb{N}$, with $u_k \to u$ in $H^s(\Omega)$. Now, as a special case, (2.7.4) implies that

$$\frac{\partial^j(\sigma_0 u_k)}{\partial n_l^j} = \sum_{\nu=0}^{j} \binom{j}{\nu} \frac{\partial^\nu(\phi \circ \theta)}{\partial n_l^\nu} \frac{\partial^{j-\nu} u_k}{\partial n_l^{j-\nu}}, \quad j = 0, \ldots, s-1.$$

Since u_k is a test function, it surely holds in particular that $|D^\beta u_k| \le C_k r^s$ for all $|\beta| \le s$. Using (2.7.7) and the boundedness of the restricted test functions $u_k|_{\Omega_0}$ and all its derivatives then follows

$$\gamma_l(\sigma_0 u_k) = \gamma_l \frac{\partial(\sigma_0 u_k)}{\partial n_l} = \cdots = \gamma_l \frac{\partial^{s-1}(\sigma_0 u_k)}{\partial n_l^{s-1}} = 0, \quad l = 1, \ldots, 4.$$

Thus, it is $\sigma_0 u_k \in H_0^s(\Omega_0)$ for all $k \in \mathbb{N}$. Combining this with (2.7.1) and the boundedness of the trace operator from $H^{s-j}(\Omega_0)$ to $H^{s-j-1/2}(\partial \Omega_0)$, $j = 0, \ldots, s-1$, we finally arrive at

$$\left\| \frac{\partial^j(\sigma_0 u)}{\partial n_l^j} \right\|_{H^{s-j-1/2}(\partial\Omega_0)} \leq \left\| \frac{\partial^j(\sigma_0(u-u_k))}{\partial n_l^j} \right\|_{H^{s-j-1/2}(\partial\Omega_0)} \lesssim \left\| \frac{\partial^j(\sigma_0(u-u_k))}{\partial n_l^j} \right\|_{H^{s-j}(\Omega_0)}$$

$$\lesssim \|\sigma_0(u-u_k)\|_{H^s(\Omega_0)} \lesssim \|u-u_k\|_{H^s(\Omega)},$$

for $j = 0, \ldots, s-1$, which tends to zero as $k \to \infty$, so that $\sigma_0 u \in H_0^s(\Omega_0)$. $\qquad\square$

Consequently, Proposition 2.8 is applicable for this particular domain decomposition.

It is also possible to decompose the L-shaped domain in such a way that smooth partitions of unity exist. Such a covering is constructed in Figure 2.6. Here any

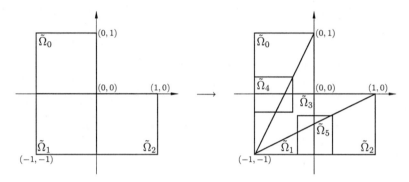

Figure 2.6: Left: Non-overlapping decomposition of the L-shaped domain with three congruent squares. Right: Overlapping covering including the non-convex polygonal patch $\tilde\Omega_3$ corresponding to the knot sequence $(0,0)$, $(0,1)$, $(-1,-1)$, $(1,0)$.

boundary segment of any patch has either empty intersection with $\partial\Omega$ or it is contained in $\partial\Omega$. However, now six patches instead of two are needed, and, in addition, the non-convex subdomain $\tilde\Omega_3$ is much harder to handle in an implementation. For instance, if one has to decide whether the supports of two wavelets in the overlapping region intersect each other, this will be quite complicated and costly if one function stems from $\tilde\Omega_3$ and the other one, say, from $\tilde\Omega_0$. Altogether, it can be expected that this decomposition will not lead to competitive performances in practice.

Summarizing, we can say that, due to the practical obligation to obtain matching grids in the overlapping region, the domain covering should not always be chosen at one's convenience. Therefore, as it has been depicted for the special case of the L-shaped domain, one may encounter situations where no arbitrary smooth partitions of unity exist. In the next chapter, where the N-term approximation of solutions of elliptic operator equations with aggregated wavelet frames will be studied, we will have to take care of this observation.

It is important to note that the above construction of a partition of unity subordinate to $\{\Omega_0, \Omega_1\}$ for the L-shaped domain can be surely generalized to more complicated two-dimensional polygonal domains with re-entrant corners. Thanks to Lemma 2.4, the choice of the practically unfavorable non-convex patches around such re-entrant corners can be avoided, which are inevitable if one aims at a smooth partition of unity. However, it should also be stated that with $\Omega = \Omega_0 \cup \Omega_1$, in view of the boundary conditions on $\partial\Omega$ incorporated in the space $H_0^t(\Omega)$, one has to choose wavelets on Ω_0 and Ω_1 satisfying homogeneous boundary conditions of order $t-1$ on $\partial\Omega_i \cap \partial\Omega$, $i = 0, 1$. In view of (2.4.5), on $\partial\Omega_i \cap \Omega$, we want to choose boundary conditions of order $d-2$. Unfortunately, both $\partial\Omega_0$ and $\partial\Omega_1$ have one face which intersects $\partial\Omega$ as well as Ω. Hence, the global smoothness of the wavelets on Ω_i near such a face cannot be chosen larger than $C^{t-1}(\Omega)$ with this covering.

In this context of choosing appropriate boundary conditions at internal boundaries $\partial\Omega_i \cap \Omega$, we also state the following remark.

Remark 2.7. In [104] the open covering is assumed to satisfy Assumption 2.2, and the local spaces are also chosen to be $H_0^t(\Omega_i)$. Moreover, it is assumed that one is equipped with a collection of non-negative functions $\{\omega_i\}_{i=0}^{m-1}$ with each ω_i being smooth on Ω_i and zero outside Ω_i such that an open covering $\Omega = \bigcup_{i=0}^{m-1} \hat{\Omega}_i$ exists with $\hat{\Omega}_i \subset \Omega_i$ and $\omega_i \approx 1$ on $\hat{\Omega}_i$. A frame for $H_0^t(\Omega)$ is then obtained by setting $\Psi := \bigcup_{i=0}^{m-1} \omega_i \Psi^{(i)}$, where $\Psi^{(i)}$ is a basis for $H_0^t(\Omega_i)$. Following this approach, if each ω_i is chosen such that it vanishes at the internal boundary $\partial\Omega_i \cap \Omega$, the boundary conditions of the local bases at $\partial\Omega_i \cap \Omega$ can be chosen at one's convenience. That means, as the local spaces one may as well choose $H_{0,\partial\Omega_i \cap \partial\Omega}^t(\Omega_i)$. In addition, if for one patch one has $\partial\Omega_i \cap \partial\Omega = \emptyset$, it is possible to work with a periodic wavelet basis on Ω_i, which is favorable because of their simple implementation and their good condition numbers. This is particularly interesting for the case of integral equations posed on a closed manifold.

Chapter 3

Nonlinear Approximation with Aggregated Wavelet Frames

An adaptive method based on a discretization with respect to a wavelet basis or frame Ψ for $H_0^t(\Omega)$ generates a sequence of approximations $u_k = (\mathbf{u}_k)^\top \Psi$, $\mathbf{u}_k \in \ell_2(\Lambda)$, $k \geq 0$, to the unknown solution u. In order to develop a benchmark for the *optimal* relation between the error $\|u - u_k\|_{H^t(\Omega)}$ and the number of degrees of freedom, i.e., $\#\operatorname{supp}\mathbf{u}_k$, one has to study the convergence rate of the best N-term approximations of u in $H_0^t(\Omega)$ with respect to Ψ, as N tends to infinity. It is known (see [55]) that for the case of a basis this rate is governed by the Besov regularity of u in a certain scale. In particular, it can be deduced from the decay rate of the *unique* expansion coefficients of u with respect to the basis, and its specific value also depends on the regularity and approximation power of the latter. We shall explain this in more detail below. The intention of this chapter is to show that the well-known results in the context of best N-term approximation with respect to classical wavelet Riesz bases carry over to the case of aggregated wavelet frames.

Basically, the idea is to consider the coefficients of u with respect to the non-canonical global dual frame developed in Proposition 2.11 and to prove that these are decaying at the same rate as the coefficients with respect to a wavelet basis with similar regularity and approximation properties would. According to (2.3.29), the (non-canonical) frame coefficients are simply composed of the expansion coefficients of $\sigma_i u$ with respect to the local lifted wavelet bases $\Psi^{(i)}$ on Ω_i, $i = 0, \ldots, m-1$. Provided that the partition of unity is smooth, so that a sufficient Besov regularity for $\sigma_i u$ can be guaranteed, the well-known results for the case of wavelet bases indeed carry over immediately by applying the classical results for $\Psi^{(i)}$. However, as §2.7 has highlighted, the smoothness assumption on the partition cannot always be assured, particularly not for domain coverings of special practical interest. For this reason, in this chapter we spend some effort treating those cases where the smoothness requirement fails to hold. For the sake of simplicity, we shall focus on the model case of the L-shaped domain made up of two rectangles as in §2.7; recall Figure 2.4.

First, we collect some basic facts about function spaces as well as interpolation and approximation spaces and their relation. Afterwards, we revisit the classical results on nonlinear approximation with classical wavelet bases which will be (partly) generalized to the case of aggregated wavelet frames subsequent to that.

3.1 Function spaces

For an $h \in \mathbb{R}^n$, let $\Delta_h f := f(\cdot + h) - f(\cdot)$, and for an integer $r > 1$ let $\Delta_h^r f := \Delta_h(\Delta_h^{r-1} f)$ be the difference operator of order r. We further define for $f \in L_p(\Omega)$, $0 < p \leq \infty$, the r-th order L_p-modulus of smoothness by

$$\omega_r(f,t)_p := \sup_{\|h\| \leq t} \|\Delta_h^r f\|_{L_p(\Omega_{h,r})}, \quad t \geq 0, \tag{3.1.1}$$

where $\Omega_{h,r} := \{x \in \Omega : x + kh \in \Omega, \ k = 0, \dots, r\}$. For $p = \infty$, $L_\infty(\Omega)$ is replaced by $C(\Omega)$, the space of uniformly continuous functions on Ω. Obviously, $\omega_r(f,t)_p$ is finite for each t. Moreover, it is a continuous and increasing function of t with $\omega_r(f,0)_p = 0$. For $0 < p, q \leq \infty$, $s > 0$, and $r := \lfloor s \rfloor + 1$, as usual, we define the *Besov space* $B_{p,q}^s(\Omega)$, measuring s orders of smoothness in $L_p(\Omega)$, as the space of all $f \in L_p(\Omega)$ for which

$$|f|_{B_{p,q}^s(\Omega)} := \begin{cases} \left(\displaystyle\int_0^\infty (t^{-s} \omega_r(f,t)_p)^q \frac{dt}{t} \right)^{1/q}, & 0 < q < \infty, \\ \displaystyle\sup_{t>0} \ t^{-s} \omega_r(f,t)_p, & q = \infty, \end{cases} \tag{3.1.2}$$

is finite. The quantity (3.1.2) defines a seminorm for $1 \leq p, q \leq \infty$, a quasi-seminorm for all other cases, and the expression $\|f\|_{B_{p,q}^s(\Omega)} := \|f\|_{L_p(\Omega)} + |f|_{B_{p,q}^s(\Omega)}$ defines a (quasi-)norm on $B_{p,q}^s(\Omega)$. For convenience we recall that a quasi-norm on a linear space X is a norm with the triangle inequality replaced by requiring

$$\|x + y\|_X \leq C(\|x\|_X + \|y\|_X), \quad \text{for all } x, y \in X, \tag{3.1.3}$$

with some constant C depending on X.

The spaces $B_{\infty,\infty}^s(\Omega)$ coincide with the classical *Hölder spaces*, when s is not an integer, thus $C^s(\Omega) = B_{\infty,\infty}^s(\Omega)$ in that case. When s is not an integer, we also denote the spaces $C^s(\Omega)$ by $W_\infty^s(\Omega)$.

We stress the following important note on the relation between Sobolev and Besov spaces.

Remark 3.1. In [112] it has been proved that $W_p^s = B_{p,p}^s$, if s is not an integer and $p \neq 2$, while for $p = 2$ this equality always holds.

Throughout this chapter, essential use will be made of embedding results concerning Sobolev and Besov spaces. Clearly, decreasing p means to measure smoothness in a weaker metric, thus the spaces $B_{p,q}^s(\Omega)$ get larger. Concerning the secondary index q, let us mention that on the one hand it is

$$B_{p,q_1}^s(\Omega) \hookrightarrow B_{p,q_2}^s(\Omega), \quad q_1 < q_2, \tag{3.1.4}$$

and on the other hand we have for any $\varepsilon > 0$,

$$B^{s+\varepsilon}_{p,q_1}(\Omega) \hookrightarrow B^s_{p,q_2}(\Omega), \quad q_1 > q_2. \tag{3.1.5}$$

We also recall the well-known *Sobolev embedding theorem*, see [2], and a similar result for Besov spaces; cf. [112].

Theorem 3.1. *Let* $1 \le p_1, p_2 \le \infty$ *and* $s_1, s_2 \ge 0$. *We have the continuous embedding* $W^{s_1}_{p_1}(\Omega) \subset W^{s_2}_{p_2}(\Omega)$, *if* $s_1 - s_2 \ge n(1/p_1 - 1/p_2) > 0$, *except for the case when* $p_2 = \infty$ *and* $s_1 - n(1/p_1 - 1/p_2)$ *is an integer, for which one has to assume that* $s_1 - s_2 > n(1/p_1 - 1/p_2)$.

Theorem 3.2. *It is* $B^{s_1}_{p_1,p_1}(\Omega) \subset B^{s_2}_{p_2,p_2}(\Omega)$, *if* $s_1 - s_2 \ge n(1/p_1 - 1/p_2) > 0$, *for all* $s_1, s_2 \ge 0$ *and* $1 \le p_1, p_2 \le \infty$, *in the sense of a continuous embedding.*

The latter result can also be extended to the case $p_1, p_2 > 0$, see [26, Corollary 3.7.1], which is of particular interest in the context of nonlinear wavelet approximation, as we will see in §3.3.2 below. It is often convenient to describe embeddings of Besov spaces $B^\alpha_{\tau,q}(\Omega)$ by means of a $(\frac{1}{\tau}, \alpha)$-DeVore/Triebel diagram (cf. [55]), in which each Besov space is identified with a point in the $(\frac{1}{\tau}, \alpha)$-plane. For instance, in Figure 3.1, Theorem 3.2 is visualized for $p_2 = 2$, $s_2 = 0$, and $p_2 = 2$, $s_2 = t$, respectively. The

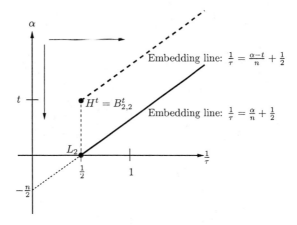

Figure 3.1: A classical DeVore/Triebel diagram for the visualization of embeddings of Besov spaces.

space $L_2(\Omega)$ is represented by the point $(\frac{1}{2}, 0)$, and $B^t_{2,2}(\Omega) = H^t(\Omega)$ by $(\frac{1}{2}, t)$. The spaces on (and above) the solid thick line, namely $B^\alpha_{\tau,\tau}(\Omega)$, $\alpha > 0$, with $\frac{1}{\tau} = \frac{\alpha}{n} + \frac{1}{2}$, are embedded in $L_2(\Omega)$, and the spaces on (and above) the thick dashed line, $B^\alpha_{\tau,\tau}(\Omega)$, $\alpha > t$, with $\frac{1}{\tau} = \frac{\alpha-t}{n} + \frac{1}{2}$, are embedded in $H^t(\Omega)$. The slope of the embedding line is always equal to the space dimension n. The vertical arrow indicates the trivial embedding $B^{s_1}_{\tau,q_1}(\Omega) \subset B^{s_2}_{\tau,q_2}(\Omega)$ for $s_1 > s_2 \geq 0$. The horizontal arrow visualizes the embedding when the metric is getting weaker (and the secondary index is not getting smaller).

For further information on the theory of function spaces, the reader is referred to [2, 99, 111–115].

3.2 Interpolation and approximation spaces

3.2.1 Interpolation spaces

We collect in the following the main features from interpolation theory needed for our purposes, where interpolation of function spaces is performed via the real method of Lions and Peetre. A detailed discussion of this topic can be found in [10] and [11]. Our presentation will be particularly restricted to the case where $(X, \| \cdot \|_X)$ and $(Y, \| \cdot \|_Y)$ form (quasi-)normed Banach function spaces such that $Y \subset X$ in the sense of a continuous and dense embedding. For this situation we define the so-called *K-functional*

$$K(f,t) := \inf_{g \in Y} \|f - g\|_X + t\|g\|_Y, \quad f \in X, \ t \geq 0. \tag{3.2.1}$$

For $\theta \in (0,1)$ and $0 < q \leq \infty$, we define the spaces $(X,Y)_{\theta,q}$ as the set of all functions $f \in X$ such that the norm (quasi-norm in case $q < 1$)

$$\|f\|_{(X,Y)_{\theta,q}} := \begin{cases} \left(\displaystyle\int_0^\infty (t^{-\theta} K(f,t))^q \frac{dt}{t} \right)^{1/q}, & 0 < q < \infty, \\ \displaystyle\sup_{t>0} \ t^{-\theta} K(f,t), & q = \infty, \end{cases} \tag{3.2.2}$$

is finite. Hence, $(X,Y)_{\theta,q}$ defines a family of intermediate (quasi-)Banach spaces, $Y \subset (X,Y)_{\theta,q} \subset X$. Note that, due to the monotonicity of $K(f,t)$ in t, one also has the equivalent discrete (quasi-)norm

$$\|f\|_{(X,Y)_{\theta,q}} \approx \|(\rho^{j\theta} K(f,\rho^{-j}))_{j \geq 0}\|_{\ell_q(\mathbb{N}_0)}, \tag{3.2.3}$$

for any fixed $\rho > 1$.

As a prominent example, we recall the most important results on interpolation of Besov spaces. In particular, we have for $0 < q_1, p \leq \infty$, $\alpha > 0$ that

$$(L_p(\Omega), B^\alpha_{p,q_1}(\Omega))_{\theta,q} = B^{\theta\alpha}_{p,q}(\Omega), \quad 0 < \theta < 1, \ 0 < q \leq \infty, \tag{3.2.4}$$

in the sense of equivalent norms. Moreover, from the *reiteration theorem* for interpolation (see [11]) it follows that if $0 < q_1, q_2, p \leq \infty$, $0 < \alpha_1 < \alpha_2$, it is

$$(B^{\alpha_1}_{p,q_1}(\Omega), B^{\alpha_2}_{p,q_2}(\Omega))_{\theta,q} = B^\alpha_{p,q}(\Omega), \quad \text{with } \alpha := (1-\theta)\alpha_1 + \theta\alpha_2, \tag{3.2.5}$$

for each $0 < \theta < 1$, $0 < q \leq \infty$. Furthermore, the following result will be of importance.

Proposition 3.1 (see [57, Corollary 6.3]). *Let $0 < q_1, q_2, p_1, p_2 \leq \infty$ and $\alpha_1, \alpha_2 > 0$. Then, for each $0 < \theta < 1$ and for $1/q = \theta/q_1 + (1-\theta)/q_2$, $1/p = \theta/p_1 + (1-\theta)/p_2$, we have*

$$(B_{p_1,q_1}^{\alpha_1}(\Omega), B_{p_2,q_2}^{\alpha_2}(\Omega))_{\theta,q} = B_{p,q}^{\alpha}(\Omega), \quad \text{with } \alpha := (1-\theta)\alpha_1 + \theta\alpha_2, \tag{3.2.6}$$

provided that $p = q$.

3.2.2 Approximation classes

Definition 3.1. Let $\{V_j\}_{j\geq 0}$ be a nested sequence of subspaces of X such that $\bigcup_{j\geq 0} V_j$ is dense in X. For $s > 0$ and $0 < q \leq \infty$, we define the approximation space $\mathcal{A}_q^s(X)$ related to the sequence V_j by

$$\mathcal{A}_q^s(X) := \left\{ f \in X : |f|_{\mathcal{A}_q^s(X)} := \|\{2^{sj} \operatorname{dist}_X(f, V_j)\}_{j\geq 0}\|_{\ell_q(\mathbb{N}_0)} < \infty \right\}. \tag{3.2.7}$$

It is readily seen (cf. [56, Chapter 7, §9]) that $\mathcal{A}_q^s(X)$ constitutes a (quasi-)Banach subspace of X with the norm

$$\|f\|_{\mathcal{A}_q^s(X)} := \|f\|_X + |f|_{\mathcal{A}_q^s(X)}, \tag{3.2.8}$$

which is again a quasi-norm for $q < 1$. Clearly, we have the continuous embedding

$$A_{q_1}^s(X) \hookrightarrow A_{q_2}^s(X), \quad q_1 < q_2, \ s > 0. \tag{3.2.9}$$

Note that for any $f \in \mathcal{A}_q^s(X)$, $0 < q \leq \infty$, we have that $\operatorname{dist}_X(f, V_j) \lesssim 2^{-sj}$, where the parameter q provides some additional information. In that case we say that f can be approximated at a *rate* s. A central problem in approximation theory is to characterize those functions f which can be approximated at a rate $s > 0$ by some smoothness condition. For this purpose, the following theorem, showing that certain approximation spaces are actually interpolation spaces, can be very helpful, as we shall see in §3.3.

Theorem 3.3 (cf. [26, Theorem 3.5.2, Theorem 4.2.1]). *Let $\{T_j\}_{j\geq 0}$ be a sequence of subsets of Y with $T_j \subset T_{j+1} \subset \cdots \subset Y \subset X$ such that for some $m > 0$ one has the Jackson-type estimate*

$$\operatorname{dist}_X(f, T_j) \lesssim 2^{-mj}\|f\|_Y, \tag{3.2.10}$$

and a Bernstein-type estimate

$$\|f\|_Y \lesssim 2^{mj}\|f\|_X, \quad f \in T_j. \tag{3.2.11}$$

In case the sets T_j do not *constitute linear subspaces, let us also assume the existence of a fixed integer a such that $T_j + T_j \subset T_{j+a}$, $j \geq 0$. Then, for the approximation spaces $\mathcal{A}_q^s(X)$ related to the sequence $\{T_j\}_{j\geq 0}$ and $s \in (0, m)$, one has the norm equivalence*

$$\|(2^{js} K(f, 2^{-mj}))_{j\geq 0}\|_{\ell_q(\mathbb{N}_0)} \eqsim \|f\|_X + \|(2^{js} \operatorname{dist}_X(f, T_j))_{j\geq 0}\|_{\ell_q(\mathbb{N}_0)}, \tag{3.2.12}$$

and thus by (3.2.3) we have the relation

$$(X, Y)_{\theta,q} = \mathcal{A}_q^s(X), \quad \text{for } s = \theta m. \tag{3.2.13}$$

3.3 The classical wavelet basis setting

In the course of this section, suppose that $\Psi = \{\psi_\lambda\}_{\lambda \in \Lambda}$ represents a classical wavelet basis for $L_2(\Omega)$ over the domain $\Omega \subset \mathbb{R}^n$ (cf. §2.2), constructed via a pair of biorthogonal multiresolution sequences $\{V_j\}_{j \geq j_0}$ and $\{\tilde{V}_j\}_{j \geq j_0}$ which are generated by compactly supported tensor product type, biorthogonal scaling functions ϕ and $\tilde{\phi}$, respectively. Let again $d > 0$ represent the order of polynomial exactness of the spaces V_j, $j \geq j_0$. Without loss of generality we may also assume $j_0 = 0$.

3.3.1 Linear approximation with wavelets in $H^t(\Omega)$

Let us assume throughout this subsection that the scaling function ϕ is contained in $H^\alpha(\Omega)$ with an $\alpha \in (0, \gamma)$, for a $\gamma \in (0, d)$. At first, we want to determine the approximation classes corresponding to *linear approximation* in $H^t(\Omega)$, $0 \leq t < \alpha$, from the linear spaces V_j.

Proposition 3.2. *Under the above assumptions, we have the equivalence*

$$\|f\|_{\mathcal{A}_2^{sn}(H^t(\Omega))} \eqsim \|f\|_{H^{sn+t}(\Omega)}, \quad 0 < s < \frac{\gamma - t}{n}. \tag{3.3.1}$$

Proof. Let $\alpha \in (0, \gamma)$. The claim follows by an application of the direct and inverse estimates from (2.2.26) and (2.2.27) and Theorem 3.3. In particular, we have $\inf_{v_j \in V_j} \|f - v_j\|_{H^t(\Omega)} \lesssim 2^{-j(\alpha - t)} \|f\|_{H^\alpha(\Omega)}$, $f \in H^\alpha(\Omega)$, and $\|f\|_{H^\alpha(\Omega)} \lesssim 2^{j(\alpha - t)} \|f\|_{H^t(\Omega)}$, $f \in V_j$. Now, applying Theorem 3.3 with $T_j = V_j$, $X = H^t(\Omega)$, $Y = H^\alpha(\Omega)$, $m = \alpha - t$, $q = 2$, and using Remark 3.1 and (3.2.5), we get

$$\mathcal{A}_2^{\theta(\alpha - t)}(H^t(\Omega)) = (H^t(\Omega), H^\alpha(\Omega))_{\theta, 2} = H^{\theta\alpha + (1-\theta)t}(\Omega) = H^{\theta(\alpha - t) + t}(\Omega),$$

for $0 < \theta < 1$. We can write $\theta(\alpha - t) = sn$, with a suitable $0 < s < \frac{\alpha - t}{n}$. This yields

$$\mathcal{A}_2^{sn}(H^t(\Omega)) = \mathcal{A}_2^{\theta(\alpha - t)}(H^t(\Omega)) = H^{\theta(\alpha - t) + t}(\Omega) = H^{sn+t}(\Omega),$$

for all $0 < s < \frac{\alpha - t}{n}$. $\qquad\square$

Noting that usually it is $\dim V_j \eqsim 2^{jn}$, we get the following corollary.

Corollary 3.1. *For $0 < s < \frac{\gamma - t}{n}$ and $f \in H^{sn+t}(\Omega)$, one has*

$$\mathrm{dist}_{H^t(\Omega)}(f, V_j) \lesssim 2^{-j(sn)} \eqsim (\dim V_j)^{-s}. \tag{3.3.2}$$

3.3.2 N-term wavelet approximation in $H^t(\Omega)$

Next, the approximation properties in $H^t(\Omega)$, $t \geq 0$, of the *nonlinear spaces*

$$S_N := \left\{ \sum_{\lambda \in \mathcal{J}} d_\lambda \psi_\lambda : \mathcal{J} \subset \Lambda, \ \#\mathcal{J} \leq N \right\}, \quad N = 1, 2, \ldots, \tag{3.3.3}$$

shall be recalled. To this end, we will first consider the subsequence

$$\Sigma_j := S_{2^{jn}}, \quad j \geq 0. \tag{3.3.4}$$

The process of approximating a function f from the spaces S_N is usually called N-*term approximation* which is a special kind of *nonlinear approximation*.

In §2.2.2 we have already seen that wavelets can have the ability to characterize L_2-Sobolev spaces by means of norm equivalences. We state here another result in this direction involving Besov spaces.

Theorem 3.4 (see [26, Theorem 3.7.7]). *For some $r \in [1, \infty]$, let us assume $\phi \in L_r(\Omega)$, and let the dual scaling function $\tilde{\phi}$ be contained in $L_{r'}(\Omega)$, $1/r + 1/r' = 1$. If $f = \sum_{\lambda \in \Lambda} c_\lambda \psi_\lambda$ is the decomposition of f into the elements of the wavelet basis Ψ, then we have the norm equivalence*

$$\|f\|_{B_{p,q}^s(\Omega)} \approx \left(\sum_{j \geq j_0} 2^{qj(s+n(1/2-1/p))} \left(\sum_{|\lambda|=j} |c_\lambda|^p \right)^{q/p} \right)^{1/q}, \tag{3.3.5}$$

if $0 < p \leq r$ and $s > 0$ such that $n(1/p - 1/r) < s < \min\{\alpha, d\}$ and $\tilde{\phi} \in B_{p,q_0}^\alpha(\Omega)$ for some q_0.

Note that this result represents a generalization of Corollary 2.1 which is (at least partly) retrieved in case $p = q = 2$. The latter theorem, in combination with the interpolation tool Theorem 3.3, gives us the desired result.

Proposition 3.3. *Assume that the space $B_{q,q}^{sn+t}(\Omega)$ admits a wavelet characterization of the kind (3.3.5) for $sn + t \in [t, \alpha']$, $t < \alpha'$, $\frac{1}{q} = s + \frac{1}{2}$. Then, for $0 < s < \frac{\alpha'-t}{n}$, $\frac{1}{q} = s + \frac{1}{2}$, we have the norm equivalence*

$$\|f\|_{B_{q,q}^{sn+t}(\Omega)} \approx \|f\|_{H^t(\Omega)} + \|(2^{jsn} \operatorname{dist}_{H^t(\Omega)}(f, \Sigma_j))_{j \geq 0}\|_{\ell_q(\mathbb{N}_0)}. \tag{3.3.6}$$

Hence, for the approximation spaces associated with the sequence of nonlinear spaces $\{\Sigma_j\}_{j \geq 0}$, we have

$$\mathcal{A}_q^{sn}(H^t(\Omega)) = B_{q,q}^{sn+t}(\Omega). \tag{3.3.7}$$

Sketch of proof. An extensive proof can be found in [26, Theorem 4.2.2]. The outline is to use the characterization (3.3.5) to verify the direct and inverse estimates from Theorem 3.3 for $X = H^t(\Omega)$ and $Y = B_{q',q'}^{s'n+t}(\Omega)$, $s'n + t = \alpha'$, $\frac{1}{q'} = s' + \frac{1}{2}$ and $m := \alpha' - t$. Then, one realizes that the interpolation space appearing in (3.2.13) is

$$B_{q,q}^{sn+t}(\Omega) = (H^t(\Omega), B_{q',q'}^{\alpha'}(\Omega))_{\theta,q}, \quad \theta = \frac{sn}{\alpha' - t}, \quad 0 < s < \frac{\alpha' - t}{n}. \tag{3.3.8}$$

Finally, the proof is concluded by an application of (3.2.12) and (3.2.13) with $\theta m = sn$ in Theorem 3.3. $\square$

Corollary 3.2. *If $f \in B_{q,q}^{sn+t}(\Omega)$, $0 < s < \frac{\alpha'-t}{n}$, $t < \alpha' < \min\{\alpha, d\}$, $\phi \in B_{q,q_0}^{\alpha}(\Omega)$ for some q_0, and $\frac{1}{q} = s + \frac{1}{2}$, then we have the estimate*

$$\mathrm{dist}_{H^t(\Omega)}(f, S_N) \lesssim N^{-s}. \tag{3.3.9}$$

Proof. We state that by the monotonicity of the sequence $\mathrm{dist}_{H^t(\Omega)}(f, S_N)$, we have the equivalence

$$\sum_{j \geq 0} (2^{jsn} \, \mathrm{dist}_{H^t(\Omega)}(f, \Sigma_j))^q \approx \sum_{N \geq 1} N^{-1}(N^s \, \mathrm{dist}_{H^t(\Omega)}(f, S_N))^q,$$

which, together with Proposition 3.3, finishes the proof. $\qquad\square$

Remark 3.2. The Corollaries 3.1 and 3.2 give sufficient conditions on f under which the error of approximation with respect to V_j or S_N is proportional to N^{-s}, where N represents the number of degrees of freedom used in the approximation process. In order to obtain the same rate, in the linear case the requirement $f \in H^{sn+t}(\Omega)$ has to be made, while in the nonlinear case the weaker requirement $f \in B_{q,q}^{sn+t}(\Omega)$ is needed, $\frac{1}{q} = s + \frac{1}{2}$. Indeed, $H^{sn+t}(\Omega)$ can be embedded into $B_{q,q}^{sn+t-\varepsilon}(\Omega)$, for an arbitrary small $\varepsilon > 0$ (cf. (3.1.5)). The difference between $H^{sn+t}(\Omega)$ and $B_{q,q}^{sn+t}(\Omega)$ is growing with s. The two relevant scales of function spaces are depicted in Figure 3.2 in an $(\alpha, \frac{1}{\tau})$-DeVore/Triebel diagram. It is important to note that for growing s, the functions in $H^{sn+t}(\Omega)$ become smooth in the classical sense, whereas $B_{q,q}^{sn+t}(\Omega)$ might still contain functions with discontinuities.

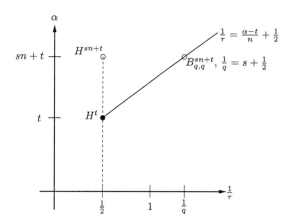

Figure 3.2: Visualization of the scales of functions spaces governing the convergence rates of linear and nonlinear wavelet approximation.

Remark 3.3. The approximation class $\mathcal{A}_{\infty}^s(H^t(\Omega))$, consisting of exactly those functions for which $\mathrm{dist}_{H^t(\Omega)}(f, S_N) \lesssim N^{-s/n}$, is slightly larger than $\mathcal{A}_q^s(H^t(\Omega))$, $0 < q < 2$.

The space $\mathcal{A}_{\infty}^s(H^t(\Omega))$ can be characterized by properties of the expansion coefficients $\mathbf{c} = \{c_\lambda\}_{\lambda \in \Lambda}$ of a function $f = \sum_{\lambda \in \Lambda} c_\lambda \psi_\lambda \in H^t(\Omega)$, as we shall now describe.

3.3.3 N-term approximation in ℓ_2

We introduce the concept of N-term approximation in ℓ_2. For $N = 1, 2, \ldots$, let this time Σ_N denote the nonlinear subspace of $\ell_2(\Lambda)$ consisting of all vectors with at most N non-zero coordinates. Given $\mathbf{v} \in \ell_2(\Lambda)$, we introduce the error of approximation

$$\sigma_N(\mathbf{v}) := \inf_{\mathbf{w} \in \Sigma_N} \|\mathbf{v} - \mathbf{w}\|_{\ell_2(\Lambda)}. \tag{3.3.10}$$

Clearly, a best approximation to $\mathbf{v}$ from Σ_N is obtained by taking a vector $\mathbf{v}_N \in \Sigma_N$ which agrees with $\mathbf{v}$ on those coordinates on which $\mathbf{v}$ takes its N largest values in modulus and which is zero everywhere else. Thus, $\mathbf{v}_N$ is called a *best N-term approximation* to $\mathbf{v}$, and it is not necessarily unique, especially not when many of the (largest) coefficients are equal in modulus.

The concept of N-term approximation in ℓ_2 is closely related to the *weak ℓ_τ-spaces* $\ell_\tau^w(\Lambda)$. Given some $0 < \tau < 2$, these are defined according to

$$\ell_\tau^w(\Lambda) := \{\mathbf{c} \in \ell_2(\Lambda) : |\mathbf{c}|_{\ell_\tau^w(\Lambda)} := \sup_{k \in \mathbb{N}} k^{1/\tau} |\gamma_k(\mathbf{c})| < \infty\}, \tag{3.3.11}$$

where $\gamma_k(\mathbf{c})$ denotes the kth largest coefficient in modulus of $\mathbf{c}$. For $\delta \in (0, 2 - \tau]$, it is $\ell_\tau(\Lambda) \hookrightarrow \ell_\tau^w(\Lambda) \hookrightarrow \ell_{\tau+\delta}(\Lambda)$, justifying the denotation as weak ℓ_τ. This is verified by

$$k^{1/\tau} |\gamma_k(\mathbf{c})| = (k|\gamma_k(\mathbf{c})|^\tau)^{1/\tau} \leq \left(\sum_{j \leq k} |\gamma_j(\mathbf{c})|^\tau \right)^{1/\tau} \leq \|\mathbf{c}\|_{\ell_\tau(\Lambda)},$$

and

$$\|\mathbf{c}\|_{\ell_{\tau+\delta}(\Lambda)}^{\tau+\delta} = \sum_{k \in \mathbb{N}} |\gamma_k(\mathbf{c})|^{\tau+\delta} k^{1+\delta/\tau} k^{-1-\delta/\tau} \leq \sum_{k \in \mathbb{N}} |\mathbf{c}|_{\ell_\tau^w(\Lambda)}^{\tau+\delta} k^{-1-\delta/\tau} \lesssim |\mathbf{c}|_{\ell_\tau^w(\Lambda)}^{\tau+\delta}.$$

The quantity $|\mathbf{c}|_{\ell_\tau^w(\Lambda)}$ defines a quasi-seminorm,

$$|\mathbf{v} + \mathbf{w}|_{\ell_\tau^w(\Lambda)} \leq C_1(\tau)(|\mathbf{v}|_{\ell_\tau^w(\Lambda)} + |\mathbf{w}|_{\ell_\tau^w(\Lambda)}), \quad \mathbf{v}, \mathbf{w} \in \ell_\tau^w(\Lambda), \tag{3.3.12}$$

with a $C_1(\tau) > 0$. In addition, we set $\|\mathbf{c}\|_{\ell_\tau^w(\Lambda)} := |\mathbf{c}|_{\ell_\tau^w(\Lambda)} + \|\mathbf{c}\|_{\ell_2(\Lambda)}$ to obtain a quasi-norm. We will often use the shorter notation $|\cdot|_{\ell_\tau^w} := |\cdot|_{\ell_\tau^w(\Lambda)}$ in case the choice of the index set is clear from the context.

The weak ℓ_τ-spaces can be obtained by interpolation. In particular, we have

$$\ell_\tau^w(\Lambda) = (\ell_{\tau_0}(\Lambda), \ell_{\tau_1}(\Lambda))_{\theta,\infty}, \tag{3.3.13}$$

where $\frac{1}{\tau} = \frac{1-\theta}{\tau_0} + \frac{\theta}{\tau_1}$, $0 < \theta < 1$; cf. [11, Theorem 5.3.1].

For some $s > 0$, let us furthermore define the approximation class

$$\mathcal{A}^s(\Lambda) := \left\{ \mathbf{v} \in \ell_2(\Lambda) : \|\mathbf{v}\|_{\mathcal{A}^s(\Lambda)} := \sup_{N \geq 0} (N+1)^s \sigma_N(\mathbf{v}) < \infty \right\}, \tag{3.3.14}$$

where $\sigma_0(\mathbf{v}) := \|\mathbf{v}\|_{\ell_2(\Lambda)}$. Obviously, $\mathcal{A}^s(\Lambda)$ consists of all vectors $\mathbf{v} \in \ell_2(\Lambda)$ which can be approximated with order $\mathcal{O}(N^{-s})$ by the elements of Σ_N. The importance of the weak ℓ_τ-spaces lies in the following well-known statement, which, for instance, can be found in [27, Proposition 3.2].

Proposition 3.4. *Given an $s > 0$, let τ be defined as $\frac{1}{\tau} = s + \frac{1}{2}$. Then, the space $\mathcal{A}^s(\Lambda)$ coincides with $\ell_\tau^w(\Lambda)$ up to equivalent (semi-)norms*

$$|\mathbf{v}|_{\ell_\tau^w(\Lambda)} \approx \sup_{N \geq 1} N^s \sigma_N(\mathbf{v}), \quad \mathbf{v} \in \ell_\tau^w(\Lambda), \tag{3.3.15}$$

with constants depending on τ only when τ tends to zero.

Using this result, one can now characterize all functions f for which (3.3.9) holds, i.e., the class $\mathcal{A}_\infty^{sn}(H^t(\Omega))$ from §3.3.2.

Corollary 3.3. *Let $\Psi = \{\psi_\lambda\}_{\lambda \in \Lambda}$ be a wavelet Riesz basis for $L_2(\Omega)$ with dual basis $\tilde{\Psi} = \{\tilde{\psi}_\lambda\}_{\lambda \in \Lambda}$ such that, as usual, $\mathbf{D}^{-1}\Psi$ is a Riesz basis for $H^t(\Omega)$, where $\mathbf{D} = \mathrm{diag}(2^{t|\lambda|})_{\lambda \in \Lambda}$. Then, for a function $f = \sum_{\lambda \in \Lambda} 2^{t|\lambda|} \langle f, \tilde{\psi}_\lambda \rangle_{L_2(\Omega)} 2^{-t|\lambda|} \psi_\lambda \in H^t(\Omega)$, $s > 0$, and $\frac{1}{\tau} = s + \frac{1}{2}$, the following are equivalent*

   (i) $\mathbf{c} = \{2^{t|\lambda|} \langle f, \tilde{\psi}_\lambda \rangle_{L_2(\Omega)}\}_{\lambda \in \Lambda} \in \ell_\tau^w(\Lambda)$,

   (ii) $\mathrm{dist}_{H^t(\Omega)}(f, S_N) \lesssim N^{-s}$.

If in addition $f \in B_{\tau,\tau}^{sn+t}(\Omega)$, $0 < s < \frac{\min\{\alpha,d\}-t}{n}$, $\phi \in B_{\tau,\tau_0}^\alpha(\Omega)$ for some τ_0, $\frac{1}{\tau} = s + \frac{1}{2}$, then (i) and (ii) hold for that range of s.

Proof. Using the Riesz basis property of $\mathbf{D}^{-1}\Psi$ in $H^t(\Omega)$, for $f = \mathbf{c}^\top \mathbf{D}^{-1}\Psi$, we get

$$\mathrm{dist}_{H^t(\Omega)}(f, S_N) = \inf_{\mathbf{d} \in \Sigma_N} \|(\mathbf{c} - \mathbf{d})^\top \mathbf{D}^{-1}\Psi\|_{H^t(\Omega)}$$

$$\approx \inf_{\mathbf{d} \in \Sigma_N} \|\mathbf{c} - \mathbf{d}\|_{\ell_2(\Lambda)} = \sigma_N(\mathbf{c}). \tag{3.3.16}$$

Hence, the first part of the claim follows by Proposition 3.4. The second part is just the assertion of Corollary 3.2. $\qquad\square$

Remark 3.4.   (i) From (3.3.16) one may now deduce that for an f as in Corollary 3.3, a best approximation (up to a constant factor) to f from S_N is obtained by taking a best N-term approximation $\mathbf{c}_N$ of $\mathbf{c} = \{2^{t|\lambda|} \langle f, \tilde{\psi}_\lambda \rangle_{L_2(\Omega)}\}_{\lambda \in \Lambda} \in \ell_\tau^w(\Lambda)$ and setting $f_N := \mathbf{c}_N^\top(\mathbf{D}^{-1}\Psi)$. Indeed, we have

$$\|f - f_N\|_{H^t(\Omega)} \approx \|\mathbf{c} - \mathbf{c}_N\|_{\ell_2(\Lambda)} = \sigma_N(\mathbf{c}) \approx \mathrm{dist}_{H^t(\Omega)}(f, S_N) \lesssim N^{-s}.$$

  (ii) An application of Theorem 3.4 yields that

$$f \in B_{\tau,\tau}^{sn+t}(\Omega) \quad \text{if and only if} \quad \mathbf{c} = \{2^{t|\lambda|} \langle f, \tilde{\psi}_\lambda \rangle_{L_2(\Omega)}\}_{\lambda \in \Lambda} \in \ell_\tau(\Lambda), \tag{3.3.17}$$

for all $0 < s < \frac{\min\{\alpha,d\}-t}{n}$, $\phi \in B_{\tau,\tau_0}^\alpha(\Omega)$ for some τ_0, $\frac{1}{\tau} = s + \frac{1}{2}$. Since $\ell_\tau(\Lambda) \hookrightarrow \ell_\tau^w(\Lambda)$, we also have an immediate proof of the last part of the statement in Corollary 3.3.

3.3.4 Spline wavelets and boundary conditions

Spline wavelets

As one important special case, let us analyze the case of tensor product type spline wavelet bases of order $d \geq 1$, as mentioned in §2.2.4 and §2.2.5, say, on a smooth image Ω of the unit cube. The properties of the elements of such bases have already been collected in §2.4. In this case, the spaces V_j contain all polynomials up to degree $d-1$, thus the order of polynomial reproduction is d. Moreover, for the piecewise smooth generator ϕ, an application of [99, Theorem 4.6.3 (ii)] shows that

$$\phi \in H^\alpha(\Omega), \quad \text{for all } \alpha < d - \frac{1}{2} =: \gamma, \tag{3.3.18}$$

and in addition that for $n \geq 2$

$$\phi \in B^\alpha_{\tau,\tau}(\Omega), \quad \text{for all } t < \alpha < \frac{n}{2(n-1)} + d - 1, \quad \frac{1}{\tau} = \frac{\alpha - t}{n} + \frac{1}{2}, \tag{3.3.19}$$

whereas in case $n = 1$ one has $\phi \in B^\alpha_{\tau,\tau}(\Omega)$ for all $\alpha > 0$, $\frac{1}{\tau} = \alpha - t + \frac{1}{2}$.

Remark 3.5. (i) Using (3.3.18), (3.3.19) and Corollary 3.3, one can show that for $n \geq 2$ and a function $f \in B^{sn+t}_{\tau,\tau}(\Omega)$, $\frac{1}{\tau} = s + \frac{1}{2}$, we get for $\mathbf{c} = \{2^{t|\lambda|}\langle f, \tilde{\psi}_\lambda\rangle_{L_2(\Omega)}\}_{\lambda \in \Lambda}$ that $\sigma_N(\mathbf{c}) \lesssim N^{-s}$ for all $0 < s < \frac{\min\{\gamma,d\}-t}{n}$ with $\gamma := \frac{n}{2(n-1)} + d - 1$ which for $n = 2$ reads as $0 < s < \frac{d-t}{2}$. For $n = 1$ we obtain the range $0 < s < d - t$.

(ii) From Proposition 3.2 and (3.3.18), for the linear case, one may infer that for a function $f \in H^{sn+t}(\Omega)$ it follows $\text{dist}_{H^t(\Omega)}(f, V_j) \lesssim N^{-s}$, $N = \dim V_j$, for all $0 < s < \frac{d-1/2-t}{n}$.

Boundary conditions

We also need to consider the case where homogeneous boundary conditions are prescribed on $\Gamma = \partial\Omega$. We therefore give adapted definitions of Besov spaces. For a more detailed discussion we refer to [26, §3.10] and to [69, Chapter 1] for this issue in the context of Sobolev spaces.

Definition 3.2. For $p, q > 0$, $\alpha > 0$, we define the Besov spaces with boundary conditions $\mathring{B}^\alpha_{p,q}(\Omega)$ as the closure in the Besov space $B^\alpha_{p,q}(\Omega)$ of all smooth functions on Ω that vanish at order $t - 1$ on Γ.

In view of the frame construction from §2.3.2 and the fact that the wavelets are assumed to be piecewise smooth and globally C^{d-2}, where $\frac{d-t}{n} \geq \frac{1}{2}$ (recall §2.4), on the patches Ω_i of an open covering $\Omega = \bigcup_{i=0}^{m-1} \Omega_i$, we also give the following definition.

Definition 3.3. Let $p, q > 0$, $\alpha > 0$.

(i) We define the Besov spaces $\mathring{B}^\alpha_{p,q}(\Omega_i)$ on Ω_i as the closure in the Besov space $B^\alpha_{p,q}(\Omega_i)$ of all smooth functions on Ω_i that vanish at order $t - 1$ on $\partial\Omega_i \cap \Gamma$ and at order $d - 2$ on $\partial\Omega_i \cap \Omega$.

(ii) Equivalently, for $\alpha - \frac{1}{p} < d - 1$, $\mathring{B}^\alpha_{p,q}(\Omega_i)$ can be defined as the closure in $B^\alpha_{p,q}(\Omega_i)$ of the space of smooth functions on Ω_i that vanish at order $t - 1$ on $\partial\Omega_i \cap \Gamma$, and whose supports intersected with Ω are contained in Ω_i.

(iii) For $\alpha - \frac{1}{p} > t - 1$, $\alpha - \frac{1}{p} \notin \{0, \ldots, d - 2\}$, another equivalent definition is

$$\mathring{B}^\alpha_{p,q}(\Omega_i) := \{f \in B^\alpha_{p,q}(\Omega_i) : \gamma_l f = 0 \text{ on } \partial\Omega \cap \partial\Omega_i \text{ for } l = 0, \ldots, t - 1,$$
$$\gamma_l f = 0 \text{ on } \partial\Omega_i \cap \Omega \text{ for } l = 0, \ldots, \min(d - 2, \lfloor \alpha - \tfrac{1}{p} \rfloor)) \}.$$
$$(3.3.20)$$

In [26, Theorem 3.10.5, Remark 3.10.6] it is shown that Theorem 3.4 can be transferred to this context. In particular, for the pair of biorthogonal wavelet bases $(\Psi^{(i)}, \tilde\Psi^{(i)})$ from (2.3.12), we have the result that for the same range of s as in (3.3.17) it is

$$f \in \mathring{B}^{sn+t}_{\tau,\tau}(\Omega_i) \quad \text{if and only if} \quad \mathbf{c} = \{2^{t|\mu|}\langle f, \tilde\psi_{i,\mu}\rangle_{L_2(\Omega)}\}_{\mu \in \Lambda_i^\square} \in \ell_\tau(\Lambda_i^\square), \quad (3.3.21)$$

provided that $sn + t - \frac{1}{\tau} \notin \{0, \ldots, d - 2\}$.

3.4 Aggregated wavelet frames

The next step is to generalize Corollary 3.3 to the case of an aggregated wavelet Gelfand frame. In this situation, for the expansion coefficients $\mathbf{c} = \{2^{t|\lambda|}\langle f, \tilde\psi_\lambda\rangle_{L_2(\Omega)}\}_{\lambda \in \Lambda}$ of f with respect to *some* dual frame, $\text{dist}_{H^t(\Omega)}(f, S_N) \lesssim \sigma_N(\mathbf{c})$ can be show to hold as in (3.3.16) using the boundedness of F^* from (2.3.21). In general, the converse estimate fails to hold. For example, it can happen that the sequence of coefficients $\mathbf{d} := \{\langle 2^{-t|\mu|}\psi_\mu, 2^{t|\lambda|}\tilde\psi_\lambda\rangle_{L_2(\Omega)}\}_{\lambda \in \Lambda}$ of a special frame element $2^{-t|\mu|}\psi_\mu$ has infinite support. In this case $\sigma_N(\mathbf{d}) > 0$, for all $N \in \mathbb{N}$, but $\text{dist}_{H^t(\Omega)}(2^{-t|\mu|}\psi_\mu, S_N) = 0$ for all $N \geq 1$.

It turns out that the dual frame given in Proposition 2.11 is an adequate choice. We intend to show in the sequel that, under the same regularity assumptions on f as in Corollary 3.3, the sequence $\mathbf{c}$ is contained in $\ell^w_\tau(\Lambda)$, so that again $\text{dist}_{H^t(\Omega)}(f, S_N) = \mathcal{O}(N^{-s})$ can be ensured.

3.4.1 Basic assumptions

For the remainder of this chapter, we consider an aggregated wavelet Gelfand frame $\Psi_{L_2} = \{\psi_\lambda\}_{\lambda \in \Lambda}$ as in Proposition 2.11 with the dual frame $\tilde{\Psi}_{L_2} = \{\tilde{\psi}_\lambda\}_{\lambda \in \Lambda}$ defined in (2.3.28). We will use exactly the same notation as in §2.3.2 and assume that the wavelets have all the properties collected in §2.4.

We further assume that all local wavelet bases are exact of order d and that the nonlinear spaces S_N are defined analogous to (3.3.3). For the sake of simplicity, let us also assume that the elements of the wavelet bases $\Psi^{(i)}$ (cf. (2.3.12)) are sufficiently smooth to ensure (3.3.21) on Ω_i for all $0 < s < \frac{d-t}{n}$.

3.4.2 Coverings permitting a smooth partition of unity

In case a smooth partition of unity is available, the above aim is easy to achieve.

Proposition 3.5. *Assume that an arbitrary smooth partition of unity $\{\sigma_i\}_{i=0}^{m-1}$ subordinate to the open covering $\Omega = \bigcup_{i=0}^{m-1} \Omega_i$ exists. Let $f \in H_0^t(\Omega) \cap B_{\tau,\tau}^{sn+t}(\Omega) = \mathring{B}_{\tau,\tau}^{sn+t}(\Omega)$, $0 < s < \frac{d-t}{n}$, $\frac{1}{\tau} = s + \frac{1}{2}$, $sn + t - \frac{1}{\tau} \notin \{0, \ldots, d-2\}$. Then, the expansion coefficients of f with respect to $\mathbf{D}\tilde{\Psi}_{L_2}$, i.e., $\mathbf{c} = (\mathbf{c}^{(0)}, \ldots, \mathbf{c}^{(n)})$ where $\mathbf{c}^{(i)} := \{2^{t|\mu|} \langle f, \sigma_i \tilde{\psi}_{i,\mu} \rangle_{L_2(\Omega_i)}\}_{\mu \in \Lambda_i^\square}$, are contained in $\ell_\tau(\Lambda) \hookrightarrow \ell_\tau^w(\Lambda)$. Moreover, it is $\operatorname{dist}_{H^t(\Omega)}(f, S_N) = \mathcal{O}(N^{-s})$.*

Proof. Since σ_i is smooth, we have $\sigma_i f \in \mathring{B}_{\tau,\tau}^{sn+t}(\Omega_i)$. By (3.3.21) we get $\mathbf{c}^{(i)} := \{2^{t|\mu|} \langle \sigma_i f, \tilde{\psi}_{i,\mu} \rangle_{L_2(\Omega_i)}\}_{\mu \in \Lambda_i^\square} = \{2^{t|\mu|} \langle f, \sigma_i \tilde{\psi}_{i,\mu} \rangle_{L_2(\Omega_i)}\}_{\mu \in \Lambda_i^\square} \in \ell_\tau(\Lambda_i^\square)$, $i = 0, \ldots, m-1$. Hence, $(\mathbf{c}^{(0)}, \ldots, \mathbf{c}^{(n)}) \in \ell_\tau(\Lambda) \hookrightarrow \ell_\tau^w(\Lambda)$. In addition, we have $\operatorname{dist}_{H^t(\Omega)}(f, S_N) \lesssim \sigma_N(\mathbf{c})$ which by (3.3.15) completes the proof. $\square$

In §2.7 we have seen that one cannot always assume to be equipped with a smooth partition of unity. We will take care of this matter for the case of two-dimensional polygonal domains in the following. In particular, the prototypical case of the L-shaped domain shall be treated.

3.4.3 The L-shaped domain

As the prototype of a non-convex two-dimensional polygonal domain, we consider again the L-shaped domain $\Omega := (-1, 1)^2 \setminus [0, 1)^2$ with the open covering given by the patches $\Omega = \Omega_0 \cup \Omega_1$, $\Omega_0 = (-1, 0) \times (-1, 1)$, $\Omega_1 = (-1, 1) \times (-1, 0)$ as in §2.7 and the associated partition $\{\sigma_0, \sigma_1\}$ constructed there. As stated at the very end of §2.7, on Ω_0 and Ω_1 we cannot choose wavelets with homogeneous boundary conditions of order higher than $t - 1$ at internal boundaries. Therefore, (3.3.21) holds except for the cases where $sn + t - \frac{1}{\tau} \notin \{0, \ldots, t-1\}$. But, since $\frac{1}{\tau} = s + \frac{1}{2}$ and $n = 2$, it is $sn + t - \frac{1}{\tau} = s + t - \frac{1}{2} > t - 1$, so that no exception occurs in the present context.

It is our basic intention to apply the results on nonlinear wavelet approximation obtained in this chapter to solutions $u \in H_0^t(\Omega)$, $t \in \mathbb{N}$, of the elliptic operator equations $\mathcal{L}u = f$ from (1.3.5). In principle, we want to proceed as in the proof of Proposition 3.5. Therefore, a Besov regularity analysis of the functions $\sigma_i u$, $i = 0, \ldots, m-1$, is mandatory. However, to obtain satisfactory results, it is necessary to impose the following assumption on u.

Assumption 3.1.   (i) Let the function u be contained in $H_0^t(\Omega) \cap H^{t+\nu}(\Omega)$ for a $\nu > 0$.

(ii) Let for all multi-indices α with $|\alpha| = j$, $j = 0, \ldots, t$, the function u satisfy

$$D^\alpha u(x,y) = \mathcal{O}(r(x,y)^{\beta_j}), \text{ for } r(x,y) \to 0, \text{ with } \beta_j > t - (j+1). \quad (3.4.1)$$

Thus, we require some L_2-Sobolev regularity that goes beyond $H^t(\Omega)$, and, furthermore, a sufficiently fast decay of some derivatives of $u(x,y)$ is assumed when (x,y) approaches the re-entrant corner at the origin. It will be carried out later that, in case u represents the solution to the Poisson or the biharmonic equation with homogeneous Dirichlet boundary data, these assumptions are actually satisfied.

Note that if u satisfies part (ii) of the assumption, then without loss of generality we may also assume

$$t - j > \beta_j, \quad j = 0, \ldots, t.$$

In fact, if a function is of the order r^β, for $r \to 0$, then it is surely also of the order $r^{\tilde{\beta}}$ for all $\tilde{\beta} < \beta$. We will make use of this observation in the proof of the next proposition where we establish Sobolev regularity for

$$u_0 := \sigma_0 u = (\phi \circ \theta)u \quad \text{and} \quad u_1 := (1 - \sigma_0)u. \quad (3.4.2)$$

Proposition 3.6. *Let Assumption 3.1 be satisfied. Then u_0 and u_1 are contained in $H^{t+\eta}(\Omega)$ for all $0 < \eta < \eta^*$, where*

$$\eta^* := \min \left\{ \nu, 1 - t + \min_{0 \le j \le t}(j + \beta_j) \right\}. \quad (3.4.3)$$

For the proof of Proposition 3.6, we shall need the following Besov space multiplier theorem [99, Theorem 2 (ii), Section 4.4.4] and a simple conclusion thereof.

Theorem 3.5. *Suppose that the following conditions are valid:*

(i) $0 < s_1 < s_2$,   (ii) $\frac{1}{p} \le \frac{1}{p_1} + \frac{1}{p_2}$,

(iii) $\frac{2}{p} - s_1 = (\frac{2}{p_1} - s_1)_+ + (\frac{2}{p_2} - s_2)_+$ *and* $\max_i(\frac{2}{p_i} - s_i) > 0$,

(iv) $s_1 + s_2 > \frac{2}{p_1} + \frac{2}{p_2} - 2$,

(v) $q \ge q_1$,   (vi) $q \ge q_2$ *if* $s_1 - \frac{2}{p} = s_2 - \frac{2}{p_2}$,

(vii) $\{i \in \{1,2\} : s_i = \frac{2}{p_i} \text{ and } q_i > 1\} \cup \{i \in \{1,2\} : s_i < \frac{2}{p_i} \text{ and } q_i > \frac{2}{\frac{2}{p_i} - s_i}\} = \emptyset$.

Then, we have the continuous embedding $B^{s_1}_{p_1,q_1}(\Omega) \cdot B^{s_2}_{p_2,q_2}(\Omega) \hookrightarrow B^{s_1}_{p,q}(\Omega).$

Corollary 3.4. *Assume that* $f \in H^\nu(\Omega)$ *for some* $0 < \nu < 1$ *and that* $g \in H^\eta(\Omega)$ *for all* $\nu < \eta < 1$. *Then* fg *is contained in* $H^{\nu-\varepsilon}(\Omega)$ *for all* $\varepsilon > 0$.

Proof. It is sufficient to prove that $fg \in B^\nu_{p,2}(\Omega)$ for all $p \in (\frac{2}{2-\nu}, 2)$, since for any sufficiently small $\varepsilon > 0$ we can pick $p = \frac{4}{2+\varepsilon} < 2$ and append the continuous embeddings

$$B^\nu_{p,2}(\Omega) \hookrightarrow B^{\nu-\varepsilon/2}_{p,p}(\Omega) \hookrightarrow H^{\nu-\varepsilon/2-2(1/p-1/2)}(\Omega) = H^{\nu-\varepsilon}(\Omega),$$

where we have used (3.1.5) in the first step and Theorem 3.2 in the second one. For any $\frac{2}{2-\nu} < p < 2$, setting $s_2 = 2 - \frac{2}{p}$, we have $\nu < s_2 < 1$ and hence $g \in H^\eta(\Omega) \hookrightarrow H^{s_2}(\Omega)$ for all $s_2 < \eta < 1$. We shall apply Theorem 3.5 for $p_1 = p_2 = q_1 = q_2 = q = 2$ and $s_1 = \nu$ to finish the proof. In fact, by $s_2 > \nu = s_1$, condition (i) is satisfied. Using $p > \frac{2}{2-\nu} > 1$, we have $\frac{1}{p} < 1 = \frac{1}{p_1} + \frac{1}{p_2}$, and hence condition (ii) holds. Since $\nu < 1$, it is $\frac{2}{p_1} - s_1 = 1 - \nu > 0$, so that condition (iii) can be verified by $(\frac{2}{p_1} - s_1)_+ + (\frac{2}{p_2} - s_2)_+ - (\frac{2}{p} - s_1) = 2 - s_2 - \frac{2}{p} = 0$. Condition (iv) holds due to $s_1 + s_2 = \nu + s_2 > 0 = \frac{2}{p_1} + \frac{2}{p_2} - 2$, and condition (v) is true by the choice $q = q_1 = 2$. Since $s_1 - \frac{2}{p} - (s_2 - \frac{2}{p_2}) = \nu - 1 < 0$, there is nothing to prove for condition (vi). Finally, the first set in condition (vii) is empty by $p_i = 2$ and $s_i < 1$, and the second set is empty due to $\frac{2}{\frac{2}{p_1}-s_1} = \frac{2}{1-\nu} > 2 = q_1$ and $\frac{2}{\frac{2}{p_2}-s_2} = \frac{2}{\frac{2}{p}-1} > \frac{2}{1-\nu} > 2 = q_2$. $\qquad\square$

Proof of Proposition 3.6. It is sufficient to show that $u_0 \in H^{t+\eta}(\Omega)$ for $0 < \eta < \eta^*$, since then the statement for u_1 follows from $u_1 = (1 - \sigma_0)u$. Given any $\alpha \in \mathbb{N}_0^n$ with $|\alpha| = t$, it will be our strategy to verify that $D^\alpha u_0 \in H^\eta(\Omega)$ by showing that $D^\alpha u_0 \in W^1_p(\Omega)$ for some $p > 1$, since then the claim follows by Theorem 3.1. By the representation (2.7.4), we can decompose $D^\alpha u_0$ into

$$D^\alpha u_0 = (\phi \circ \theta)D^\alpha u + \sum_{\substack{0 \le \beta \le \alpha \\ \beta \ne 0}} \binom{\alpha}{\beta} D^\beta(\phi \circ \theta)D^{\alpha-\beta}u. \tag{3.4.4}$$

We shall treat both addends in (3.4.4) separately. For the first one, note that $D^\alpha u \in H^\nu(\Omega)$ by Assumption 3.1 (i). Moreover, using (2.7.7), we clearly have

$$\left\| \frac{\partial(\phi \circ \theta)}{\partial x_j} \right\|^p_{L_p(\Omega)} = \int_\Omega \left| h_{e_j} \circ \theta \right|^p r^{-p} \, \mathrm{d}x_1 \, \mathrm{d}x_2 \lesssim \int_0^{\sqrt{2}} r^{1-p} \, \mathrm{d}r,$$

which is finite for all $1 \le p < 2$. Hence, by the Sobolev embedding theorem, it is $\phi \circ \theta \in H^\eta(\Omega)$ for all $0 < \eta < 1$, so that Corollary 3.4 implies $(\phi \circ \theta)D^\alpha u \in H^\eta(\Omega)$ for all $0 < \eta < \min\{\nu, 1\}$.

It remains to study the rightmost sum in (3.4.4). For one single addend, (2.7.7), $|\alpha| = t$ and Assumption 3.1 (ii) imply that for $k = 1, 2$

$$\left\| \frac{\partial (D^\beta (\phi \circ \theta) D^{\alpha - \beta} u)}{\partial x_k} \right\|_{L_p(\Omega)}^p \lesssim \int_\Omega |(h_{\beta + e_k} \circ \theta) D^{\alpha - \beta} u|^p r^{-(|\beta| + 1)p} \, dx_1 \, dx_2$$

$$+ \int_\Omega |(h_\beta \circ \theta) D^{\alpha + e_k - \beta} u|^p r^{-|\beta| p} \, dx_1 \, dx_2$$

$$\lesssim \int_0^{\sqrt{2}} (r^{\beta_{t - |\beta|} - |\beta| - 1} + r^{\beta_{t + 1 - |\beta|} - |\beta|})^p r \, dr$$

$$\lesssim \int_0^{\sqrt{2}} r^{p \min\{\beta_{t - |\beta|} - |\beta| - 1, \beta_{t + 1 - |\beta|} - |\beta|\} + 1} \, dr < \infty,$$

if $p \min\{\beta_{t - |\beta|} - |\beta| - 1, \beta_{t + 1 - |\beta|} - |\beta|\} > -2$. Since $\beta_j < t - j$ for all $j \leq t$, the sufficient condition on p is equivalent to $\frac{2}{p} > \max\{|\beta| + 1 - \beta_{t - |\beta|}, |\beta| - \beta_{t + 1 - |\beta|}\} > 1$. Consequently, the rightmost sum in (3.4.4) is contained in $W_p^1(\Omega)$ for

$$\frac{2}{p} > \max_{1 \leq j \leq t} \max\{j + 1 - \beta_{t - j}, j - \beta_{t + 1 - j}\} = \max_{0 \leq j \leq t} (t - j + 1 - \beta_j). \qquad (3.4.5)$$

Using the Sobolev embedding $W_p^1(\Omega) \hookrightarrow H^{2 - 2/p}(\Omega)$, it follows for the whole sum (3.4.4) that $D^\alpha u_0 \in H^\eta(\Omega)$ whenever

$$\eta < \min\{\nu, 1, 2 - \max_{0 \leq j \leq t} (t - j + 1 - \beta_j)\} = \min\left\{\nu, 1, 1 - t + \min_{0 \leq j \leq t} (j + \beta_j)\right\} = \eta^*,$$

where we have used once more that $\beta_j < t - j$ in the last step. Note that η^* is strictly positive by $\beta_j > t - (j + 1)$. $\qquad \square$

We are now in a position to formulate the main result of this section.

Theorem 3.6. *Let Assumption 3.1 be satisfied. For some $s > t$ and $\delta \in (0, s - t)$, let $u \in B_{\tau, \tau}^s(\Omega)$, where $\frac{1}{\tau} = \frac{s - (t + \delta)}{2} + \frac{1}{2}$. Then, the sequence of frame coefficients $(2^{t|\mu|} \langle u, \sigma_i \tilde{\psi}_{i, \mu} \rangle)_{\mu \in \Lambda_i^\square, i = 0, 1}$ belongs to the space $\ell_{\tau_0}(\Lambda)$, where $\frac{1}{\tau_0} = \frac{s' - t}{2} + \frac{1}{2}$, for all $t < s' < \min\{d, \frac{\eta^* s + t - 1}{t + \eta^* - 1} + t - 1\} =: s^*$, where η^* is given by (3.4.3).*

Here we have assumed that u is contained in a scale of Besov spaces which lies slightly above the classical scale of spaces which govern the convergence rates for best N-term approximation in $H^t(\Omega)$, see the DeVore/Triebel diagram in Figure 3.3. Below we will show that, under mild regularity assumptions on the right-hand side f, also this requirement is satisfied for the weak solutions of the Poisson and the biharmonic equation with homogeneous Dirichlet boundary conditions.

The proof of Theorem 3.6 will be based on the embedding theorems for Sobolev spaces and on interpolation arguments between Sobolev and Besov spaces. The first step is now to establish Besov regularity for u_0 and u_1. Before this, we analyze σ_0.

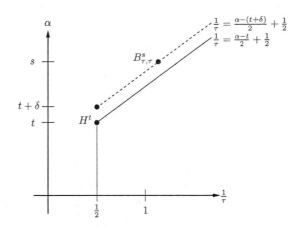

Figure 3.3: Classical scale of Besov spaces governing best N-term approximation in $H^t(\Omega)$ w.r.t. wavelet bases (solid line), and the relevant scale for aggregated wavelet frames in Theorem 3.6 (dashed line).

Lemma 3.1. *For any $s \geq 1 - \varepsilon$, the function $\sigma_0 = \phi \circ \theta$ is contained in $B^s_{\tau,\tau}(\Omega)$, for $\frac{1}{\tau} = \frac{s-(1-\varepsilon)}{2} + \frac{1}{2}$, where ε may be chosen arbitrarily small.*

Proof. As we have seen in the proof of Proposition 3.6, $\sigma_0 = \phi \circ \theta \in H^\alpha(\Omega)$ for all $\alpha \in (0,1)$. Furthermore, analogously to the proof of [33, Theorem 2.3], one can show that $\sigma_0 \in B^{s'}_{\tau,\tau}(\Omega)$, $\frac{1}{\tau} = \frac{s'}{2} + \frac{1}{2}$ for any $s' > 0$. The asserted statement then immediately follows by employing interpolation between the latter Besov spaces and $H^\alpha(\Omega)$, $\alpha \in (0,1)$, i.e., by an application of Proposition 3.1. $\qquad\square$

Using Lemma 3.1 and Theorem 3.5, we obtain the following result.

Lemma 3.2. *Let $s > t$. If u is contained in $B^s_{\tau,\tau}(\Omega)$, for a $\delta \in (0, s - t)$ and $\frac{1}{\tau} = \frac{s-(t+\delta)}{2} + \frac{1}{2}$, then $u_0 = \sigma_0 u \in B^{s'}_{\tau',\tau'}(\Omega)$, $\frac{1}{\tau'} = \frac{s'-(1-\varepsilon)}{2} + \frac{1}{2}$ for any $s' \in [1 - \varepsilon, s)$ and for any $\varepsilon \in (0, 2 + \delta)$.*

Proof. Lemma 3.1 tells us that $\sigma_0 \in B^{s'}_{\tau',\tau'}(\Omega)$, $\frac{1}{\tau'} = \frac{s'-(1-\varepsilon)}{2} + \frac{1}{2}$ for any $s' > 0$. It is therefore sufficient to verify that $B^{s'}_{\tau',\tau'}(\Omega) \cdot B^s_{\tau,\tau}(\Omega) \hookrightarrow B^{s'}_{\tau',\tau'}(\Omega)$, for $s > s'$, by an application of Theorem 3.5. To this end, let us denote $s_2 = s$, $s_1 = s'$, $p_1 = q_1 = p = q = \tau'$ and $p_2 = q_2 = \tau$. Since $s > s' > 0$, condition (i) holds, and the validity of condition (ii) follows from $p = p_1$ and $p_2 > 0$. We clearly have $\frac{2}{p_1} - s_1 = \frac{2}{\tau'} - s' = \varepsilon > 0$, so that condition (iii) is verified by $\frac{2}{p_2} - s_2 = -\delta - t + 1 < 0$. Furthermore, we have $s_1 + s_2 = \frac{2}{p_1} + \frac{2}{p_2} - \varepsilon + \delta + t - 1 > \frac{2}{p_1} + \frac{2}{p_2} - 2$, for any $0 < \varepsilon < \delta + 2$, and since $\frac{2}{p_2} - s_2 = -\delta - t + 1 \neq \varepsilon = \frac{2}{p_1} - s_1$ both conditions (iv) and (vi) are satisfied. For condition (v), there is nothing to prove due to $q = q_1 = \tau'$. Moreover, the set $\{i \in \{1,2\} : s_i = \frac{2}{p_i}$ and $q_i = p_i > 1\}$ is clearly empty. We have

77

$s_2 > \frac{2}{p_2}$ and $\frac{2}{s_1+\varepsilon} = p_1 < \frac{2}{\frac{2}{p_1}-s_1} = \frac{2}{\varepsilon}$, for $\varepsilon \in (0, 2+\delta)$, ensuring the validity of condition (vii). Therefore, we can apply Theorem 3.5 and conclude the proof. $\qquad\square$

We are now prepared to prove the main result of this section.

Proof of Theorem 3.6. Lemma 3.2 implies $u_i = \sigma_i u \in B^{s'}_{\tau',\tau'}(\Omega)$, $\frac{1}{\tau'} = \frac{s'-(1-\varepsilon)}{2} + \frac{1}{2}$ for any $s' \in [1-\varepsilon, s)$ and for any $\varepsilon \in (0, 2+\delta)$. Moreover, Proposition 3.6 gives $u_i \in H^{t+\eta}(\Omega)$ for each $0 < \eta < \eta^*$, and by using interpolation between $H^{t+\eta}(\Omega)$ and $B^{s'}_{\tau',\tau'}(\Omega)$, it can be easily checked that $u_i \in B^{t_0}_{\tau_0,\tau_0}(\Omega)$, where $\frac{1}{\tau_0} = \frac{t_0-t}{2} + \frac{1}{2}$ and $t_0 = (t+\eta+\varepsilon-1)^{-1}((\eta+1)(s'+\varepsilon-1)-(s'-t-\eta))+t-1$. Indeed, t_0 is obtained by computing the intersection of the bold-faced and the dashed line in the DeVore/Triebel diagram in Figure 3.4. For $\varepsilon \to 0$ and $s' \to s$, t_0 tends to $\frac{\eta s+t-1}{t+\eta-1}+t-1$. Consequently, since ε may be chosen arbitrarily small, and since s' may be chosen arbitrarily close to s, it follows that $u_i \in B^{\breve{s}}_{\breve{\tau},\breve{\tau}}(\Omega)$, where $\frac{1}{\breve{\tau}} = \frac{\breve{s}-t}{2}+\frac{1}{2}$, for all $t < \breve{s} < \frac{\eta s+t-1}{t+\eta-1}+t-1$. Since $s > t \geq 1$, the fraction $\frac{\eta s+t-1}{t+\eta-1}$ is strictly monotonically increasing in η in case $t > 1$, so that we may replace η by η^*, and it is equal to s independent of η if $t = 1$. Finally, from Lemma 2.4 and the fact that the local systems $\Psi^{(i)}$ are Riesz bases of order d with dual bases $\tilde{\Psi}^{(i)}$, it can be inferred from (3.3.21) that the sequence of wavelet coefficients $(2^{t|\mu|}\langle u, \sigma_i \tilde{\psi}_{i,\mu}\rangle)_{\mu\in\Lambda_i^{\square},i=0,1}$ belongs to the space $\ell_{\breve{\tau}}(\Lambda)$ for $\frac{1}{\breve{\tau}} = \frac{\breve{s}-t}{2}+\frac{1}{2}$ for all $t < \breve{s} < \min\{d, \frac{\eta^* s+t-1}{t+\eta^*-1}+t-1\}$. $\qquad\square$

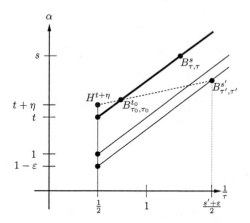

Figure 3.4: DeVore/Triebel diagram corresponding to the proof of Theorem 3.6.

In the following it will be pointed out that for two prominent examples of elliptic problems, i.e., the Poisson and the biharmonic equation, Assumption 3.1 is satisfied and that the additional Besov regularity of u assumed in Theorem 3.6 is guaranteed. We thus can apply Theorem 3.6 to these special cases.

The Poisson equation

Let us first consider the Poisson equation with homogeneous boundary conditions in the L-shaped domain, $t = 1$. First of all, we state the following well-known theorem from [70, Sections 2.4, 2.7] which characterizes the weak solution to problem (1.3.7) in an arbitrary polygonal domain.

Theorem 3.7. *Let $\Omega \subset \mathbb{R}^2$ be a polygonal domain with vertices S_l, $l = 1 \ldots, M$. Let the measures of the inner angles at S_l be denoted with ω_l. With respect to polar coordinates (r_l, θ_l) in the vicinity of each vertex S_l, we introduce the functions*

$$\mathcal{S}_{l,m}(r_l, \theta_l) = \zeta_l(r_l) r_l^{\lambda_{l,m}} \sin(m\pi\theta_l/\omega_l), \text{ when } \lambda_{l,m} = m\pi/\omega_l \notin \mathbb{Z}, \tag{3.4.6}$$

$$\mathcal{S}_{l,m}(r_l, \theta_l) = \zeta_l(r_l) r_l^{\lambda_{l,m}} [\log r_l \sin(m\pi\theta_l/\omega_l) + \theta_l \cos(m\pi\theta_l/\omega_l)] \text{ otherwise.} \tag{3.4.7}$$

Here ζ_l denotes a suitable C^∞ truncation function, and $m \geq 1$ is an integer. Then, for a given $f \in H^s(\Omega)$, $s \geq -1$, the corresponding variational solution to (1.3.7) has an expansion $u = u_R + u_S$, where $u_R \in H^{s+2}(\Omega)$ and

$$u_S - \sum_{l=1}^{M} \sum_{0 < \lambda_{l,m} < s+1} c_{l,m} \mathcal{S}_{l,m}, \tag{3.4.8}$$

provided that no $\lambda_{l,m}$ is equal to $s + 1$.

Usually u_S and u_R are called *singular* and *regular part* of the solution, respectively. For the special case of the L-shaped domain, we denote the re-entrant corner with S_1, where the measure of the inner angle is $\frac{3}{2}\pi$, and thus $\lambda_{1,m} = \frac{2}{3}m$, while at all other vertices S_l, $l \neq 1$, we have $\omega_l = \frac{\pi}{2}$ and $\lambda_{l,m} = 2m$. The following lemma shows that Assumption 3.1 is satisfied.

Lemma 3.3. *Let u be the variational solution to (1.3.7) with a right-hand side $f \in H^\mu(\Omega)$ for some $\mu > 0$. Then, there exists a decomposition $u = u_S + u_R$ such that $u_S \in H^\alpha(\Omega)$, for any $\alpha < 5/3$ and $u_R \in H^{\alpha'}(\Omega)$ for all $\alpha' < 2 + \min\{\frac{1}{3}, \mu\}$. Moreover, $u_R(x, y) = \mathcal{O}(r(x, y))$, and $\nabla u_R(x, y) = \mathcal{O}(1)$, for $r \to 0$, whereas the singular part satisfies $u_S(x, y) = \mathcal{O}(r(x, y)^{2/3})$, as well as $\nabla u_S(x, y) = \mathcal{O}(r(x, y)^{-1/3})$, for $r \to 0$.*

Proof. Let $0 < \mu' < \min\{\frac{1}{3}, \mu\}$. Surely, $f \in H^{\mu'}(\Omega)$. We apply Theorem 3.7 for $s = \mu'$ and notice that $u_S = c_{1,1}\mathcal{S}_{1,1} \in H_0^1(\Omega)$. Hence, we have $u_R \in H_0^1(\Omega) \cap H^{2+\mu'}(\Omega)$. For u_S, using [70, Theorem 1.2.18], it can be easily inferred that it is contained in $H^\alpha(\Omega)$, for any $\alpha < 5/3$. The rest of the statements on u_S immediately follow from (3.4.6) in Theorem 3.7 and $\lambda_{1,1} = \frac{2}{3}$. In order to prove the remaining statements for u_R, it will be convenient to show that $u_R \in \mathring{W}_p^1(\Omega)$ for all $2 \leq p < \infty$. Indeed, since $u_R \in H^{2+\mu'}(\Omega)$, the Sobolev embedding theorem implies $u_R \in W_p^1(\Omega)$, for all $2 \leq p < \infty$. Furthermore, using $u_R \in H_0^1(\Omega)$ and taking the trace in the H-scale, i.e. for $p = 2$, we get $\|\gamma u_R\|_{H^{\frac{1}{2}}} = 0$. Since $u_R \in H^{2+\mu'}(\Omega)$, and thus for sure $u_R \in C^0(\overline{\Omega})$, any trace operator applied to u_R, i.e. for arbitrary $2 \leq p < \infty$, gives the same result, namely $\gamma u_R = 0$. Hence, $u_R \in \overset{\circ}{W}_p^1(\Omega)$, for all $2 \leq p < \infty$.

Now Theorem 2.7 tells us that $\rho^{-1+|\alpha|} D^\alpha u_R \in L_p(\Omega)$ for all $|\alpha| \leq 1$, and in addition also

$$\|\rho^{-1+|\alpha|} D^\alpha u_R\|_{L_p(\Omega)} \lesssim \|u_R\|_{W_p^1(\Omega)}, \quad \text{for all } |\alpha| \leq 1, \tag{3.4.9}$$

and all $2 \leq p < \infty$, with a constant that does not depend on p; recall (2.7.3). Therefore the task is to prove that the right-hand side in (3.4.9) can be uniformly bounded as p approaches infinity, because in that case it follows that $\rho^{-1+|\alpha|} D^\alpha u \in L_\infty(\Omega)$, $|\alpha| \leq 1$; cf. Lemma 3.4 below. Together with $r^{-1} \leq \rho^{-1}$, this yields $u_R(x, y) = \mathcal{O}(r(x, y))$ and $\nabla u_R(x, y) = \mathcal{O}(1)$, and thus the proof would be completed.

To this end, we will again make use of the fact that $u_R \in H^{2+\mu'}(\Omega)$. Note that by the continuity of the embeddings $H^{2+\mu'}(\Omega) = B_{2,2}^{2+\mu'}(\Omega) \hookrightarrow B_{\infty,\infty}^{1+\mu'}(\Omega) \hookrightarrow C^1(\overline{\Omega})$, we may infer $u_R \in C^1(\overline{\Omega})$, so that the boundedness of the first order derivative of u_R is anyway guaranteed. Hence,

$$\|u_R\|_{W_p^1(\Omega)} \leq \sum_{|\gamma| \leq 1} \|D^\gamma u_R\|_{L_p(\Omega)} \leq |\Omega|^{1/p} \sum_{|\gamma| \leq 1} \|D^\gamma u_R\|_{L_\infty(\Omega)},$$

where the upper bound stays bounded as p tends to infinity. $\qquad\square$

Lemma 3.4. *If $f \in L_p(\Omega)$ for all $1 \leq p < \infty$ with $\|f\|_{L_p(\Omega)} \leq C_1$, then $f \in L_\infty(\Omega)$ with $\|f\|_{L_\infty(\Omega)} \leq C_1$.*

Proof. Although this Lemma is well-known, cf. [58], for the reader's convenience we state its proof here. Let us suppose that the statement is wrong. Then, there exists a set A of positive measure, $\mu(A) > 0$, and a constant $C_2 > C_1$ such that $|f(x)| \geq C_2$ in A. Hence, $\int_\Omega |f(x)|^p dx \geq \int_A |f(x)|^p dx \geq \mu(A) C_2^p$. Therefore, since $\mu(A)^{1/p}$ tends to 1 as p tends to infinity, it holds $\|f\|_{L_p(\Omega)} > C_1$ for a sufficiently large p, which is a contradiction. $\qquad\square$

Finally, we are now in a position to prove the following special result which is an application of Theorem 3.6.

Corollary 3.5. *Let u be the variational solution to (1.3.7). Let the right-hand side f be contained in $H^\mu(\Omega)$ for some $\mu > 0$. Then, the sequence of frame coefficients $(2^{|\mu|} \langle u, \sigma_i \tilde{\psi}_{i,\mu} \rangle)_{\mu \in \Lambda_i^\Box, i=0,1}$ belongs to the space $\ell_{\tau_0}(\Lambda)$, where $\frac{1}{\tau_0} = \frac{s-1}{2} + \frac{1}{2}$, for all $1 < s < \min\{d, \mu + 2\}$.*

Proof. We want to apply Theorem 3.6 for $t = 1$ and $\mathcal{L} = -\Delta : H_0^1(\Omega) \to H^{-1}(\Omega)$. Note first that Lemma 3.3 shows that Assumption 3.1 is satisfied. Moreover, [33, Theorem 2.4] implies that the variational solution u to (1.3.7) is contained in $B_{\tau,\tau}^s(\Omega)$, for all $\frac{3}{2} < s < \mu + 2$, $\frac{1}{\tau} = \frac{s-3/2}{2} + \frac{1}{2} = \frac{s-(t+1/2)}{2} + \frac{1}{2}$, if $f \in H^\mu(\Omega)$, for $\mu > -1/2$. Therefore, the assumption on the Besov regularity of u in Theorem 3.6 is fulfilled for

$\delta = \frac{1}{2}$ and any $\frac{3}{2} < s < \mu + 2$. Consequently, Theorem 3.6 tells us that the sequence of coefficients $(2^{t|\mu|}\langle u, \sigma_i\tilde{\psi}_{i,\mu}\rangle)_{\mu \in \Lambda_i^\square, i=0,1}$ is contained in the space $\ell_{\tau_0}(\Lambda)$, where $\frac{1}{\tau_0} = \frac{s'-t}{2} + \frac{1}{2}$, for all $t < s' < \min\{d, \frac{\eta^* s + t - 1}{t + \eta^* - 1} + t - 1\} =: s^*$. Note that with $t = 1$ we have $s^* = \min\{d, \frac{\eta^* s + 1 - 1}{1 + \eta^* - 1} + 1 - 1\} = \min\{d, s\}$ independent of η^*, completing the proof. □

Remark 3.6.   (i) It is important to stress the fact that in the case of a wavelet basis with similar regularity and approximation properties, for the unique expansion coefficients, the analogous statement holds under the only slightly milder requirement $f \in H^\mu(\Omega)$, $\mu > -1/2$.

(ii) The crucial result Theorem 3.7 can also be generalized to the case of elliptic second order operators with variable coefficients by imposing rather mild regularity assumptions on the coefficients; see [69, Chapter 5.2].

Example 3.1. *Consider the case $u = u_S = \mathcal{S}_{1,1}$, and $f = -\Delta u_S$. In [33] it has been shown that for any $\alpha > 0$ each function $\mathcal{S}_{l,m}$, and thus also u_S, is contained in $B_{\tau,\tau}^\alpha(\Omega)$, $\frac{1}{\tau} = \frac{\alpha}{2} + \frac{1}{2}$. Together with the fact that $u_S \in H^\omega(\Omega)$, with $\omega < 5/3$, this implies that in this case the assumption of Theorem 3.6 is satisfied for any $s > 1$. Moreover, the right-hand side f vanishes in a vicinity of the re-entrant corner and is contained in $C^\infty(\overline{\Omega})$. Consequently, the sequence of frame coefficients $(2^{|\mu|}\langle u, \sigma_i\tilde{\psi}_{i,\mu}\rangle)_{\mu \in \Lambda_i^\square, i=0,1}$ belongs to $\ell_\tau(\Lambda)$, and thus to $\ell_\tau^w(\Lambda)$, for $\frac{1}{\tau} = \frac{\gamma-1}{2} + \frac{1}{2}$ and for all $1 < \gamma < d$. This means that the convergence order of the best N-term frame approximation is $\mathcal{O}(N^{-(\gamma-1)/2})$, for any $1 < \gamma < d$, so that in principle the order of convergence is only limited by the order d of the wavelets. For the case of linear approximation, Corollary 3.1 yields that the rate is strictly limited by $(5/3-1)/2 = 1/3$, no matter how large d is chosen. Hence, in this situation, the nonlinear approximation is really superior to the linear case.*

The biharmonic equation

As another important application of Theorem 3.6 we study the biharmonic problem, $t = 2$,

$$
\begin{aligned}
\Delta^2 u &= f \text{ in } \Omega, \\
u = \frac{\partial u}{\partial n_l} &= 0 \text{ on } \Gamma_l, \ l = 1, \ldots, 6,
\end{aligned}
\tag{3.4.10}
$$

where the Γ_l denote the open segments of $\partial\Omega$ and clearly n_l the outward normal at Γ_l. We denote, as above, with S_l, $l = 1, \ldots, 6$, the vertices of the L-shaped domain and declare S_1 to be the re-entrant corner. The counterpart of Theorem 3.7 for the biharmonic equation reads as follows; cf. [70, Theorem 3.4.4].

Theorem 3.8. *Let $u \in H_0^2(\Omega)$ be the solution of (3.4.10) with f given in $L_2(\Omega)$. Then, there exists a function $u_R \in H^4(\Omega)$ and constants c_l, $l = 0, \ldots, 7$, such that u can be written as $u = u_R + u_S$ with*

$$u_S = \zeta_1(r_1)(c_0 r_1^{1+z_0} v_0(\theta_1) + c_1 r_1^{1+z_1} v_1(\theta_1) + c_2 r_1^2 v_2(\theta_1)) + \sum_{l=2}^{6} c_{l+1} \zeta_l(r_l) r_l^2 v_{l+1}(\theta_l),$$

(3.4.11)

where $z_0 \approx 0.5445$, $z_1 \approx 0.9085$. Clearly, (r_l, θ_l) denote polar coordinates with respect to the corners S_l, $l = 1, \ldots, 6$, whereas ζ_l are suitable smooth cut-off functions. Moreover, the functions v_k are smooth, $k = 0, \ldots, 7$, with $v_k(0) = v_k(\frac{3}{2}\pi) = v_k'(0) = v_k'(\frac{3}{2}\pi) = 0$, $k = 0, 1, 2$, and $v_k(0) = v_k(\frac{\pi}{2}) = v_k'(0) = v_k'(\frac{\pi}{2}) = 0$, $k = 3, \ldots, 7$.

From (3.4.11) and the boundary conditions of the functions v_k, we may also infer that u_S and therefore also u_R is contained in $H_0^2(\Omega)$. Another application of [70, Theorem 1.2.18] shows that u_S is at least contained in $H^{2.54}(\Omega)$, so that Assumption 3.1 (i) for u_S is satisfied for $\nu = 0.54$, provided that $f \in L_2(\Omega)$.

Lemma 3.5. *Let u be the variational solution to (3.4.10) with a right-hand side $f \in L_2(\Omega)$. Then, Assumption 3.1 (ii) is satisfied. In particular, it holds $D^\alpha u_S(x, y) = \mathcal{O}(r(x, y)^{1+0.54-j})$ and $D^\alpha u_R(x, y) = \mathcal{O}(r(x, y)^{2-j})$, $|\alpha| = j$, $j = 0, 1, 2$, $r \to 0$.*

By (3.4.11), the statement for u_S is obvious. The proof of the result for the regular part u_R works in a completely analogous way to the proof of Lemma 3.3 and can therefore be omitted.

Finally, an application of Theorem 3.6 yields the following result.

Corollary 3.6. *Let u be the variational solution to (3.4.10) with a right-hand side $f \in L_2(\Omega)$. Then, the sequence of frame coefficients $(2^{2|\mu|} \langle u, \sigma_i \tilde{\psi}_{i,\mu} \rangle)_{\mu \in \Lambda_i^\square, i=0,1}$ belongs to the space $\ell_{\tau_0}(\Lambda)$, where $\frac{1}{\tau_0} = \frac{s'-2}{2} + \frac{1}{2}$, for all $2 < s' < \min\{d, 3.05\}$.*

Proof. In order to be able to apply Theorem 3.6, it remains to establish Besov regularity of u in the scale $\frac{1}{\tau} = \frac{s-(2+\delta)}{2} + \frac{1}{2}$ for a $\delta > 0$ and an $s > 2 + \delta$. To this end, note first that following the lines of [33], it can be shown that the singular part satisfies $u_S \in B^\alpha_{\tau_1, \tau_1}(\Omega)$, for any $\alpha > 0$, $\frac{1}{\tau_1} = \frac{\alpha}{2} + \frac{1}{2}$. Then, by interpolation between $H^{2.54}(\Omega)$ and $B^\alpha_{\tau_1, \tau_1}(\Omega)$, $\frac{1}{\tau_1} = \frac{\alpha}{2} + \frac{1}{2}$ (Proposition 3.1), we obtain $u_S \in B^s_{\tau, \tau}(\Omega)$, for all $s > 2 + \delta$, where $\frac{1}{\tau} = \frac{s-(2+\delta)}{2} + \frac{1}{2}$ and $\delta \in (0, 0.54)$ is fixed. Using the embedding $H^4(\Omega) = B^4_{2,2}(\Omega) \hookrightarrow B^4_{\tau,2}(\Omega) \hookrightarrow B^s_{\tau, \tau}(\Omega)$, we get $u_R \in B^s_{\tau, \tau}(\Omega)$, for $2 + \delta < s < 4$, $\frac{1}{\tau} = \frac{s-(2+\delta)}{2} + \frac{1}{2}$ (cf. (3.1.5)). Hence, the same holds for $u = u_S + u_R$. Inserting the values for β_j from Lemma 3.5 into Proposition 3.6, we obtain $\eta^* = \min\{0.54, -1 + \min_{0 \leq j \leq 2}(j + \beta_j)\} = 0.54$. Hence, an application of Theorem 3.6 with $t = 2$, and $2 + \delta < s < 4$ implies that the sequence of coefficients $(2^{2|\mu|} \langle u, \sigma_i \tilde{\psi}_{i,\mu} \rangle)_{\mu \in \Lambda_i^\square, i=0,1}$ is contained in $\ell_{\tau_0}(\Lambda)$, where $\frac{1}{\tau_0} = \frac{s'-2}{2} + \frac{1}{2}$, for all $2 < s' < \min\{d, \frac{\eta^* s + t - 1}{t + \eta^* - 1} + t - 1\} = \min\{d, \frac{0.54s+1}{1.54} + 1\}$. The claim follows by choosing s close to 4. $\qquad\square$

Remark 3.7. For wavelets of order $d \leq 3$, Corollary 3.6 gives the desired decay rate $s < \frac{d-2}{2}$. The outline to deduce more general results, addressing the case of smoother right hand sides $f \in H^\mu(\Omega)$, $\mu > 0$, is to apply generalizations of Theorem 3.8, as they can be found in the book of Grisvard, cf. [69, Theorem 7.2.2.3].

Theorem 3.7 and Theorem 3.8 show that for the case of polygonal domains, the solutions of the Poisson and the biharmonic equation exhibit a limited L_2-Sobolev regularity, even if the right-hand side is a smooth function, due to the singularities appearing near the corners. In this situation, the convergence rate of linear wavelet approximation as in §3.3.1 is strictly limited. But in the above analysis we have frequently used the fact that in the scale of Besov spaces being relevant for N-term approximation, the singular parts are arbitrary smooth, so that the convergence rates may increase with the smoothness of the right-hand side.

Chapter 4

An Adaptive Steepest Descent Wavelet Frame Algorithm

This chapter is dedicated to the development of an *adaptive* numerical wavelet frame method for the iterative solution of elliptic operator equations. The method will be based on a steepest descent iteration for the solution of the consistent linear system $\mathbf{Au} = \mathbf{f}$ from (2.5.4). The fundamental aim is to design the method in such a way that it realizes the same convergence rate as the best N-term wavelet frame approximations for u in linear time. The adaptive solution of infinite-dimensional systems using a steepest descent method has also been studied in [20], however, there the results were restricted to the case of a Riesz basis, whereas we are concerned with frames.

In [104] it has been shown that the discrete operator equation (2.5.4) can be solved by applying a damped Richardson iteration using appropriate implementable numerical routines for the finite realization of the involved matrix-vector operations; see also [38, 97, 117]. The basic deficiency of this method though is the need of a good estimate of the *optimal relaxation parameter* $2/(\|\mathbf{A}\| + \|\mathbf{A}^{\dagger}\|^{-1})$ where $\mathbf{A}^{\dagger}$ represents the *Moore-Penrose pseudo-inverse* of $\mathbf{A}$ and $\|\mathbf{A}\|$ its spectral norm. Particularly, $\|\mathbf{A}^{\dagger}\|$ is very hard to access; see [117] for a detailed discussion of this matter. This is the main motivation to develop in this chapter an algorithm based on the steepest descent method, in which the relaxation parameter is, as usual, adapted in each iteration, and it can be directly computed from available quantities.

In a first step, the optimality requirement for an adaptive algorithm will be particularized. Afterwards, the convergence of the exact iterative scheme as a solver in $\ell_2(\Lambda)$ is discussed, including the case when the descent direction and the descent parameter are perturbed up to some admissible tolerance. Then, an implementable adaptive scheme is developed, and its convergence and optimality in the above sense is proved. At the end of this chapter, in §4.4, the results of numerical experiments will be presented, showing that the adaptive scheme based on the steepest descent approach may outperform the damped Richardson iteration in case the relaxation parameter within the latter is not optimally chosen.

4.1 Optimality of adaptive algorithms

The results in the preceding chapter have shown that in case u exhibits singularities, as it is the case for certain elliptic operator equations on non-smooth domains, the use of adaptive algorithms really pays off. Indeed, note that the results on linear wavelet approximation from §3.3.1 correspond to such numerical algorithms where the discretization is uniformly refined. In this context, the singularities often deteriorate the rate of convergence which can be achieved (cf. Example 3.1), whereas this is possibly not happening in the nonlinear case, as we have learned in §3.3.2 and §3.4.

Recall that the rate of convergence $s > 0$ in the estimate (3.3.9) clearly describes the trade-off between the accuracy of the approximation and its complexity in the case of N-term approximation. Note further that the largest s for which this estimate can be guaranteed represents the natural benchmark for the best convergence rate a sequence of adaptive approximations generated by any numerical algorithm can exhibit. We specify this observation.

Definition 4.1. Let Ψ_{L_2} be an aggregated wavelet frame for $(H_0^t(\Omega), L_2(\Omega), H^{-t}(\Omega))$ and set $\Psi := \mathbf{D}^{-1}\Psi_{L_2}$. Moreover, let for some $s > 0$ the solution $u \in H_0^t(\Omega)$ to problem (1.3.4) have a representation $u = \tilde{\mathbf{u}}^\top \Psi$ with $\tilde{\mathbf{u}} \in \ell_\tau^w(\Lambda)$, $\frac{1}{\tau} = s + \frac{1}{2}$. Then, a numerical algorithm, producing for any $\varepsilon > 0$ a $\mathbf{u}_\varepsilon \in \ell_2(\Lambda)$ with

$$\|u - \mathbf{u}_\varepsilon^\top \Psi\|_{H^t(\Omega)} \lesssim \varepsilon \quad \text{and} \quad \# \operatorname{supp} \mathbf{u}_\varepsilon \lesssim \varepsilon^{-1/s} |\tilde{\mathbf{u}}|^{1/s}_{\ell_\tau^w(\Lambda)}, \tag{4.1.1}$$

taking a number of operations and storage locations to compute $\mathbf{u}_\varepsilon$ that can also be bounded by the expression $\varepsilon^{-1/s}|\tilde{\mathbf{u}}|^{1/s}_{\ell_\tau^w(\Lambda)}$, is called *(quasi-)optimal*, or to be of *asymptotically optimal complexity*.

Remark 4.1. This definition is further justified by the following observation. Suppose that we had best N-term approximations $\tilde{\mathbf{u}}_N$ to $\tilde{\mathbf{u}}$ available, and let $N = N(\tilde{\mathbf{u}}, \varepsilon)$ be the smallest number for which $\|\tilde{\mathbf{u}} - \tilde{\mathbf{u}}_N\| \leq \varepsilon$. By (3.3.15) we have $\|\tilde{\mathbf{u}} - \tilde{\mathbf{u}}_N\| = \sigma_N(\tilde{\mathbf{u}}) \leq CN^{-s}|\tilde{\mathbf{u}}|_{\ell_\tau^w}$. Then, setting $\bar{N} := \lceil C^{1/s}\varepsilon^{-1/s}|\tilde{\mathbf{u}}|^{1/s}\rceil$, we get $\sigma_{\bar{N}}(\tilde{\mathbf{u}}) \leq \varepsilon$, thus $N \leq \bar{N} = \lceil C^{1/s}\varepsilon^{-1/s}|\tilde{\mathbf{u}}|^{1/s}_{\ell_\tau^w}\rceil$, so that $N \lesssim \varepsilon^{-1/s}|\tilde{\mathbf{u}}|^{1/s}_{\ell_\tau^w}$. Moreover, in view of the upper bound for the number of operations needed to compute $\mathbf{u}_\varepsilon$ in Definition 4.1, we may infer that an optimal method is of *linear complexity* $\mathcal{O}(\# \operatorname{supp} \mathbf{u}_\varepsilon)$.

For the solution of elliptic operator equations as introduced in §1.3, an *adaptive wavelet Galerkin algorithm* has been developed in [27] which is based on an *a posteriori error estimator* designed in [34]. The method in [27] was proved to be optimal in the above sense. However, one has to strive for a different strategy when dealing with frames.

Remark 4.2. Using a frame instead of a Riesz basis, a classical Galerkin approach cannot be the method of choice, because a selected finite collection of frame elements can actually be only almost linearly dependent. The respective stiffness matrix may then possess a very small but non-zero smallest eigenvalue, resulting in a large spectral condition number.

4.2 The exact steepest descent scheme

In the following, we consider the bi-infinite dimensional linear system of equations $\mathbf{A}\mathbf{u} = \mathbf{f}$ from (2.5.4) where we assume that $\mathbf{A}$ and $\mathbf{f}$ represent the discretizations of $\mathcal{L}$ and f from (1.3.5) with respect to the aggregated wavelet Gelfand frame Ψ_{L_2} from Proposition 2.11. Again we set $\Psi := \mathbf{D}^{-1}\Psi = \{2^{-t|\lambda|}\psi_\lambda\}_{\lambda \in \Lambda}$ to obtain a frame for $H_0^t(\Omega)$, $t \in \mathbb{N}$. Thus, $\mathbf{A}$ is a positive semi-definite, symmetric matrix, $\inf_{0 \neq \mathbf{v} \in \mathrm{ran}(\mathbf{A})} \langle \mathbf{A}\mathbf{v}, \mathbf{v} \rangle_{\ell_2(\Lambda)} > 0$, $\mathbf{A} = \mathbf{A}^*$, inducing the orthogonal decomposition $\ell_2(\Lambda) = \mathrm{ran}(\mathbf{A}) \oplus \mathrm{ker}(\mathbf{A})$.

In the following, we write $\|\cdot\|$ and $\langle \cdot, \cdot \rangle$ for $\|\cdot\|_{\ell_2(\Lambda)}$, or $\|\cdot\|_{\ell_2(\Lambda) \to \ell_2(\Lambda)}$, and $\langle \cdot, \cdot \rangle_{\ell_2(\Lambda)}$, respectively. We set $\langle\!\langle \cdot, \cdot \rangle\!\rangle := \langle \mathbf{A}\cdot, \cdot \rangle$ and define the natural energy seminorm $\|\cdot\| := \langle\!\langle \cdot, \cdot \rangle\!\rangle^{\frac{1}{2}}$. With $\mathbf{Q} : \ell_2(\Lambda) \to \mathrm{ran}(\mathbf{A})$ denoting again the orthogonal projector onto the range of $\mathbf{A}$, for any $\mathbf{v} \in \ell_2(\Lambda)$ we have

$$\|\mathbf{A}^\dagger\|^{-\frac{1}{2}}\|\mathbf{Q}\mathbf{v}\| \leq \|\mathbf{v}\| \leq \|\mathbf{A}\|^{\frac{1}{2}}\|\mathbf{Q}\mathbf{v}\|, \quad \|\mathbf{A}^\dagger\|^{-\frac{1}{2}}\|\mathbf{v}\| \leq \|\mathbf{A}\mathbf{v}\| \leq \|\mathbf{A}\|^{\frac{1}{2}}\|\mathbf{v}\|. \quad (4.2.1)$$

For example, the left equivalence can be verified as follows. It is $\|\mathbf{v}\|^2 = \langle \mathbf{A}\mathbf{v}, \mathbf{v} \rangle = \langle \mathbf{A}\mathbf{Q}\mathbf{v}, \mathbf{Q}\mathbf{v} \rangle \leq \|\mathbf{A}\|\|\mathbf{Q}\mathbf{v}\|^2$. For the proof of the lower estimate, note that the pseudo inverse of $\mathbf{A}$ can be written as $\mathbf{A}^\dagger = (\mathbf{A}|_{\mathrm{ran}(\mathbf{A})})^{-1}\mathbf{Q}$. Thus, it is $\langle \mathbf{A}\mathbf{Q}\mathbf{v}, \mathbf{Q}\mathbf{v} \rangle = \langle \mathbf{A}|_{\mathrm{ran}(\mathbf{A})}\mathbf{Q}\mathbf{v}, \mathbf{Q}\mathbf{v} \rangle \geq \|(\mathbf{A}|_{\mathrm{ran}(\mathbf{A})})^{-1}\|^{-1}\|\mathbf{Q}\mathbf{v}\|^2$. By $\|\mathbf{A}^\dagger\| = \|(\mathbf{A}|_{\mathrm{ran}(\mathbf{A})})^{-1}\|$, the left part in (4.2.1) is proved.

Using this notation, the steepest descent scheme and its error-reduction in one iterative step read as follows.

Proposition 4.1. *Let $\mathbf{w}$ be an approximation for $\mathbf{u}$ with $\mathbf{r} := \mathbf{f} - \mathbf{A}\mathbf{w} \neq 0$. Then, with $\kappa(\mathbf{A}) := \|\mathbf{A}\|\|\mathbf{A}^\dagger\|$, for*

$$\bar{\mathbf{w}} := \mathbf{w} + \frac{\langle \mathbf{r}, \mathbf{r} \rangle}{\langle \mathbf{A}\mathbf{r}, \mathbf{r} \rangle}\mathbf{r}, \quad (4.2.2)$$

we have $\|\mathbf{u} - \bar{\mathbf{w}}\| \leq \frac{\kappa(\mathbf{A})-1}{\kappa(\mathbf{A})+1}\|\mathbf{u} - \mathbf{w}\|$.

The proof of this proposition is obtained by using standard arguments on the convergence of iterative descent methods (see [75]), for which reason we need not carry out the details here.

Remark 4.3. The fact that $\mathbf{A}$ has a nontrivial kernel in case Ψ is not a Riesz basis does not affect the convergence of the iterative scheme specified by (4.2.2). If a starting vector $\mathbf{w}$ is chosen from $\mathrm{ran}(\mathbf{A})$, then all subsequent iterates will also have this property. Hence, for the convergence analysis of the exact steepest descent iteration, $\mathbf{A}$ can as well be replaced by $\mathbf{A}|_{\mathrm{ran}(\mathbf{A})}$ which is a positive definite operator on $\mathrm{ran}(\mathbf{A})$, and the well-known convergence theory for finite dimensional matrices directly carries over.

It is clear that (4.2.2) cannot be implemented directly since infinite sequences and a bi-infinite matrix are involved. Therefore, the challenging task is to transform (4.2.2) into an implementable version. To this end, one has to replace the infinite sequences by finite ones without destroying the overall convergence of the scheme. One basic tool for this purpose will be the following perturbation result.

Proposition 4.2. *Let again* $\mathbf{r} := \mathbf{f} - \mathbf{Aw}$. *For any* $\lambda \in (\frac{\kappa(\mathbf{A})-1}{\kappa(\mathbf{A})+1}, 1)$, *there exists a* $\delta = \delta(\lambda) > 0$ *small enough, such that if* $\|\tilde{\mathbf{r}} - \mathbf{r}\| \leq \delta\|\tilde{\mathbf{r}}\|$ *and* $\|\mathbf{z} - \mathbf{A}\tilde{\mathbf{r}}\| \leq \delta\|\tilde{\mathbf{r}}\|$, *then with*

$$\tilde{\mathbf{w}} := \mathbf{w} + \frac{\langle \tilde{\mathbf{r}}, \tilde{\mathbf{r}} \rangle}{\langle \mathbf{z}, \tilde{\mathbf{r}} \rangle} \tilde{\mathbf{r}}, \tag{4.2.3}$$

we have $\|\mathbf{u} - \tilde{\mathbf{w}}\| \leq \lambda\|\mathbf{u} - \mathbf{w}\|$ *and* $|\frac{\langle \tilde{\mathbf{r}}, \tilde{\mathbf{r}} \rangle}{\langle \mathbf{z}, \tilde{\mathbf{r}} \rangle}| \lesssim 1$. *If, for some* $\eta > 0$, *in addition* $\|\tilde{\mathbf{r}} - \mathbf{r}\| \leq \eta$, *then* $\|(\mathbf{I} - \mathbf{Q})(\tilde{\mathbf{w}} - \mathbf{w})\| \leq C_3\eta$, *with some absolute constant* $C_3 > 0$.

Proof. Equation (4.2.1) implies that $\langle \mathbf{Ar}, \mathbf{r} \rangle \eqsim \|\mathbf{r}\|^2$. The first step is to show that, for a sufficiently small $\bar{\delta}$ and any $0 < \delta \leq \bar{\delta}$,

$$\|\tilde{\mathbf{r}}\| \eqsim \|\mathbf{r}\| \qquad \text{and} \qquad \langle \mathbf{z}, \tilde{\mathbf{r}} \rangle \eqsim \|\tilde{\mathbf{r}}\|^2 \tag{4.2.4}$$

hold. It is $\|\mathbf{r}\| \leq \|\mathbf{r} - \tilde{\mathbf{r}}\| + \|\tilde{\mathbf{r}}\| \leq (\delta + 1)\|\tilde{\mathbf{r}}\|$ and $\|\tilde{\mathbf{r}}\|^2 = \langle \tilde{\mathbf{r}} - \mathbf{r} + \mathbf{r}, \tilde{\mathbf{r}} \rangle \leq (\|\tilde{\mathbf{r}} - \mathbf{r}\| + \|\mathbf{r}\|)\|\tilde{\mathbf{r}}\| \leq (\delta\|\tilde{\mathbf{r}}\| + \|\mathbf{r}\|)\|\tilde{\mathbf{r}}\|$, from which the first equivalence follows. Further, we have $\langle \mathbf{z}, \tilde{\mathbf{r}} \rangle = \langle \mathbf{z} - \mathbf{A}\tilde{\mathbf{r}} + \mathbf{A}\tilde{\mathbf{r}}, \tilde{\mathbf{r}} \rangle \leq \|\mathbf{z} - \mathbf{A}\tilde{\mathbf{r}}\|\|\tilde{\mathbf{r}}\| + \|\mathbf{A}\tilde{\mathbf{r}}\|\|\tilde{\mathbf{r}}\| \leq (\delta + \|\mathbf{A}\|)\|\tilde{\mathbf{r}}\|^2$. Thus, $\langle \mathbf{z}, \tilde{\mathbf{r}} \rangle \lesssim \|\tilde{\mathbf{r}}\|^2$. In addition, $\|\mathbf{Q}\tilde{\mathbf{r}}\|^2 \eqsim \langle \mathbf{A}\tilde{\mathbf{r}}, \tilde{\mathbf{r}} \rangle = \langle \mathbf{A}\tilde{\mathbf{r}} - \mathbf{z} + \mathbf{z}, \tilde{\mathbf{r}} \rangle \leq \|\mathbf{A}\tilde{\mathbf{r}} - \mathbf{z}\|\|\tilde{\mathbf{r}}\| + \langle \mathbf{z}, \tilde{\mathbf{r}} \rangle \leq \delta\|\tilde{\mathbf{r}}\|^2 + \langle \mathbf{z}, \tilde{\mathbf{r}} \rangle$. And $\|\tilde{\mathbf{r}} - \mathbf{r}\|^2 = \|\mathbf{Q}\tilde{\mathbf{r}} - \mathbf{r} + (\mathbf{I} - \mathbf{Q})\tilde{\mathbf{r}}\|^2 = \|\mathbf{Q}\tilde{\mathbf{r}} - \mathbf{r}\|^2 + \|(\mathbf{I} - \mathbf{Q})\tilde{\mathbf{r}}\|^2 \leq \delta^2\|\tilde{\mathbf{r}}\|^2$. Hence, $\|(\mathbf{I} - \mathbf{Q})\tilde{\mathbf{r}}\|^2 \leq \delta^2\|\tilde{\mathbf{r}}\|^2$. We arrive at $\|\tilde{\mathbf{r}}\|^2 = \|\mathbf{Q}\tilde{\mathbf{r}}\|^2 + \|(\mathbf{I} - \mathbf{Q})\tilde{\mathbf{r}}\|^2 \lesssim \delta\|\tilde{\mathbf{r}}\|^2 + \langle \mathbf{z}, \tilde{\mathbf{r}} \rangle + \delta^2\|\tilde{\mathbf{r}}\|^2$, so that also $\|\tilde{\mathbf{r}}\|^2 \lesssim \langle \mathbf{z}, \tilde{\mathbf{r}} \rangle$. From (4.2.4) we want to infer that

$$\left| \frac{\langle \tilde{\mathbf{r}}, \tilde{\mathbf{r}} \rangle}{\langle \mathbf{z}, \tilde{\mathbf{r}} \rangle} - \frac{\langle \mathbf{r}, \mathbf{r} \rangle}{\langle \mathbf{Ar}, \mathbf{r} \rangle} \right| \lesssim \delta. \tag{4.2.5}$$

Indeed, it is

$$\frac{\langle \tilde{\mathbf{r}}, \tilde{\mathbf{r}} \rangle}{\langle \mathbf{z}, \tilde{\mathbf{r}} \rangle} - \frac{\langle \mathbf{r}, \mathbf{r} \rangle}{\langle \mathbf{Ar}, \mathbf{r} \rangle} = \frac{\langle \tilde{\mathbf{r}}, \tilde{\mathbf{r}} \rangle - \langle \mathbf{r}, \mathbf{r} \rangle}{\langle \mathbf{z}, \tilde{\mathbf{r}} \rangle} + \langle \mathbf{r}, \mathbf{r} \rangle \left[\frac{1}{\langle \mathbf{z}, \tilde{\mathbf{r}} \rangle} - \frac{1}{\langle \mathbf{Ar}, \mathbf{r} \rangle} \right]$$

$$= \frac{\langle \tilde{\mathbf{r}}, \tilde{\mathbf{r}} \rangle - \langle \mathbf{r}, \mathbf{r} \rangle}{\langle \mathbf{z}, \tilde{\mathbf{r}} \rangle} + \frac{\langle \mathbf{r}, \mathbf{r} \rangle}{\langle \mathbf{z}, \tilde{\mathbf{r}} \rangle \langle \mathbf{Ar}, \mathbf{r} \rangle} [\langle \mathbf{Ar}, \mathbf{r} \rangle - \langle \mathbf{z}, \tilde{\mathbf{r}} \rangle]$$

$$= \frac{\langle 2(\tilde{\mathbf{r}} - \mathbf{r}), \mathbf{r} \rangle + \|\tilde{\mathbf{r}} - \mathbf{r}\|^2}{\langle \mathbf{z}, \tilde{\mathbf{r}} \rangle}$$

$$+ \frac{\langle \mathbf{r}, \mathbf{r} \rangle}{\langle \mathbf{z}, \tilde{\mathbf{r}} \rangle \langle \mathbf{Ar}, \mathbf{r} \rangle} [\langle \mathbf{Ar}, \mathbf{r} - \tilde{\mathbf{r}} \rangle + \langle \mathbf{A}\tilde{\mathbf{r}} - \mathbf{z}, \tilde{\mathbf{r}} \rangle + \langle \mathbf{A}(\mathbf{r} - \tilde{\mathbf{r}}), \tilde{\mathbf{r}} \rangle].$$

Therefore, we obtain

$$
\left| \frac{\langle \tilde{\mathbf{r}}, \tilde{\mathbf{r}} \rangle}{\langle \mathbf{z}, \tilde{\mathbf{r}} \rangle} - \frac{\langle \mathbf{r}, \mathbf{r} \rangle}{\langle \mathbf{Ar}, \mathbf{r} \rangle} \right| \lesssim \frac{2\|\tilde{\mathbf{r}} - \mathbf{r}\|\|\tilde{\mathbf{r}}\| + \|\tilde{\mathbf{r}} - \mathbf{r}\|^2}{\|\tilde{\mathbf{r}}\|^2}
$$
$$
+ \frac{\|\tilde{\mathbf{r}}\|^2 \left[\|\tilde{\mathbf{r}}\|\|\mathbf{r} - \tilde{\mathbf{r}}\| + \|\mathbf{A}\tilde{\mathbf{r}} - \mathbf{z}\|\|\tilde{\mathbf{r}}\| + \|\mathbf{r} - \tilde{\mathbf{r}}\|\|\tilde{\mathbf{r}}\| \right]}{\|\tilde{\mathbf{r}}\|^2 \|\tilde{\mathbf{r}}\|^2}
$$
$$
\leq \frac{2\delta\|\tilde{\mathbf{r}}\|^2 + \delta^2\|\tilde{\mathbf{r}}\|^2}{\|\tilde{\mathbf{r}}\|^2} + \frac{3\delta\|\tilde{\mathbf{r}}\|^2}{\|\tilde{\mathbf{r}}\|^2} \lesssim \delta.
$$

Writing

$$
\frac{\langle \tilde{\mathbf{r}}, \tilde{\mathbf{r}} \rangle}{\langle \mathbf{z}, \tilde{\mathbf{r}} \rangle} \tilde{\mathbf{r}} - \frac{\langle \mathbf{r}, \mathbf{r} \rangle}{\langle \mathbf{Ar}, \mathbf{r} \rangle} \mathbf{r} = \left[\frac{\langle \tilde{\mathbf{r}}, \tilde{\mathbf{r}} \rangle}{\langle \mathbf{z}, \tilde{\mathbf{r}} \rangle} - \frac{\langle \mathbf{r}, \mathbf{r} \rangle}{\langle \mathbf{Ar}, \mathbf{r} \rangle} \right] \mathbf{r} + \frac{\langle \tilde{\mathbf{r}}, \tilde{\mathbf{r}} \rangle}{\langle \mathbf{z}, \tilde{\mathbf{r}} \rangle} [\tilde{\mathbf{r}} - \mathbf{r}], \tag{4.2.6}
$$

we find that

$$
\left\| \frac{\langle \tilde{\mathbf{r}}, \tilde{\mathbf{r}} \rangle}{\langle \mathbf{z}, \tilde{\mathbf{r}} \rangle} \tilde{\mathbf{r}} - \frac{\langle \mathbf{r}, \mathbf{r} \rangle}{\langle \mathbf{Ar}, \mathbf{r} \rangle} \mathbf{r} \right\| \lesssim \delta\|\mathbf{r}\| \lesssim \delta\|\mathbf{u} - \mathbf{w}\|. \tag{4.2.7}
$$

Now, let $\bar{\mathbf{w}} := \mathbf{w} + \frac{\langle \mathbf{r}, \mathbf{r} \rangle}{\langle \mathbf{Ar}, \mathbf{r} \rangle} \mathbf{r}$. We obtain

$$
\|\mathbf{u} - \tilde{\mathbf{w}}\| \leq \|\mathbf{u} - \bar{\mathbf{w}}\| + \|\bar{\mathbf{w}} - \tilde{\mathbf{w}}\| \leq \frac{\kappa(\mathbf{A}) - 1}{\kappa(\mathbf{A}) + 1}\|\mathbf{u} - \mathbf{w}\| + \|\bar{\mathbf{w}} - \tilde{\mathbf{w}}\|
$$
$$
\leq \frac{\kappa(\mathbf{A}) - 1}{\kappa(\mathbf{A}) + 1}\|\mathbf{u} - \mathbf{w}\| + \|\mathbf{A}\|^{1/2}\|\mathbf{Q}\|\|\bar{\mathbf{w}} - \tilde{\mathbf{w}}\|
$$
$$
\leq \max\left\{ \frac{\kappa(\mathbf{A}) - 1}{\kappa(\mathbf{A}) + 1}, C\|\mathbf{A}\|^{1/2}\delta \right\}\|\mathbf{u} - \mathbf{w}\|,
$$

where $C > 0$ is the constant hidden in (4.2.7). The first statement now follows by choosing $\delta \leq \lambda C^{-1}\|\mathbf{A}\|^{-1/2}$. From (4.2.6) and $(\mathbf{I} - \mathbf{Q})\mathbf{r} = 0$, we have $(\mathbf{I} - \mathbf{Q})(\tilde{\mathbf{w}} - \mathbf{w}) = \frac{\langle \tilde{\mathbf{r}}, \tilde{\mathbf{r}} \rangle}{\langle \mathbf{z}, \tilde{\mathbf{r}} \rangle}(\mathbf{I} - \mathbf{Q})(\tilde{\mathbf{r}} - \mathbf{r})$, which by $|\frac{\langle \tilde{\mathbf{r}}, \tilde{\mathbf{r}} \rangle}{\langle \mathbf{z}, \tilde{\mathbf{r}} \rangle}| \lesssim 1$ and $\|\mathbf{I} - \mathbf{Q}\| \leq 1$ completes the proof of the second statement. $\qquad\square$

4.3 Development of the adaptive solver

In order to develop an implementable adaptive scheme, three basic routines will be required. In particular, we need a numerical routine for the realization of the matrix-vector products that have to be performed in (4.2.2) and another one for the approximation of $\mathbf{f}$. Moreover, in order to obtain an optimal balance between the number of non-zero coefficients of the iterates (and thus the computational cost) and the accuracy of the approximation, we shall use a thresholding routine to throw away an adequate amount of small coefficients. For an extensive discussion of these building blocks we also refer to [7, 27, 104].

Assumption 4.1. For some $s > 0$, there exists a solution $\mathbf{u}$ to (2.5.4) that is contained in $\ell_\tau^w(\Lambda)$, in which here and in the remainder of this chapter, τ and s are always related according to $\frac{1}{\tau} = s + \frac{1}{2}$.

For some s^* larger than any s for which Assumption 4.1 can be expected, we assume the existence of the following three routines:

- **APPLY**$[\mathbf{A}, \mathbf{w}, \varepsilon] \to \mathbf{z}_\varepsilon$. Determines, for an $\varepsilon > 0$ and a finitely supported $\mathbf{w}$, a finitely supported $\mathbf{z}_\varepsilon$ with

$$\|\mathbf{A}\mathbf{w} - \mathbf{z}_\varepsilon\| \le \varepsilon. \tag{4.3.1}$$

  Moreover, for any $s < s^*$, it is

$$\#\mathrm{supp}\, \mathbf{z}_\varepsilon \lesssim \varepsilon^{-1/s} |\mathbf{w}|_{\ell_\tau^w}^{1/s}, \tag{4.3.2}$$

  where the number of arithmetic operations and storage locations used by this call is bounded by some absolute multiple of

$$\varepsilon^{-1/s} |\mathbf{w}|_{\ell_\tau^w}^{1/s} + \#\mathrm{supp}\, \mathbf{w} + 1. \tag{4.3.3}$$

- **RHS**$[\mathbf{f}, \varepsilon] \to \mathbf{f}_\varepsilon$. Determines, for an $\varepsilon > 0$, a finitely supported $\mathbf{f}_\varepsilon$ with $\|\mathbf{f} - \mathbf{f}_\varepsilon\| \le \varepsilon$. Moreover, for any $s < s^*$, we require

$$\#\mathrm{supp}\, \mathbf{f}_\varepsilon \lesssim \varepsilon^{-1/s} |\mathbf{u}|_{\ell_\tau^w}^{1/s}, \tag{4.3.4}$$

  where the number of arithmetic operations and storage locations used by the call is bounded by some absolute multiple of

$$\varepsilon^{-1/s} |\mathbf{u}|_{\ell_\tau^w}^{1/s} + 1. \tag{4.3.5}$$

- **COARSE**$[\mathbf{w}, \varepsilon] \to \mathbf{w}_\varepsilon$. Determines, for an $\varepsilon > 0$ and a finitely supported $\mathbf{w}$, a finitely supported $\mathbf{w}_\varepsilon$ such that

$$\|\mathbf{w} - \mathbf{w}_\varepsilon\| \le \varepsilon. \tag{4.3.6}$$

  Moreover, $\#\mathrm{supp}\, \mathbf{w}_\varepsilon \lesssim \inf\{N : \sigma_N(\mathbf{w}) \le \varepsilon\}$, and **COARSE** can be arranged to take a number of arithmetic operations and storage locations that is bounded by an absolute multiple of

$$\#\mathrm{supp}\, \mathbf{w} + \max\{\log(\varepsilon^{-1}\|\mathbf{w}\|), 1\}. \tag{4.3.7}$$

We shall see that these assumptions are indeed sufficient to guarantee optimality of an adaptive solver using these three routines if the tolerances with which they are called are appropriately steered. For instance, we will have to make sure that the log-term in (4.3.7), and the constant terms in (4.3.3), (4.3.5), and (4.3.7), can be bounded by $\varepsilon^{-1/s}|\mathbf{u}|_{\ell_\tau^w}^{1/s}$, which will turn out to be uncritical.

Clearly, whether **APPLY** and **RHS** can really be arranged this ways depends on the properties of $\mathbf{A}$ and $\mathbf{f}$.

4.3.1 The adaptive matrix-vector product

In case the matrix $\mathbf{A}$ is sufficiently sparse, it admits the development of a suitable routine **APPLY**. In particular, we assume $\mathbf{A}$ to be s^*-*compressible*.

Definition 4.2. The matrix $\mathbf{A}$ is called s^*-*compressible if for each* $j \in \mathbb{N}_0$ *there exists an infinite matrix* $\mathbf{A}_j$, *constructed by dropping entries from* $\mathbf{A}$ *and having in each column at most* $\mathcal{O}(2^j)$ *non-zero entries, such that for any* $\tilde{s} < s^*$ *it is* $\|\mathbf{A} - \mathbf{A}_j\| \le C2^{-j\tilde{s}} =: C_j$, *for a* $C > 0$.

The basic idea for the realization of **APPLY** is to detect significant columns of $\mathbf{A}$ in the linear combination $\mathbf{Aw} = \sum_\lambda \mathbf{A}^{(\lambda)} \mathbf{w}_\lambda$ by extracting those columns $\mathbf{A}^{(\lambda)}$ that are multiplied with large coefficients (in modulus) $\mathbf{w}_\lambda$ and by computing finite approximations to each of the chosen columns within a tolerance depending on the size of the corresponding coefficient and the prescribed accuracy ε. Following this idea, an *adaptive* routine is obtained, by which we may speak of an adaptive scheme whenever **APPLY** is used in a numerical algorithm. More precisely, for s^*-compressible $\mathbf{A}$ we will make use of the following routine.

APPLY$[\mathbf{A}, \mathbf{w}, \bar{\varepsilon}] \to \mathbf{z}_\varepsilon$:

- $q := \lceil \log((\# \operatorname{supp} \mathbf{w})^{1/2} \|\mathbf{w}\| \|\mathbf{A}\| 2/\varepsilon) \rceil$.

- *Divide the elements of* $\mathbf{w}$ *into sets* $V_0, \ldots, V_q$, *where for* $0 \le i \le q - 1$, V_i *contains the elements with modulus in* $(2^{-i-1}\|\mathbf{w}\|, 2^{-i}\|\mathbf{w}\|]$, *and possible remaining elements are put into* V_q.

- *For* $k = 0, 1, \ldots$, *generate vectors* $\mathbf{w}_{[k]}$ *by subsequently extracting* $2^k - \lfloor 2^{k-1} \rfloor$ *elements from* $\bigcup_i V_i$, *starting from* V_0 *and when it is empty continuing with* V_1 *and so forth, until for some* $k = l$ *either* $\bigcup_i V_i$ *becomes empty or*

$$\|\mathbf{A}\| \left\| \mathbf{w} - \sum_{k=0}^l \mathbf{w}_{[k]} \right\| \le \varepsilon/2. \tag{4.3.8}$$

In both cases $\mathbf{w}_{[l]}$ *may contain less than* $2^l - \lfloor 2^{l-1} \rfloor$ *elements.*

- *Compute the smallest* $j \ge l$ *such that*

$$\sum_{k=0}^l C_{j-k} \|\mathbf{w}_{[k]}\| \le \varepsilon/2. \tag{4.3.9}$$

- *For* $k = 0, \ldots, l$, *compute the non-zero entries in the matrices* $\mathbf{A}_{j-k}$ *which have a column index in common with one of the entries of* $\mathbf{w}_{[k]}$, *and compute*

$$\mathbf{z}_\varepsilon := \sum_{k=0}^l \mathbf{A}_{j-k} \mathbf{w}_{[k]}. \tag{4.3.10}$$

Note that in the second step of **APPLY**, the entries of $\mathbf{w}$ are only *quasi-sorted*, which requires a linear amount of work, whereas a complete sorting would take $\mathcal{O}((\#\operatorname{supp}\mathbf{w}) \times \log(\#\operatorname{supp}\mathbf{w}))$ operations, with which (4.3.3) would be violated. Such an approach is also know as a *binary binning strategy*; see [7, 90].

Assuming that all entries of $\mathbf{A}$ can be exactly computed at unit cost, it can be shown that this method has the desired properties; see [104, Proposition 3.8] for a rigorous proof. In Chapter 5 will shall see that $\mathbf{A}$ is s^*-compressible for an s^* greater than any s for which a solution $\mathbf{u}$ to (2.5.4) can be expected to be contained in $\ell_\tau^w(\Lambda)$, i.e., $s^* > \frac{d-t}{n}$. In general, however, numerical quadrature is needed for the approximation of the entries of $\mathbf{A}$, meaning that the above assumption is rather simplistic. In particular, reconsidering §2.6, in the overlapping regions of the patches Ω_i, due to possibly non-matching grids, the quadrature is a demanding problem. Yet, in Chapter 5 we shall see that for the case of aggregated wavelet frames the matrix $\mathbf{A}$ can be well approximated by computable sparse matrices. That means, for some suitably large $s^* > 0$, it is s^*-*computable* in the following sense.

Definition 4.3. The matrix $\mathbf{A}$ is called s^*-*computable* if for each $j \in \mathbb{N}_0$ one can construct a matrix $\mathbf{A}_j^*$, having in each column $\mathcal{O}(2^j)$ entries whose joint computation takes $\mathcal{O}(2^j)$ operations, such that for any $\tilde{s} < s^*$ it is $\|\mathbf{A} - \mathbf{A}_j^*\| \lesssim 2^{-\tilde{s}j}$.

In case $\mathbf{A}$ has this property, the linear complexity of **APPLY** can be retained.

Using that $\mathbf{A} : \ell_2(\Lambda) \to \ell_2(\Lambda)$ is bounded, the properties of **APPLY** and **RHS** imply the following.

Proposition 4.3. *For any $s \in (0, s^*)$, $\mathbf{A} : \ell_\tau^w(\Lambda) \to \ell_\tau^w(\Lambda)$ is bounded. For $\mathbf{z}_\varepsilon :=$* **APPLY**$[\mathbf{A}, \mathbf{w}, \varepsilon]$ *and* $\mathbf{f}_\varepsilon := $ **RHS**$[\mathbf{f}, \varepsilon]$, *we have*

$$|\mathbf{z}_\varepsilon|_{\ell_\tau^w} \lesssim |\mathbf{w}|_{\ell_\tau^w} \tag{4.3.11}$$

and

$$|\mathbf{f}_\varepsilon|_{\ell_\tau^w} \lesssim |\mathbf{u}|_{\ell_\tau^w}, \tag{4.3.12}$$

uniformly over $\varepsilon > 0$ and all finitely supported $\mathbf{w}$.

Proof. We first show that for $s \in (0, s^*)$, $\mathbf{A} : \ell_\tau^w(\Lambda) \to \ell_\tau^w(\Lambda)$ is bounded. Let $C > 0$ be a constant such that for $\mathbf{z}_\varepsilon := $ **APPLY**$[\mathbf{A}, \mathbf{w}, \varepsilon]$, $\#\operatorname{supp}\mathbf{z}_\varepsilon \leq C\varepsilon^{-1/s}|\mathbf{w}|_{\ell_\tau^w}^{1/s}$. Let $\mathbf{v} \in \ell_\tau^w(\Lambda)$ and $N \in \mathbb{N}$ be given. Let $\mathbf{v}_N$ be a best N-term approximation of $\mathbf{v}$. For $\bar{\varepsilon} := C^s|\mathbf{v}_N|_{\ell_\tau^w}N^{-s}$, let $\mathbf{z}_{\bar{\varepsilon}} := $ **APPLY**$[\mathbf{A}, \mathbf{v}_N, \bar{\varepsilon}]$. Then, by (3.3.15),

$$\|\mathbf{A}\mathbf{v} - \mathbf{z}_{\bar{\varepsilon}}\| \leq \|\mathbf{A}\mathbf{v}_N - \mathbf{z}_{\bar{\varepsilon}}\| + \|\mathbf{A}\|\|\mathbf{v} - \mathbf{v}_N\|$$
$$\lesssim C^s|\mathbf{v}_N|_{\ell_\tau^w}N^{-s} + \|\mathbf{A}\|N^{-s}|\mathbf{v}|_{\ell_\tau^w} \lesssim N^{-s}|\mathbf{v}|_{\ell_\tau^w}.$$

Since $\#\operatorname{supp}\mathbf{z}_{\bar{\varepsilon}} \leq C\bar{\varepsilon}^{-1/s}|\mathbf{v}_N|_{\ell_\tau^w}^{1/s} \leq N$, using (3.3.15), we infer that $|\mathbf{A}\mathbf{v}|_{\ell_\tau^w} \approx \sup_N N^s\sigma_N(\mathbf{A}\mathbf{v}) \leq \sup_N N^s\|\mathbf{A}\mathbf{v} - \mathbf{z}_{\bar{\varepsilon}}\| \lesssim |\mathbf{v}|_{\ell_\tau^w}$, thus $\mathbf{A}$ is bounded on

$\ell_\tau^w(\Lambda)$. The next step is to prove (4.3.11) and (4.3.12). For any $\mathbf{v} \in \ell_\tau^w(\Lambda)$, and any finitely supported $\mathbf{z}$, by [27, Lemma 4.11] or [104, Proposition 3.4] we have

$$|\mathbf{z}|_{\ell_\tau^w} \lesssim |\mathbf{v}|_{\ell_\tau^w} + (\#\operatorname{supp}\mathbf{z})^s \|\mathbf{v} - \mathbf{z}\|. \tag{4.3.13}$$

From this and (4.3.2) we infer that for finitely supported $\mathbf{w}$, $\varepsilon > 0$, and with $\mathbf{z}_\varepsilon :=$ **APPLY**$[\mathbf{A}, \mathbf{w}, \varepsilon]$, one gets $|\mathbf{z}_\varepsilon|_{\ell_\tau^w} \lesssim |\mathbf{Aw}|_{\ell_\tau^w} + (\#\operatorname{supp}\mathbf{z}_\varepsilon)^s \varepsilon \leq |\mathbf{Aw}|_{\ell_\tau^w} + C^s |\mathbf{w}|_{\ell_\tau^w} \lesssim |\mathbf{w}|_{\ell_\tau^w}$. Similarly, with (4.3.4), for $\mathbf{f}_\varepsilon := \mathbf{RHS}[\mathbf{f}, \varepsilon]$, we obtain $|\mathbf{f}_\varepsilon|_{\ell_\tau^w} \lesssim |\mathbf{Au}|_{\ell_\tau^w} + (\#\operatorname{supp}\mathbf{f}_\varepsilon)^s \varepsilon \lesssim |\mathbf{u}|_{\ell_\tau^w}$. $\qquad\square$

4.3.2 Coarsening

One way to realize the thresholding routine **COARSE** would be to sort the entries of the given vector $\mathbf{w}$ by their absolute values and to throw away the smallest coefficients in modulus as long as still the remaining coefficients approximate $\mathbf{w}$ up to accuracy ε. However, the sorting would again require $\mathcal{O}((\#\operatorname{supp}\mathbf{w}) \times \log(\#\operatorname{supp}\mathbf{w}))$ operations. Fortunately, the routine **COARSE** can also be implemented using binary binning with which (4.3.7) can be guaranteed [7, 90].

COARSE$[\mathbf{w}, \varepsilon] \to \mathbf{w}_\varepsilon$:

- $q := \lceil \log((\#\operatorname{supp}\mathbf{w})^{1/2} \|\mathbf{w}\|/\varepsilon) \rceil$.

- *Divide the elements of* $\mathbf{w}$ *into sets* $V_0, \ldots, V_q$*, where for* $0 \leq i \leq q-1$*,* V_i *contains the elements with modulus in* $(2^{-i-1}\|\mathbf{w}\|, 2^{-i}\|\mathbf{w}\|]$*, and possible remaining elements are put into* V_q*.*

- *Create* $\mathbf{w}_\varepsilon$ *by extracting elements first from* V_0 *and when it is empty from* V_1 *and so forth, until* $\|\mathbf{w} - \mathbf{w}_\varepsilon\| \leq \varepsilon$*.*

Thanks to the properties of **COARSE** we have the following result.

Proposition 4.4. *Let* $\mu > 1$ *and* $s > 0$*. Then, for any* $\varepsilon > 0$*,* $\mathbf{v} \in \ell_\tau^w(\Lambda)$*, and finitely supported* $\mathbf{w}$ *with* $\|\mathbf{v} - \mathbf{w}\| \leq \varepsilon$*, for* $\overline{\mathbf{w}} := \mathbf{COARSE}[\mathbf{w}, \mu\varepsilon]$ *we have that*

$$\#\operatorname{supp}\overline{\mathbf{w}} \lesssim \varepsilon^{-1/s} |\mathbf{v}|_{\ell_\tau^w}^{1/s}.$$

Obviously $\|\mathbf{v} - \overline{\mathbf{w}}\| \leq (1+\mu)\varepsilon$*, and*

$$|\overline{\mathbf{w}}|_{\ell_\tau^w} \leq C_2(\tau)|\mathbf{v}|_{\ell_\tau^w}, \tag{4.3.14}$$

for some $C_2(\tau) > 0$ *independent of* ε*.*

Proof. Let N be the smallest integer such that $\|\mathbf{v}_N - \mathbf{v}\| \leq (\mu-1)\varepsilon$ for a best N-term approximation $\mathbf{v}_N$ of $\mathbf{v}$. Then, $\#\operatorname{supp}\mathbf{v}_N \lesssim \varepsilon^{-1/s} |\mathbf{v}|_{\ell_\tau^w}^{1/s}$. Furthermore $\|\mathbf{v}_N - \mathbf{w}\| \leq \|\mathbf{v}_N - \mathbf{v}\| + \|\mathbf{v} - \mathbf{w}\| \leq (\mu - 1 + 1)\varepsilon = \mu\varepsilon$. From $\|\overline{\mathbf{w}} - \mathbf{w}\| \leq \mu\varepsilon$ and the required minimality of $\#\operatorname{supp}\overline{\mathbf{w}}$ we infer $\#\operatorname{supp}\overline{\mathbf{w}} \lesssim \#\operatorname{supp}\mathbf{v}_N$. The last statement follows by an application of (4.3.13) with $\mathbf{z}$ replaced by $\overline{\mathbf{w}}$. $\qquad\square$

4.3.3 Adaptive approximation of the right-hand side

Concerning the function **RHS**, we state that, if for some $s < s^*$, $\mathbf{u} \in \ell^w_\tau(\Lambda)$, then Proposition 4.3 shows that $\mathbf{f} = \mathbf{A}\mathbf{u} \in \ell^w_\tau(\Lambda)$ with $|\mathbf{f}|_{\ell^w_\tau} \lesssim |\mathbf{u}|_{\ell^w_\tau}$. From (3.3.15) one infers that for any $\varepsilon > 0$, there exists an $\mathbf{f}_\varepsilon$ with $\|\mathbf{f} - \mathbf{f}_\varepsilon\| \leq \varepsilon$ and $\#\mathrm{supp}\,\mathbf{f}_\varepsilon \lesssim \varepsilon^{-1/s}|\mathbf{u}|^{1/s}_{\ell^w_\tau}$. In case the routine **RHS** computes such an $\mathbf{f}_\varepsilon$ in $\mathcal{O}(\varepsilon^{-1/s}|\mathbf{u}|^{1/s}_{\ell^w_\tau} + 1)$ operations, we call it *(quasi-)optimal*.

The question whether it can effectively be arranged in that way cannot be answered in general, as it depends on the particular right-hand side at hand. For the practical realization, some a priori information on the smooth and singular parts of f (in case it is a function) has to be used, from which the location of the large coefficients in $\mathbf{f}$ can be deduced, and suitable quadrature is needed. We shall assume that such techniques are at our disposal.

4.3.4 Computation of an approximate residual

Now we are in the position to formulate our inexact adaptive steepest descent scheme. But before we proceed, we add the following important remark.

Remark 4.4. If the iterative steps in (4.2.2) are performed only inexactly, it can no longer be guaranteed that the iterates are contained in $\mathrm{ran}(\mathbf{A})$, as it is the case for the unperturbed iteration. Unfortunately, components in $\mathrm{ker}(\mathbf{A})$, once occurred in an iterate and thus in the error, will never be reduced by subsequent steps of the iteration. On the one hand, this does not affect the convergence of the adaptive scheme on the continuous level, because $\mathrm{ker}(\mathbf{A}) = \mathrm{ker}(F^* \mathbf{D}^{-1})$ (recall Lemma 2.3). On the other hand, the redundant parts in the iterates clearly influence the cost of the routine **APPLY**, i.e., they may cause an unbounded increase of $|\mathbf{w}|_{\ell^w_\tau}$. Therefore, the most challenging task in the development of adaptive frame methods is the control of the kernel contributions in the iterates.

The following crucial routine computes an approximate residual of the current approximation $\mathbf{w}$ for $\mathbf{u}$ within a sufficiently small tolerance ζ such that either, in view of Proposition 4.2, the relative error in this approximate residual is below some prescribed tolerance δ, or the residual itself, being a measure of the error in $\mathbf{w}$, is below some other prescribed tolerance ε. In view of controlling the components of the approximations in $\mathrm{ker}(\mathbf{A})$, the tolerance ζ should be in any case below some third input parameter ξ.

$\mathbf{RES}[\mathbf{w}, \xi, \delta, \varepsilon] \to [\tilde{\mathbf{r}}, \nu]$:
$\zeta := 2\xi$
do $\zeta := \zeta/2$
    $\tilde{\mathbf{r}} := \mathbf{RHS}[\mathbf{f}, \zeta/2] - \mathbf{APPLY}[\mathbf{A}, \mathbf{w}, \zeta/2]$
until $\nu := \|\tilde{\mathbf{r}}\| + \zeta \leq \varepsilon$ or $\zeta \leq \delta\|\tilde{\mathbf{r}}\|$

Theorem 4.1. *Let* $\mathbf{r} := \mathbf{f} - \mathbf{A}\mathbf{w}$. *The routine* **RES** *has the following properties.*

(i) $[\tilde{\mathbf{r}}, \nu] = \mathbf{RES}[\mathbf{w}, \xi, \delta, \varepsilon]$ *terminates with* $\nu \geq \|\mathbf{r}\|$, $\nu \gtrsim \min\{\xi, \varepsilon\}$ *and* $\|\mathbf{r} - \tilde{\mathbf{r}}\| \leq \xi$.

(ii) *If, for* $s \leq \breve{s} < s^*$, *with* $\tau = (\frac{1}{2} + s)^{-1}$ *and* $\breve{\tau} = (\frac{1}{2} + \breve{s})^{-1}$, $\mathbf{u} \in \ell_\tau^w(\Lambda)$, *then*

$$\#\mathrm{supp}\, \tilde{\mathbf{r}} \lesssim \min\{\xi, \nu\}^{-1/s} |\mathbf{u}|_{\ell_\tau^w}^{1/s} + \min\{\xi, \nu\}^{-1/\breve{s}} |\mathbf{w}|_{\ell_{\breve{\tau}}^w}^{1/\breve{s}}, \quad (4.3.15)$$

$$\min\{\xi, \nu\}^{(\breve{s}/s) - 1} |\tilde{\mathbf{r}}|_{\ell_{\breve{\tau}}^w} \lesssim |\mathbf{u}|_{\ell_\tau^w}^{\breve{s}/s} + \min\{\xi, \nu\}^{(\breve{s}/s) - 1} |\mathbf{w}|_{\ell_{\breve{\tau}}^w}, \quad (4.3.16)$$

and the number of arithmetic operations and storage locations required by the call is bounded by some absolute multiple of

$$\min\{\xi, \nu\}^{-1/s} |\mathbf{u}|_{\ell_\tau^w}^{1/s} + \min\{\xi, \nu\}^{-1/\breve{s}} [|\mathbf{w}|_{\ell_{\breve{\tau}}^w}^{1/\breve{s}} + \xi^{1/\breve{s}}(\#\mathrm{supp}\,\mathbf{w} + 1)].$$

(iii) *In addition, if* **RES** *terminates with* $\nu > \varepsilon$, *then* $\|\mathbf{r} - \tilde{\mathbf{r}}\| \leq \delta\|\tilde{\mathbf{r}}\|$, $\nu \leq (1+\delta)\|\tilde{\mathbf{r}}\|$, *and* $\nu \leq \frac{1+\delta}{1-\delta}\|\mathbf{r}\|$.

Proof. Let us prove (i) first. If at evaluation of the until-case, $\zeta > \delta\|\tilde{\mathbf{r}}\|$, then

$$\|\tilde{\mathbf{r}}\| + \zeta < (\delta^{-1} + 1)\zeta. \quad (4.3.17)$$

Consequently, since ζ is halved in each iteration, we infer that, if not by $\zeta \leq \delta\|\tilde{\mathbf{r}}\|$, **RES** will terminate by $\|\tilde{\mathbf{r}}\| + \zeta \leq \varepsilon$.

Clearly, after any computation of $\tilde{\mathbf{r}}$ inside the algorithm, it is $\|\tilde{\mathbf{r}} - \mathbf{r}\| \leq \zeta$. Hence, $\|\mathbf{r}\| \leq \|\tilde{\mathbf{r}}\| + \|\mathbf{r} - \tilde{\mathbf{r}}\| \leq \|\tilde{\mathbf{r}}\| + \zeta = \nu$.

If the do-loop terminates in the first iteration, it is $\nu = \|\tilde{\mathbf{r}}\| + \xi$, thus $\nu \geq \xi$, and $\|\mathbf{r} - \tilde{\mathbf{r}}\| \leq \zeta = \xi$. Otherwise, let $\tilde{\mathbf{r}}^{\mathrm{old}} := \mathbf{RHS}[\mathbf{f}, \zeta] - \mathbf{APPLY}[\mathbf{A}, \mathbf{w}, \zeta]$. We have $\|\tilde{\mathbf{r}}^{\mathrm{old}}\| + 2\zeta > \varepsilon$ and $2\zeta > \delta\|\tilde{\mathbf{r}}^{\mathrm{old}}\|$, so that, using (4.3.17) with ζ replaced by 2ζ and $\tilde{\mathbf{r}}$ by $\tilde{\mathbf{r}}^{\mathrm{old}}$, we get

$$\nu \geq \zeta > \frac{\|\tilde{\mathbf{r}}^{\mathrm{old}}\| + 2\zeta}{2\delta^{-1} + 2} > \frac{\delta\varepsilon}{2 + 2\delta}.$$

Hence, $\nu \gtrsim \varepsilon$, and clearly, $\|\mathbf{r} - \tilde{\mathbf{r}}\| \leq \zeta < \xi$, so that the proof of (i) is finished.

The next step is to establish part (ii). For any finitely supported $\mathbf{v}$, we have

$$|\mathbf{v}|_{\ell_{\breve{\tau}}^w} \leq (\#\mathrm{supp}\,\mathbf{v})^{\breve{s} - s} |\mathbf{v}|_{\ell_\tau^w}. \quad (4.3.18)$$

Indeed,

$$|\mathbf{v}|_{\ell_{\breve{\tau}}^w} = \sup_{k \geq 1} k^{1/\breve{\tau}} |\gamma_k(\mathbf{v})| = \sup_{k \geq 1} k^{1/\tau} k^{1/\breve{\tau} - 1/\tau} |\gamma_k(\mathbf{v})|$$

$$\leq \sup_{k \geq 1} k^{1/\tau} (\#\,\mathrm{supp}\,\mathbf{v})^{1/\breve{\tau} - 1/\tau} |\gamma_k(\mathbf{v})| = (\#\,\mathrm{supp}\,\mathbf{v})^{\breve{s} - s} |\mathbf{v}|_{\ell_\tau^w}.$$

So for $\mathbf{g} := \mathbf{RHS}[\mathbf{f}, \zeta]$, from $\#\mathrm{supp}\,\mathbf{g} \lesssim \zeta^{-1/s}|\mathbf{u}|_{\ell_{\tau}^{w}}^{1/s}$ and $|\mathbf{g}|_{\ell_{\tau}^{w}} \lesssim |\mathbf{u}|_{\ell_{\tau}^{w}}$, using (4.3.18), we get

$$\zeta^{(\breve{s}/s)-1}|\mathbf{g}|_{\ell_{\tau}^{w}} \leq \zeta^{(\breve{s}/s)-1}(\#\,\mathrm{supp}\,\mathbf{g})^{\breve{s}-s}|\mathbf{g}|_{\ell_{\tau}^{w}} \leq \zeta^{(\breve{s}/s)-1}(\zeta^{-1/s}|\mathbf{u}|_{\ell_{\tau}^{w}}^{1/s})^{\breve{s}-s}|\mathbf{g}|_{\ell_{\tau}^{w}}$$
$$= |\mathbf{u}|_{\ell_{\tau}^{w}}^{(\breve{s}/s)-1}|\mathbf{g}|_{\ell_{\tau}^{w}} \lesssim |\mathbf{u}|_{\ell_{\tau}^{w}}^{\breve{s}/s}.$$

With ζ, $\tilde{\mathbf{r}}$, and ν having their values at termination, the properties of **APPLY**, cf. Proposition 4.3, now show that

$$\#\mathrm{supp}\,\tilde{\mathbf{r}} \lesssim \zeta^{-1/s}|\mathbf{u}|_{\ell_{\tau}^{w}}^{1/s} + \zeta^{-1/\breve{s}}|\mathbf{w}|_{\ell_{\tau}^{w}}^{1/\breve{s}},$$

and

$$\zeta^{(\breve{s}/s)-1}|\tilde{\mathbf{r}}|_{\ell_{\tau}^{w}} \lesssim |\mathbf{u}|_{\ell_{\tau}^{w}}^{\breve{s}/s} + \zeta^{(\breve{s}/s)-1}|\mathbf{w}|_{\ell_{\tau}^{w}}.$$

Therefore, (4.3.15) and (4.3.16) follow from these expressions once we have shown that $\zeta \gtrsim \min\{\xi, \nu\}$. When the do-loop terminates in the first iteration, we have $\zeta \gtrsim \xi$. In the other case, with $\tilde{\mathbf{r}}^{\mathrm{old}}$ as above, we have $\delta\|\tilde{\mathbf{r}}^{\mathrm{old}}\| < 2\zeta$, and so from $\|\tilde{\mathbf{r}} - \tilde{\mathbf{r}}^{\mathrm{old}}\| \leq \|\tilde{\mathbf{r}} - \mathbf{r}\| + \|\mathbf{r} - \tilde{\mathbf{r}}^{\mathrm{old}}\| \leq \zeta + 2\zeta$, we infer $\|\tilde{\mathbf{r}}\| \leq \|\tilde{\mathbf{r}}^{\mathrm{old}}\| + 3\zeta < 2\zeta\delta^{-1} + 3\zeta = (2\delta^{-1} + 3)\zeta$, so that $\nu = \|\tilde{\mathbf{r}}\| + \zeta < (2\delta^{-1} + 3)\zeta + \zeta = (2\delta^{-1} + 4)\zeta$, thus $\zeta \gtrsim \nu$.

To complete the proof of (ii), it remains to estimate the number of arithmetic operations. We start with the cost of **APPLY**. Let $K \geq 1$ be the number of calls of **APPLY** that were made. Note that a geometric sum argument clearly shows

$$\sum_{k=1}^{K} \left(\frac{\xi}{2^{k}}\right)^{-1/\breve{s}} = \xi^{-1/\breve{s}} \left(\frac{(2^{1/\breve{s}})^{K+1} - 1}{2^{1/\breve{s}} - 1} - 1\right) \lesssim \left(\frac{\xi}{2^{K+1}}\right)^{-1/\breve{s}}$$
$$\eqsim (\xi 2^{-K})^{-1/\breve{s}} = \zeta^{-1/\breve{s}}, \tag{4.3.19}$$

with ζ having its value at termination of **RES**. Using (4.3.3), we infer that the total cost of the calls of **APPLY** is bounded by some multiple of $\zeta^{-1/\breve{s}}|\mathbf{w}|_{\ell_{\tau}^{w}}^{1/\breve{s}} + K(\#\mathrm{supp}\,\mathbf{w} + 1)$. With the analogous argument for **RHS** and (4.3.5), we deduce that the total cost of **RES** can be bounded by some multiple of $\zeta^{-1/s}|\mathbf{u}|_{\ell_{\tau}^{w}}^{1/s} + \zeta^{-1/\breve{s}}|\mathbf{w}|_{\ell_{\tau}^{w}}^{1/\breve{s}} + K(\#\mathrm{supp}\,\mathbf{w} + 1)$. Taking into account its initial value, and the geometric decrease of ζ inside the algorithm, we have $K\xi^{-1/\breve{s}} = K(\zeta 2^{K})^{-1/\breve{s}} = K2^{-K/\breve{s}}\zeta^{-1/\breve{s}} \lesssim \zeta^{-1/\breve{s}}$. Hence, $K(\#\mathrm{supp}\,\mathbf{w} + 1) = K\xi^{-1/\breve{s}}\xi^{1/\breve{s}}(\#\mathrm{supp}\,\mathbf{w} + 1) \lesssim \zeta^{-1/\breve{s}}\xi^{1/\breve{s}}(\#\mathrm{supp}\,\mathbf{w} + 1)$. Since we have already shown that $\zeta \gtrsim \min\{\xi, \nu\}$, this completes the proof of (ii).

Finally, let us check (iii). Suppose that **RES** terminates with $\nu > \varepsilon$, and thus with $\zeta \leq \delta\|\tilde{\mathbf{r}}\|$. Then, obviously, $\|\mathbf{r} - \tilde{\mathbf{r}}\| \leq \delta\|\tilde{\mathbf{r}}\|$.

From $\|\tilde{\mathbf{r}}\| \leq \|\mathbf{r} - \tilde{\mathbf{r}}\| + \|\mathbf{r}\| \leq \delta\|\tilde{\mathbf{r}}\| + \|\mathbf{r}\|$, we have $\|\tilde{\mathbf{r}}\| \leq \frac{\|\mathbf{r}\|}{1-\delta}$, and so we arrive at $\nu = \|\tilde{\mathbf{r}}\| + \zeta \leq (1 + \delta)\|\tilde{\mathbf{r}}\| \leq \frac{1+\delta}{1-\delta}\|\mathbf{r}\|$. $\qquad\square$

4.3.5 The adaptive steepest descent method

In view of Remark 4.4, in order to be able to prove the optimality of the method, we make the following assumption. We shall append a couple of comments on its validity in §4.3.6 below.

Assumption 4.2. For any $s \in (0, s^*)$, $\mathbf{Q}$ is bounded on $\ell_\tau^w(\Lambda)$.

The routine **RES** is the basic building block for our fundamental algorithm which reads as follows.

Algorithm 1. SD_SOLVE$[\omega, \varepsilon] \to \mathbf{w}$:

% Let $\omega \geq \|\mathbf{Qu}\|$.

% Let λ and $\delta = \delta(\lambda)$ be constants as in Proposition 4.2.

% Fix some constants $\mu > 1$, $\beta \in (0, 1)$.

% Let K, M be the smallest integers with $\beta^K \omega \leq \varepsilon$, $\lambda^M \leq \frac{1-\delta}{1+\delta} \frac{\beta}{(1+3\mu)\kappa(\mathbf{A})}$, respectively.

$\mathbf{w}_0 := 0$; $\omega_0 := \omega$

for $i := 1$ to K do

    $\bar{\mathbf{w}}_i := \mathbf{w}_{i-1}$; $\omega_i := \beta\omega_{i-1}$; $\xi_i := \frac{\omega_i}{(1+3\mu)C_3 M}$ *% C_3 from Proposition 4.2*

    while *with* $[\tilde{\mathbf{r}}_i, \nu_i] := \mathbf{RES}[\bar{\mathbf{w}}_i, \xi_i, \delta, \frac{\omega_i}{(1+3\mu)\|\mathbf{A}^\dagger\|}]$, $\nu_i > \frac{\omega_i}{(1+3\mu)\|\mathbf{A}^\dagger\|}$ *do*

        $\mathbf{z}_i := \mathbf{APPLY}[\mathbf{A}, \tilde{\mathbf{r}}_i, \delta\|\tilde{\mathbf{r}}_i\|]$

        $\bar{\mathbf{w}}_i := \bar{\mathbf{w}}_i + \frac{\langle \tilde{\mathbf{r}}_i, \tilde{\mathbf{r}}_i \rangle}{\langle \mathbf{z}_i, \tilde{\mathbf{r}}_i \rangle}\tilde{\mathbf{r}}_i$

    enddo

    $\mathbf{w}_i := \mathbf{COARSE}[\bar{\mathbf{w}}_i, \frac{3\mu\omega_i}{1+3\mu}]$

endfor

$\mathbf{w} := \mathbf{w}_K$

It turns out that Algorithm 1 is indeed optimal in the sense of Definition 4.1. This is confirmed by the following theorem which is the main result of this chapter.

Theorem 4.2.   (i) *If $\omega \geq \|\mathbf{Qu}\|$, then $\mathbf{w} := \mathbf{SD\_SOLVE}[\omega, \varepsilon]$ terminates with $\|\mathbf{Q}(\mathbf{u} - \mathbf{w})\| \leq \varepsilon$.*

  (ii) *For any $\eta \in (0, s^*)$, let $\breve{s} = s^* - \frac{\eta}{2}$, $\breve{\tau} = (\breve{s} + \frac{1}{2})^{-1}$, and let the constant β inside* **SD_SOLVE** *satisfy*

$$\beta < \min\{1, [C_1(\breve{\tau})C_2(\breve{\tau})\|\mathbf{I} - \mathbf{Q}\|_{\ell_{\breve{\tau}}^w \to \ell_{\breve{\tau}}^w}]^{-2(s^*-\eta)/\eta}\}, \tag{4.3.20}$$

where $C_1(\breve{\tau})$ and $C_2(\breve{\tau})$ are the constants from (3.3.12) and (4.3.14), respectively. Let $\varepsilon \lesssim \omega \lesssim \|\mathbf{u}\|$. If for some $s \in (0, s^* - \eta]$ it is $\mathbf{u} \in \ell_\tau^w(\Lambda)$, then

$\#\operatorname{supp} \mathbf{w} \lesssim \varepsilon^{-1/s} |\mathbf{u}|_{\ell_\tau^w}^{1/s}$ *and the number of arithmetic operations and storage locations required by the call is bounded by some absolute multiple of the same expression.*

Proof. The first step is to prove (i). Let us consider the ith iteration of the for-loop. Assume that

$$\|\mathbf{Q}(\mathbf{u} - \mathbf{w}_{i-1})\| \leq \omega_{i-1}, \tag{4.3.21}$$

which holds by assumption for $i = 1$. The inner loop terminates after not more than $M + 1$ calls of **RES**. Indeed, suppose that this was not the case, then the first $M + 1$ calls of **RES** do not terminate because the first condition in the until-clause is satisfied, and so Theorem 4.1 (iii), (4.2.1), Proposition 4.2 (applied M times for each step in the inner loop) and assumption (4.3.21) show that the $(M + 1)$th call outputs a ν_i with

$$\begin{aligned}
\nu_i &\leq \frac{1+\delta}{1-\delta}\|\mathbf{f} - \mathbf{A}\bar{\mathbf{w}}_i\| = \frac{1+\delta}{1-\delta}\|\mathbf{A}(\mathbf{u} - \bar{\mathbf{w}}_i)\| \leq \frac{1+\delta}{1-\delta}\|\mathbf{A}\|^{\frac{1}{2}}\|\mathbf{u} - \bar{\mathbf{w}}_i\| \\
&\leq \frac{1+\delta}{1-\delta}\|\mathbf{A}\|^{\frac{1}{2}}\lambda^M\|\mathbf{u} - \mathbf{w}_{i-1}\| \leq \frac{1+\delta}{1-\delta}\|\mathbf{A}\|^{\frac{1}{2}}\lambda^M\|\mathbf{A}\|^{\frac{1}{2}}\|\mathbf{Q}(\mathbf{u} - \mathbf{w}_{i-1})\| \\
&\leq \frac{\omega_i}{(1+3\mu)\|\mathbf{A}^\dagger\|},
\end{aligned}$$

where the definition of M has been used in the last step, and which is a contradiction.

With $\hat{\mathbf{w}}_i$ denoting $\bar{\mathbf{w}}_i$ at termination of the inner loop, we have by (4.2.1) and the properties of **RES**

$$\|\mathbf{Q}(\mathbf{u} - \hat{\mathbf{w}}_i)\| \leq \|\mathbf{A}^\dagger\|^{\frac{1}{2}}\|\mathbf{u} - \hat{\mathbf{w}}_i\| \leq \|\mathbf{A}^\dagger\|\|\mathbf{A}(\mathbf{u} - \hat{\mathbf{w}}_i)\| \leq \|\mathbf{A}^\dagger\|\nu_i \leq \frac{\omega_i}{1+3\mu}, \tag{4.3.22}$$

so that, by the properties of **COARSE**,

$$\|\mathbf{Q}(\mathbf{u} - \mathbf{w}_i)\| \leq \|\mathbf{Q}(\mathbf{u} - \hat{\mathbf{w}}_i)\| + \|\mathbf{Q}\|\|\hat{\mathbf{w}}_i - \mathbf{w}_i\| \leq \frac{\omega_i}{1+3\mu} + \frac{3\mu\omega_i}{1+3\mu} = \omega_i,$$

showing convergence, and by definition of K completing the proof of the first statement.

The proof of (ii) follows the lines of the proof of [104, Theorem 3.12]. In our case where $\mathbf{A}$ has possibly a non-trivial kernel, generally, due to the errors in $\operatorname{ran}(\mathbf{I} - \mathbf{Q})$, we have no convergence of $\hat{\mathbf{w}}_i$ to $\mathbf{u}$ for $i \to \infty$, and as a consequence, we are not able to bound $|\mathbf{w}_i|_{\ell_\tau^w}$ by some absolute multiple of $|\mathbf{u}|_{\ell_\tau^w}$. Instead we shall first prove the weaker result (4.3.28). Afterwards, we will show that (4.3.28) actually suffices to conclude optimal computational complexity.

By part (i) of Theorem 4.1, Proposition 4.2 and the definition of the ξ_i, we have

$$\|(\mathbf{I} - \mathbf{Q})(\hat{\mathbf{w}}_i - \mathbf{w}_{i-1})\| \leq C_3 M \xi_i = \frac{\omega_i}{1+3\mu}. \tag{4.3.23}$$

Further, $\mathbf{Q}$ is bounded on ℓ_2, and by Assumption 4.2, it is bounded on $\ell_{\check{\tau}}^w$. By [55, (4.24)], it is $\ell_\tau^w = (\ell_2, \ell_{\check{\tau}}^w)_{\theta,\infty}$, with $\frac{1}{\tau} = \frac{1-\theta}{2} + \frac{\theta}{\check{\tau}}$, for $\tau \in [\check{\tau}, 2]$. A classical interpolation argument shows that $\mathbf{Q}$ is bounded on ℓ_τ^w, uniformly in $\tau \in [\check{\tau}, 2]$. Let N_i be the smallest integer such that

$$\|\mathbf{Qu} - (\mathbf{Qu})_{N_i}\| \le \frac{\omega_i}{1+3\mu}, \tag{4.3.24}$$

where $(\mathbf{Qu})_N$ denotes the best N-term approximation for $\mathbf{Qu}$. Then, using the assumption $\mathbf{u} \in \ell_\tau^w(\Lambda)$, (3.3.15) shows that

$$N_i \lesssim \omega_i^{-1/s} |\mathbf{Qu}|_{\ell_\tau^w}^{1/s} \lesssim \omega_i^{-1/s} |\mathbf{u}|_{\ell_\tau^w}^{1/s},$$

and therefore, using (4.3.18), we get

$$
\begin{aligned}
\omega_i^{(\check{s}/s)-1} |(\mathbf{Qu})_{N_i}|_{\ell_{\check{\tau}}^w} &\lesssim \omega_i^{(\check{s}/s)-1} N_i^{\check{s}-s} |(\mathbf{Qu})_{N_i}|_{\ell_\tau^w} \\
&\lesssim \omega_i^{(\check{s}/s)-1} \omega_i^{-(1/s)(\check{s}-s)} |\mathbf{u}|_{\ell_\tau^w}^{(1/s)(\check{s}-s)} |(\mathbf{Qu})_{N_i}|_{\ell_\tau^w} \\
&= |\mathbf{u}|_{\ell_\tau^w}^{(\check{s}/s)-1} |(\mathbf{Qu})_{N_i}|_{\ell_\tau^w} \lesssim |\mathbf{u}|_{\ell_\tau^w}^{(\check{s}/s)-1} |\mathbf{Qu}|_{\ell_\tau^w} \\
&\lesssim |\mathbf{u}|_{\ell_\tau^w}^{\check{s}/s}.
\end{aligned} \tag{4.3.25}
$$

From (4.3.22), (4.3.23) and (4.3.24), we get

$$\|(\mathbf{Qu})_{N_i} + (\mathbf{I} - \mathbf{Q})\mathbf{w}_{i-1} - \hat{\mathbf{w}}_i\| \le \frac{3\omega_i}{1+3\mu}.$$

From Proposition 4.4, with $\mathbf{v}$ reading as $(\mathbf{Qu})_{N_i} + (\mathbf{I} - \mathbf{Q})\mathbf{w}_{i-1}$ and by using that $\mu > 1$, it follows that $\mathbf{w}_i := \mathbf{COARSE}[\hat{\mathbf{w}}_i, \frac{3\mu\omega_i}{1+3\mu}]$ satisfies

$$
\begin{aligned}
|\mathbf{w}_i|_{\ell_{\check{\tau}}^w} &\le C_2(\check{\tau}) |(\mathbf{Qu})_{N_i} + (\mathbf{I} - \mathbf{Q})\mathbf{w}_{i-1}|_{\ell_{\check{\tau}}^w} \\
&\le C_1(\check{\tau})C_2(\check{\tau}) |(\mathbf{Qu})_{N_i}|_{\ell_{\check{\tau}}^w} + C_1(\check{\tau})C_2(\check{\tau}) |(\mathbf{I} - \mathbf{Q})|_{\ell_{\check{\tau}}^w \to \ell_{\check{\tau}}^w} |\mathbf{w}_{i-1}|_{\ell_{\check{\tau}}^w}
\end{aligned}
$$

by (3.3.12). And so by (4.3.25), we obtain

$$\omega_i^{(\check{s}/s)-1} |\mathbf{w}_i|_{\ell_{\check{\tau}}^w} \le C|\mathbf{u}|_{\ell_\tau^w}^{\check{s}/s} + C_1(\check{\tau})C_2(\check{\tau}) |(\mathbf{I}-\mathbf{Q})|_{\ell_{\check{\tau}}^w \to \ell_{\check{\tau}}^w} \beta^{(\check{s}/s)-1} \omega_{i-1}^{(\check{s}/s)-1} |\mathbf{w}_{i-1}|_{\ell_{\check{\tau}}^w}, \tag{4.3.26}$$

for some absolute constant $C > 0$. The assumption on β made in (4.3.20) shows that

$$C_1(\check{\tau})C_2(\check{\tau}) |(\mathbf{I} - \mathbf{Q})|_{\ell_{\check{\tau}}^w \to \ell_{\check{\tau}}^w} \beta^{(\check{s}/s)-1} < 1. \tag{4.3.27}$$

For convenience, we give a short justification of the latter. Let us abbreviate $X := C_1(\check{\tau})C_2(\check{\tau}) |(\mathbf{I} - \mathbf{Q})|_{\ell_{\check{\tau}}^w \to \ell_{\check{\tau}}^w}$. If $X \le 1$, (4.3.27) is obvious. For $X > 1$, we get

$$X\beta^{(\check{s}/s)-1} < X^{((\check{s}/s)-1)(-2(s^*-\eta)/\eta)+1} = X^{((\check{s}-s)/s)(-2(s^*-\eta)/\eta)+1} \le 1,$$

because

$$\frac{2(s^*-\eta)}{\eta} = \frac{s^*-\eta}{s^* - \frac{\eta}{2} - s^* + \eta} = \frac{s^*-\eta}{\check{s} - (s^*-\eta)} \ge \frac{s}{\check{s}-s} > 0.$$

Hence, from (4.3.26) we may further conclude by a geometric series argument that

$$\omega_i^{(\check{s}/s)-1}|\mathbf{w}_i|_{\ell_\tau^w} \lesssim |\mathbf{u}|_{\ell_\tau^w}^{\check{s}/s},\tag{4.3.28}$$

which, as it should be emphasized here, holds uniformly in i. Moreover, knowing this, Proposition 4.4 and (4.3.25) show that

$$\begin{aligned}\#\mathrm{supp}\,\mathbf{w}_i &\lesssim \omega_i^{-1/\check{s}}|(\mathbf{Qu})_{N_i} + (\mathbf{I}-\mathbf{Q})\mathbf{w}_{i-1}|_{\ell_\tau^w}^{1/\check{s}}\\ &\lesssim \omega_i^{-1/s}\big(\omega_i^{(\check{s}/s)-1}\big[|(\mathbf{Qu})_{N_i}|_{\ell_\tau^w} + |\mathbf{I}-\mathbf{Q}|_{\ell_\tau^w\to\ell_\tau^w}|\mathbf{w}_{i-1}|_{\ell_\tau^w}\big]\big)^{1/\check{s}}\\ &\lesssim \omega_i^{-1/s}|\mathbf{u}|_{\ell_\tau^w}^{1/s},\end{aligned}\tag{4.3.29}$$

again uniformly in i. By our choice of K, for $K \geq 1$, it is $\omega_K \leq \varepsilon < \omega_{K-1} = \beta^{-1}\omega_K$, hence, in particular,

$$\omega_K \gtrsim \varepsilon,\tag{4.3.30}$$

which in case $K = 0$ holds by assumption. Plugging this into (4.3.29) shows the first part of (ii).

For any computed ν_i in the inner loop, Theorem 4.1 (i) shows that $\frac{\omega_i}{(1+3\mu)\|\mathbf{A}^\dagger\|} \lesssim \nu_i$. At termination of the inner loop we have $\nu_i \lesssim \omega_i$, whereas for any evaluation of **RES** that does not lead to termination, Theorem 4.1 (iii) and Proposition 4.2 show that

$$\nu_i \leq \frac{1+\delta}{1-\delta}\|\mathbf{f}-\mathbf{A}\bar{\mathbf{w}}_i\| \lesssim \|\mathbf{u}-\bar{\mathbf{w}}_i\| \leq \|\mathbf{u}-\mathbf{w}_{i-1}\| \lesssim \omega_{i-1} \approx \omega_i.$$

We conclude that

$$\nu_i \approx \omega_i,$$

uniformly in i and over all computations of ν_i in the inner loop.

Inside the body of the inner loop, we have that the tolerance for the call of **APPLY** satisfies $\delta\|\tilde{\mathbf{r}}_i\| \geq \frac{\delta\nu_i}{1+\delta}$, because $(1+\delta)\|\tilde{\mathbf{r}}_i\| \geq \nu_i$ by Theorem 4.1 (iii), thus

$$\delta\|\tilde{\mathbf{r}}_i\| \gtrsim \omega_i.\tag{4.3.31}$$

From Proposition 4.2 we learn that

$$\left|\frac{\langle\tilde{\mathbf{r}}_i,\tilde{\mathbf{r}}_i\rangle}{\langle\mathbf{z}_i,\tilde{\mathbf{r}}_i\rangle}\right| \lesssim 1.\tag{4.3.32}$$

By (4.3.28) and the fact that the number of iterations of the inner loop is uniformly bounded, Theorem 4.1 (ii) shows that

$$\omega_i^{(\check{s}/s)-1}|\tilde{\mathbf{r}}_i|_{\ell_\tau^w} \lesssim |\mathbf{u}|_{\ell_\tau^w}^{\check{s}/s}, \quad \omega_i^{(\check{s}/s)-1}|\bar{\mathbf{w}}_i|_{\ell_\tau^w} \lesssim |\mathbf{u}|_{\ell_\tau^w}^{\check{s}/s},\tag{4.3.33}$$

where for the right estimate we have used (4.3.32). With this result and (4.3.29), Theorem 4.1 (ii), the properties of **APPLY** (with s reading as $\check{s}$) and (4.3.31) show that

$$\#\mathrm{supp}\,\tilde{\mathbf{r}}_i \lesssim \omega_i^{-1/s}|\mathbf{u}|_{\ell_\tau^w}^{1/s}, \quad \#\mathrm{supp}\,\mathbf{z}_i \lesssim \omega_i^{-1/s}|\mathbf{u}|_{\ell_\tau^w}^{1/s}, \quad \#\mathrm{supp}\,\bar{\mathbf{w}}_i \lesssim \omega_i^{-1/s}|\mathbf{u}|_{\ell_\tau^w}^{1/s}.$$

By using these results concerning the lengths of the supports and the ℓ_τ^w-norms, again Theorem 4.1 (ii) and the properties of **APPLY** and **COARSE** show that the number of arithmetic operations and storage locations required for the computation of $\mathbf{w}_i$ starting from $\mathbf{w}_{i-1}$ is bounded by an absolute multiple of

$$\omega_i^{-1/s}|\mathbf{u}|_{\ell_\tau^w}^{1/s} + \max\{\log(\omega_i^{-1}\|\hat{\mathbf{w}}_i\|),1\}. \tag{4.3.34}$$

Note that $\log(\omega_i^{-1}\|\hat{\mathbf{w}}_i\|) \lesssim \omega_i^{-1/\check{s}}\|\hat{\mathbf{w}}_i\|^{1/\check{s}} \lesssim \omega_i^{-1/\check{s}}|\hat{\mathbf{w}}_i|_{\ell_*^w}^{1/\check{s}} \lesssim \omega_i^{-1/s}|\mathbf{u}|_{\ell_\tau^w}^{1/s}$, where in the last estimate we have used (4.3.33) again. Moreover, we have $1 \lesssim \omega^{-1/s}\|\mathbf{u}\|^{1/s} \lesssim \omega_i^{-1/s}|\mathbf{u}|_{\ell_\tau^w}^{1/s}$, by the assumption $\omega \lesssim \|\mathbf{u}\|$. Finally, from (4.3.34), the geometric decrease of the ω_i and (4.3.30), we infer that the proof is completed. $\qquad\square$

Remark 4.5. We emphasize that from $\|\mathbf{Q}(\mathbf{u}-\mathbf{w})\| \le \varepsilon$ in Theorem 4.2 (i), one gets $\|u - \mathbf{w}^\top\Psi\|_{H^t(\Omega)} = \|F^*\mathbf{D}^{-1}\mathbf{Q}(\mathbf{u}-\mathbf{w})\|_{H^t(\Omega)} \lesssim \|\mathbf{Q}(\mathbf{u}-\mathbf{w})\| \le \varepsilon$.

Remark 4.6. In view of the results in Chapter 3, even for $u = \mathbf{u}^\top\Psi \in C^\infty(\Omega)$, $\mathbf{u} \in \ell_\tau^w(\Lambda)$, $\frac{1}{\tau} = s + \frac{1}{2}$, can only be expected for $0 < s < \frac{d-t}{n}$. In order to obtain Theorem 4.2 (ii) for all $s < \frac{d-t}{n}$, one has to assume $s^* > \frac{d-t}{n}$. Indeed, suppose $s^* = \frac{d-t}{n}$. Then, one has to choose η arbitrary small in order to get s arbitrary close to $\frac{d-t}{n}$ in the theorem. Hence, the exponent $-2(s^* - \eta)/\eta$ in (4.3.20) tends to $-\infty$ and, generally, β to zero. This shows that for $s^* = \frac{d-t}{n}$ Theorem 4.2 would give a suboptimal result.

4.3.6 Assumption 4.2 and ways to circumvent it

In the proof of Theorem 4.2, essential use has been made of Assumption 4.2. Unfortunately, so far this assumption has only been verified for the special case that $H = L_2(\Omega)$, and that the aggregated wavelet frame is made up of local $L_2(\Omega)$-orthonormal bases, where at the internal boundaries a proper damping is applied to the wavelets, as mentioned in Remark 2.7; see [104, §4.3] for details. The general proof is still a difficult open problem.

Let $\Psi_{L_2} = \{\psi_\lambda\}_{\lambda\in\Lambda}$ be the aggregated wavelet Gelfand frame considered in Proposition 2.11. Recall from equation (2.5.7) that for $\Psi = \{2^{-t|\lambda|}\psi_\lambda\}_{\lambda\in\Lambda}$, which is a frame for $H = H_0^t(\Omega)$, denoting the $H_0^t(\Omega)$-*canonical* dual frame with $\{\tilde{\eta}_\lambda\}_{\lambda\in\Lambda}$, $\mathbf{Q}$ has the matrix representation

$$\mathbf{Q} = \{\langle 2^{-t|\lambda|}\psi_\lambda, \tilde{\eta}_\mu\rangle_{H_0^t(\Omega)\times H^{-t}(\Omega)}\}_{\mu,\lambda\in\Lambda}. \tag{4.3.35}$$

We emphasize again that the canonical dual frame $\{\tilde{\eta}_\lambda\}_{\lambda\in\Lambda}$ of $\Psi = \{2^{-t|\lambda|}\psi_\lambda\}_{\lambda\in\Lambda}$ is interpreted as a frame for $H^{-t}(\Omega)$ as outlined in §2.1.4. Unfortunately, for most cases, the collection $\{\tilde{\eta}_\lambda\}_{\lambda\in\Lambda}$ is not explicitly given. In principle, Proposition 4.3 represents an outline for the verification of Assumption 4.2. There, from the compressibility and with that the existence of a suitable routine **APPLY**, the boundedness of $\mathbf{A}$ on ℓ^w_τ could be inferred. As it will be presented in detail in the upcoming Chapter 5, compressibility of $\mathbf{A}$ is a consequence of the regularity, vanishing moment properties, and the locality of the wavelets. However, since in general we have no access to the canonical dual frame $\{\tilde{\eta}_\lambda\}_{\lambda\in\Lambda}$ and its properties, we can not deduce compressibility of the matrix in (4.3.35) by a similar technique.

This theoretical bottleneck has already been detected in [104]. There, the strategy to circumvent the assumption is to consider a modified version of the adaptive (damped Richardson) algorithm, where regularly, after a fixed number of iterations, a bounded projector $\mathbf{P} : \ell_2(\Lambda) \to \ell_2(\Lambda)$ is approximately applied to the current iterate, using the routine **APPLY**, where $\ker(\mathbf{P}) = \ker(\mathbf{A})$, $\mathbf{P}$ is sufficiently compressible, and thus bounded on the respective space ℓ^w_τ. That means $\mathbf{P}$ is actively used in the algorithm to reduce error components in $\ker(\mathbf{A})$. In particular, with $\rho := \|(\mathbf{I} - \alpha\mathbf{A})|_{\mathrm{ran}(\mathbf{A})}\| = \frac{\kappa(\mathbf{A})-1}{\kappa(\mathbf{A})+1} < 1$ for $\alpha = 2/(\|\mathbf{A}\| + \|\mathbf{A}^\dagger\|^{-1})$, the original adaptive Richardson method in [104], *not* using this technique, reads as follows.

Algorithm 2. R_SOLVE$[\varepsilon] \to \mathbf{u}_\varepsilon$:

% Let $\theta < 1/3$ and $K \in \mathbb{N}$ be fixed such that $3\rho^K < \theta$.

% $i := 0$, $\mathbf{v}^{(0)} := 0$, $\varepsilon_0 := \|(\mathbf{A}|_{\mathrm{ran}(\mathbf{A})})^{-1}\|\|\mathbf{f}\|_{\ell_2(\Lambda)}$

while $\varepsilon_i > \varepsilon$ do

   $i := i + 1$

   $\varepsilon_i := 3\rho^K \varepsilon_{i-1}/\theta$

   $\mathbf{f}^{(i)} := \mathbf{RHS}[\mathbf{f}, \frac{\theta\varepsilon_i}{6\alpha K}]$

   $\mathbf{v}^{(i,0)} := \mathbf{v}^{(i-1)}$

   for $j = 1, \ldots K$ do

       $\mathbf{v}^{(i,j)} := \mathbf{v}^{(i,j-1)} - \alpha(\mathbf{APPLY}[\mathbf{A}, \mathbf{v}^{(i,j-1)}, \frac{\theta\varepsilon_i}{6\alpha K}] - \mathbf{f}^{(i)})$

   endfor

   $\mathbf{v}^{(i)} := \mathbf{COARSE}[\mathbf{v}^{(i,K)}, (1 - \theta)\varepsilon_i]$

enddo

$\mathbf{u}_\varepsilon := \mathbf{v}^{(i)}$

The verification of its optimality still requires Assumption 4.2, whereas for the following modified algorithm this is not the case.

Algorithm 3. R_SOLVE$*[\varepsilon] \to \mathbf{u}_\varepsilon$:

% Let $\theta < 1/3$ and $K \in \mathbb{N}$ be fixed such that $3\rho^K \|\mathbf{P}\| < \theta$.
% $i := 0$, $\mathbf{u}^{(0)} := 0$, $\varepsilon_0 := \|\mathbf{P}\| \|(\mathbf{A}|_{\mathrm{ran}(\mathbf{A})})^{-1}\| \|\mathbf{f}\|_{\ell_2(\Lambda)}$
`while` $\varepsilon_i > \varepsilon$ `do`
   $i := i + 1$
   $\varepsilon_i := 3\rho^K \|\mathbf{P}\| \varepsilon_{i-1}/\theta$
   $\mathbf{v}^{(i,0)} := \mathbf{u}^{(i-1)}$
   `for` $j = 1, \ldots K$ `do`
      $\mathbf{v}^{(i,j)} := \mathbf{v}^{(i,j-1)} - \alpha(\mathbf{APPLY}[\mathbf{A}, \mathbf{v}^{(i,j-1)}, \frac{\rho^j \varepsilon_{i-1}}{2\alpha K}] - \mathbf{RHS}[\mathbf{f}, \frac{\rho^j \varepsilon_{i-1}}{2\alpha K}])$
   `endfor`
   $\mathbf{z}^{(i)} := \mathbf{APPLY}[\mathbf{P}, \mathbf{v}^{(i,K)}, \frac{\theta \varepsilon_i}{3}]$
   $\mathbf{u}^{(i)} := \mathbf{COARSE}[\mathbf{z}^{(i)}, (1-\theta)\varepsilon_i]$
`enddo`
$\mathbf{u}_\varepsilon := \mathbf{u}^{(i)}$

The operator $\mathbf{P}$ can be constructed by replacing in (4.3.35) the canonical dual $\{\tilde{\eta}_\lambda\}_{\lambda \in \Lambda}$ by the non-canonical dual collection $\{2^{t|\mu|}\sigma_i \tilde{\psi}_{i,\mu}\}_{(i,\mu) \in \Lambda}$ from (2.3.28).

Proposition 4.5. *The bi-infinite matrix*

$$\mathbf{P} := \{2^{-t(|\lambda|-|\mu|)} \langle \psi_{i,\lambda}, \sigma_j \tilde{\psi}_{j,\mu} \rangle_{H_0^t(\Omega) \times H^{-t}(\Omega)}\}_{(j,\mu),(i,\lambda) \in \Lambda}. \tag{4.3.36}$$

represents a bounded projector $\mathbf{P} : \ell_2(\Lambda) \to \ell_2(\Lambda)$ *with* $\ker(\mathbf{P}) = \ker(\mathbf{A})$.

Proof. It is immediate to see that $\mathbf{P} = \mathbf{D}\tilde{F}F^*\mathbf{D}^{-1}$. Here, $\tilde{F}$ represents the operator from (2.3.21) associated to the L_2-dual frame from (2.3.28). Then, Proposition 2.9 (ii) shows that $\mathbf{P}$ is a bounded projector with $\ker(\mathbf{P}) = \ker(F^*\mathbf{D}^{-1}) = \ker(\mathbf{A})$. $\qquad \square$

The advantage is now that the functions $\tilde{\psi}_{j,\mu}$ are just lifted versions of the dual basis functions of the reference bases on the unit cube. In case the frame construction is based on the spline wavelets as mentioned in §2.2.4, the local dual bases $\tilde{\Psi}^{(i)}$ are explicitly given and one has control on their smoothness and approximation properties. For this situation, the compressibility of $\mathbf{P}$ has been investigated in [104, §4.5]. As for the stiffness matrix $\mathbf{A}$, besides the compressibility of $\mathbf{P}$ also its computability in the sense of Definition 4.3 has to be clarified before one can consider the above approach to be a real alternative. Even for smooth σ_j, a direct computation of the inner products in (4.3.36) by numerical quadrature might be difficult, due to the fact that the local dual wavelets $\tilde{\psi}_{j,\mu}$ are usually no spline functions. Moreover, their smoothness on Ω_i is usually smaller than the one of the primal functions. Anyway, although Assumption 4.2 can be dropped when incorporating the application of $\mathbf{P}$

into the method, the resulting alternative scheme can though be expected to have slightly worse quantitative properties. In Chapter 6, in the context of an additive Schwarz domain decomposition method, we shall demonstrate an alternative way for the application of $\mathbf{P}$, avoiding the computation of L_2-inner products of primal and weighted local dual wavelets.

One might now think of the development of a suitable variant of **SD_SOLVE**, using $\mathbf{P}$, for which optimality could then probably be shown without Assumption 4.2. However, we decide not to follow this path, because, on the one hand, the numerical results presented below indicate the asymptotically optimal complexity of **SD_SOLVE** in its present form. On the other hand, in Chapter 6, we want to consider domain decomposition solvers with which we have even more convenient instruments to eliminate Assumption 4.2. In fact, we shall develop algorithms with an inherent reduction of contributions from the kernel of $\mathbf{A}$. Basically, those schemes will be constructed in such a way that in the overlapping regions of the subdomains as few wavelets contribute to the approximation as possible. This feature can already be incorporated into the unperturbed infinite dimensional iteration, so that no additional errors are introduced. Using such a strategy will also enable us to prove optimality of the method without Assumption 4.2.

Remark 4.7. Not restricting the discussion to the case of (aggregated) wavelet frames and the treatment of differential equations, we mention that there exist frames, e.g., time-frequency localized Gabor frames (and more generally all intrinsically polynomially localized frames [38, 63]), for which the boundedness of the corresponding $\mathbf{Q}$ has been rigorously proven; see [36, 38].

4.4 Numerical examples

By now we have seen all essential ingredients needed for the development of adaptive wavelet (frame) methods. It is now time to investigate their practical properties. The intention of this section is to numerically confirm the optimal computational complexity stated in Theorem 4.2. Moreover, a quantitative comparison with the adaptive Richardson method (Algorithm 2) will be given. Another important question clearly is how the solution u of an operator equation is composed of contributions from the different subdomains Ω_i, i.e., of wavelet expansions with respect to a local basis $\Psi^{(i)}$ for $H_0^t(\Omega_i)$. In other words, we want to investigate which particular frame representation of the solution u is generated by the adaptive frame methods. The results of numerical tests for the one- and two-dimensional Poisson equation with homogeneous Dirichlet boundary data will be presented. The model problems will be chosen in such a way that the exact solutions exhibit singularities either induced by the right-hand side (§4.4.1) or solely by the shape of the domain (§4.4.2), so that adaptive schemes qualitatively outperform methods based on uniform discretizations (recall the discussion at the beginning of §4.1).

4.4.1 The Poisson equation in an interval

As our first model model problem, let us consider the variational formulation of the Poisson equation in the unit interval with prescribed homogeneous boundary conditions

$$-u'' = f \quad \text{in } \Omega, \quad u(0) = u(1) = 0. \tag{4.4.1}$$

Thus, in this case we have $n = 1$, and $t = 1$. The right-hand side f shall be chosen as the functional defined by $f(v) := 4v(\frac{1}{2}) + \int_0^1 g(x)v(x)\mathrm{d}x$, where

$$g(x) = -9\pi^2 \sin(3\pi x) - 4.$$

The solution is consequently given by

$$u(x) = -\sin(3\pi x) + \begin{cases} 2x^2 & , \ x \in [0, \frac{1}{2}) \\ 2(1-x)^2, & x \in [\frac{1}{2}, 1] \end{cases}, \tag{4.4.2}$$

and it is shown in Figure 4.1. This function is contained in the Besov space $B_{\tau,\tau}^{s+1}(\Omega)$, for all $s > 0$, $\tau^{-1} = s + \frac{1}{2}$, but it is only contained in $H^\alpha(\Omega)$ for $\alpha < \frac{3}{2}$, due to the singularity at 0.5. A justification of these statements can also be obtained from [99, Theorem 4.6.3 (ii)], similar to (3.3.18) and (3.3.19) (for the case $n = 1$). The right-hand side is contained in $H^\beta(\Omega)$ for $\beta \in [-1, -\frac{1}{2})$.

We choose the overlapping domain decomposition $\Omega = \Omega_0 \cup \Omega_1 = (0, 0.7) \cup (0.3, 1)$, and, associated to this covering, we construct a wavelet frame by aggregating local wavelet bases on Ω_0 and Ω_1, respectively. In particular, as building blocks in this simple construction procedure, we use the biorthogonal spline wavelet bases developed in [95] lifted to the subdomains. Recall that all basic properties of these bases have been collected in §2.4. In particular, we use wavelets with primal spline order $d = 2, 3$ and 4, having $\tilde{d} = 2, 3$ and 6 vanishing moments, respectively. The primal functions vanish on $\partial\Omega$ at order $t - 1$, whereas at the internal boundaries (here 0.3 or 0.7) they vanish at order $d - 2$. No boundary conditions are imposed on the dual basis functions, so that all (primal) wavelets have $\tilde{d}$ vanishing moments. From Proposition 3.5 we know that u has a representation $u = \tilde{\mathbf{u}}^T \Psi$ with for all $s < d - 1$, $\tilde{\mathbf{u}} \in \ell_\tau^w(\Lambda)$, $\tau^{-1} = s + \frac{1}{2}$.

For the tests below, the various constants appearing in the algorithm have been chosen in such a manner that the best performance is obtained. For instance, it has been assumed that $\|\mathbf{A}\| = \|\mathbf{A}^\dagger\| = 1$, although from a purely theoretical point of view this may be too optimistic. Our experiments have also taught us that any call of the routine **APPLY** should be immediately followed by a call of **COARSE**. Otherwise one observes rapidly growing numbers of degrees of freedom, resulting in an unsatisfactory performance. If such an additional coarsening is included, the thresholding after exiting the inner loop can be dropped. It is important to note that, without an analogous modification, Algorithm 2 (or the method introduced in [28]) is not performing very well either. This holds for the case of frames as well as for the

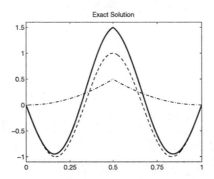

Figure 4.1: Exact solution of the one-dimensional test problem (solid line) composed of a sinusoidal (dashed line) and piecewise polynomial part (dash-dotted line).

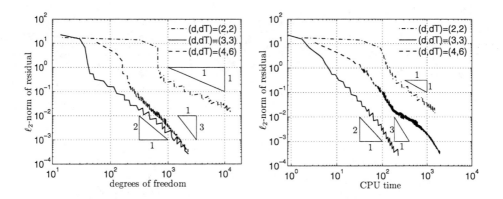

Figure 4.2: ℓ_2-norms of the residuals of $\bar{\mathbf{w}}_i$ versus $\#\operatorname{supp} \bar{\mathbf{w}}_i$ (left) and versus CPU time (right) for the one-dimensional test problem.

basis case. The numerical results given below will show that this modification has neither spoiled the convergence with optimal rates, nor the linear complexity.

In Figure 4.2 the convergence histories of **SD_SOLVE** with respect to the number of degrees of freedom and the computing time are depicted. Because of the equivalence $\|\mathbf{f} - \mathbf{A}\mathbf{w}\| \approx \|u - \mathbf{w}^\top \Psi\|_{H^t(\Omega)}$ we have computed approximations to the ℓ_2-norms of the exact discrete residuals $\mathbf{f} - \mathbf{A}\bar{\mathbf{w}}_i$ with high precision to measure the continuous H^1-errors. Obviously, the algorithm converges with the optimal order $d - 1$, where, quantitatively, the best performance is obtained for the case of piecewise quadratic wavelets, i.e., $d = 3$. Hence, Theorem 4.2 is fully confirmed by these results. Although for $d = 4$ the highest convergence rate can be observed, in this case significantly more iterations have been performed compared to $d = 3$, which explains that the algorithm performs much better for $d = 3$. This is a consequence of the worse H_0^1-Riesz constants of the underlying interval bases for $d = 4$. We also mention that for $d = 2$, from the first to the second iteration, only a small reduction of the error is created, but the support size jumps from 15 to the value of 320. This explains the almost horizontal course of the error curve in the pre-asymptotic range in this case.

A comparison of the steepest descent method with the adaptive Richardson iteration from Algorithm 2 for $d = 3$ is given in Figure 4.3. The Richardson iteration has been tested with descent parameters $\alpha = 0.4$ which is close to the optimal parameter and $\alpha = 0.15$. For $\alpha = 0.4$, **R_SOLVE** performs slightly better than **SD_SOLVE**. With the suboptimal choice $\alpha = 0.15$, however, **SD_SOLVE** outperforms **R_SOLVE**. The detection of an α close to the optimal parameter has only been possible by considering the descent parameters calculated in **SD_SOLVE** (see Figure 4.4).

For $d = 3$ the amount of 43 iterations in the inner loop of **SD_SOLVE** has been performed, while within **R_SOLVE** with $\alpha = 0.4$ altogether 186 inner iterations had to be spent which is more than four times as much. On the other hand, the total number of calls of the routine **APPLY** in the steepest descent method, including those performed in **RES**, has been 221. With $\alpha = 0.15$ in **R_SOLVE** 329 iterations and thus also calls of **APPLY** were needed. We learn that, for the quantitative performance, the number of calls of **APPLY** is crucial, rather than the number of (inner) iterations. This is due to the fact that these calls actually represent the most expensive building block.

Some more insight on the specialty of the frame approximation of the solution can be gained by looking at the distribution of degrees of freedom on the different subdomains. This particular issue is addressed in Figure 4.5, whereas the corresponding contributions from $H_0^t(\Omega_i)$ are depicted in Figure 4.6. We have to state that in case $d = 2$ the sparsity of the approximation is not really satisfactory. For the higher orders this is significantly better. On both subdomains the singularity of the solution at 0.5 is detected (Figure 4.5). Particularly for $d = 4$, it becomes visible that in the overlapping region $(0.3, 0.7)$ also away from the singularity, where the solution is smooth, considerably more coefficients appear than outside the overlapping region. Obviously, in this case a rather redundant approximation has been generated.

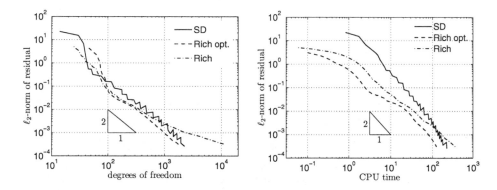

Figure 4.3: Comparison of **SD_SOLVE** and **R_SOLVE** for $d = 3$. **R_SOLVE** is tested with $\alpha = 0.4$ (dashed line) and with $\alpha = 0.15$ (dash-dotted line).

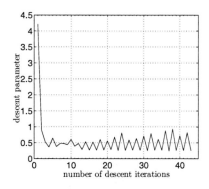

Figure 4.4: Descent parameters calculated in **SD_SOLVE** for $d = 3$.

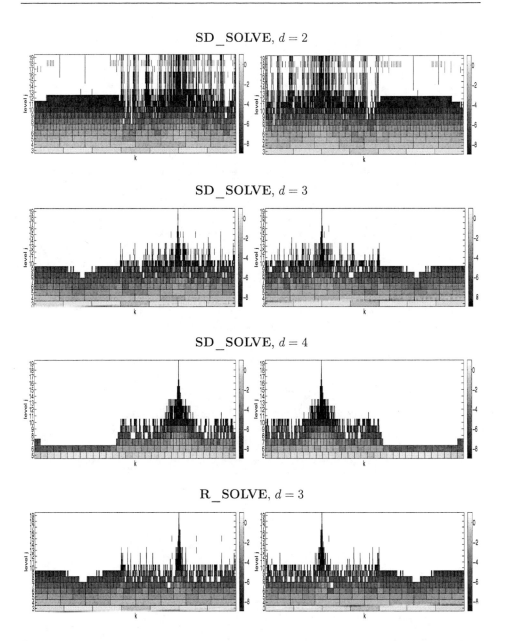

Figure 4.5: Distribution of wavelet coefficients of the final approximation produced by **SD_SOLVE** or **R_SOLVE** on Ω_0 (left) and Ω_1 (right) for different spline orders d.

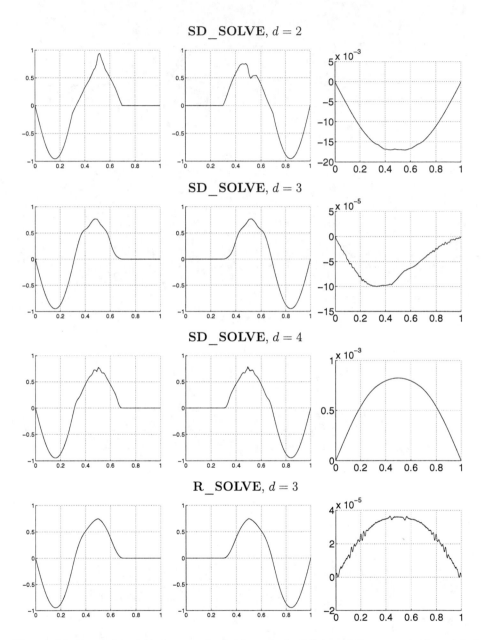

Figure 4.6: Local approximations on Ω_0 (left), on Ω_1 (middle), and the corresponding global pointwise errors (right).

Looking at the local approximations in Figure 4.6, we see that for each setup *different* local components are generated, all of them summing up to (approximately) the same function, namely u. Let us emphasize again that, although the exact global solution is known, a priori it is not clear which particular local approximation will be produced. Moreover, their shapes may change even if only the tolerances in the algorithm are steered differently.

4.4.2 The Poisson equation in an L-shaped domain

We now consider the variational form of the Poisson equation with homogeneous Dirichlet boundary conditions

$$-\Delta u = f \text{ in } \Omega, \quad u|_{\partial\Omega} = 0, \tag{4.4.3}$$

in the L-shaped domain $\Omega = (-1,1)^2 \backslash [0,1)^2$. In this case it is $n = 2$, and $t = 1$. As the exact solution of this model problem we choose the singularity function $\mathcal{S}(r,\theta) := \mathcal{S}_{1,1}(r,\theta)$ from (3.4.6), i.e.,

$$\mathcal{S}(r,\theta) := \zeta(r)r^{2/3}\sin(\frac{2}{3}\theta), \tag{4.4.4}$$

with (r,θ) denoting polar coordinates with respect to the re-entrant corner at the origin, and where ζ is a smooth function on $[0,1]$ that is identically 1 on $[0,r_0]$ and vanishes on $[r_1,1]$, for some $0 < r_0 < r_1 < 1$. The function $\mathcal{S}$, together with the corresponding right-hand side $-\Delta\mathcal{S}$, is shown in Figure 4.7. It is known that $\mathcal{S} \in B^{2s+1}_{\tau,\tau}(\Omega)$, for all $s > 0$, $\tau^{-1} = s + \frac{1}{2}$ (see [33]) but only $\mathcal{S} \in H^\alpha(\Omega)$, for all $\alpha < 5/3$, although the right-hand side is arbitrary smooth and vanishes in a neighborhood of the re-entrant corner; see [69] for details. As in §2.7, we decompose Ω into two overlapping rectangles $\Omega_0 = (-1,1)\times(-1,0)$, $\Omega_1 = (-1,0)\times(-1,1)$. Again the wavelet frames used for the discretization of our model problem are constructed by simply lifting wavelet bases on the unit cube to the subdomains and collecting the resulting local bases into a global system of elements. The reference bases are once more chosen to consist of tensor products of the scaling functions and wavelets constructed in [95]. Note that in §2.3.2, §2.3.3, and §2.7 it has been shown that indeed by this procedure a proper wavelet Gelfand frame is obtained. Moreover, in §3.4.3, Corollary 3.5, we have proved that the solution u has a representation $u = \tilde{\mathbf{u}}^\top \Psi$ with $\tilde{\mathbf{u}} \in \ell^w_\tau(\Lambda)$, for any $s < \frac{d-1}{2}$, which is now the benchmark for the optimal convergence rate of our adaptive method; recall also Example 3.1.

The relation between the degrees of freedom/CPU time and the attained accuracy is analyzed in Figure 4.8. Again the best performance is obtained for $d = 3$.

For a comparison we have tested **R_SOLVE** for $d = 3$ with the relaxation parameter $\alpha = 0.25$, being close to the optimal parameter, and with $\alpha = 0.05$ as an example of a too pessimistic guess, as one might be equipped with in practice. Again, the estimate of α has been obtained from the experienced descent parameters in

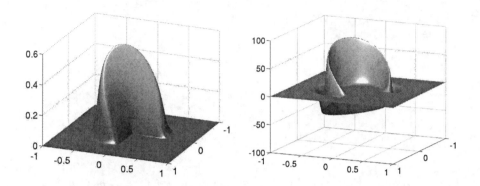

Figure 4.7: Exact solution (left) and right-hand side of the two-dimensional test problem.

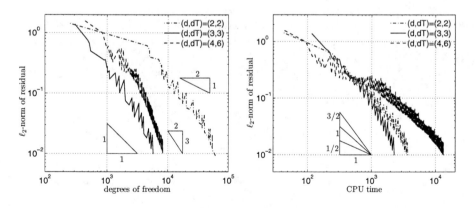

Figure 4.8: ℓ_2-norms of the residuals of $\bar{\mathbf{w}}_i$ versus $\#\operatorname{supp}\bar{\mathbf{w}}_i$ (left) and versus CPU time (right) for the two-dimensional test problem.

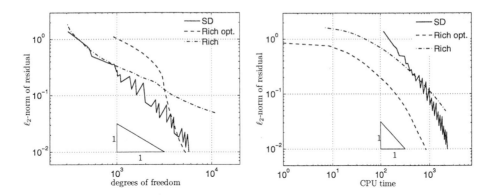

Figure 4.9: Comparison of **SD_SOLVE** (solid line) and **R_SOLVE** for $d = 3$. **R_SOLVE** is tested with $\alpha = 0.25$ (dashed line) and with $\alpha = 0.05$ (dash-dotted line).

SD_SOLVE. Figure 4.9 reveals that **R_SOLVE**, as **SD_SOLVE**, converges with the optimal rate $\frac{d-1}{2} = 1$, it is more efficient when $\alpha = 0.25$ is applied, but it results in a much worse performance for $\alpha = 0.05$. This confirms that indeed the steepest descent method is superior to the Richardson iteration.

In **SD_SOLVE**, for $d = 3$, all in all 38 inner iterations and 146 calls of **APPLY** have been performed. Within **R_SOLVE** with $\alpha = 0.25$ only 61 but with $\alpha = 0.05$ even 137 iterations were spent, although in the latter case the algorithm has been stopped at much larger accuracy.

Figures 4.10–4.12 visualize the active wavelet coefficients at termination of **SD_SOLVE** for the different spline orders. There is one column for each refinement level and one row for each type of tensor product wavelet, i.e., one row for *generator* × *generator*, *generator* × *wavelet* and so on. Each colored patch refers to one wavelet. The darker a patch is, the larger is the corresponding coefficient. In Figure 4.13 the decomposition of the final global continuous iterates into functions from $H_0^t(\Omega_i)$ and the respective pointwise errors are collected. It is remarkable that on the lower levels the degrees of freedom are spread over the hole subdomain Ω_i, $i = 0, 1$, especially for $d = 2$ and $d = 3$. Consequently, the local approximations happen to be non-zero in an area where the true global solution does actually vanish, but they cancel each other out. Thus, we have discovered the redundant parts in the approximation. On the higher levels, especially for $d = 3$, the plotted sparsity pattern nicely reflects the shape of the exact solution with its singularity at the origin

4.4.3 Conclusion and motivation for further improvement

Summarizing, we can say that the optimal convergence rates of the adaptive frame schemes have been practically confirmed. The above tests have shown that

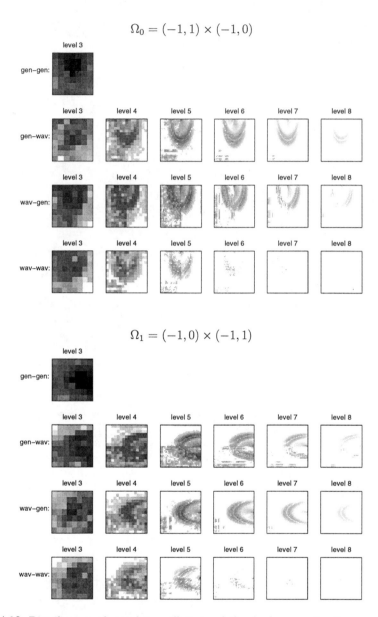

Figure 4.10: Distribution of wavelet coefficients of the final approximation produced by **SD_SOLVE** on Ω_0 (upper part) and Ω_1 (lower part) for $d = 2$.

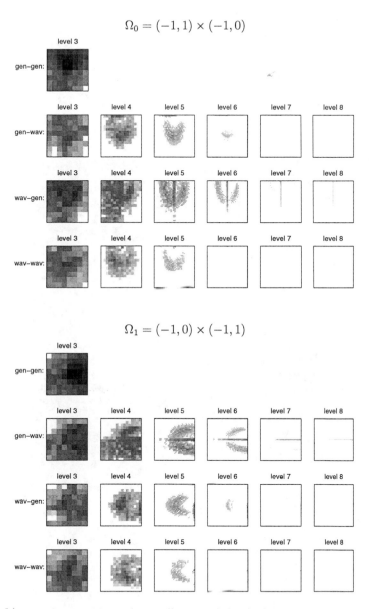

Figure 4.11: Distribution of wavelet coefficients of the final approximation produced by **SD_SOLVE** on Ω_0 (upper part) and Ω_1 (lower part) for $d = 3$.

115

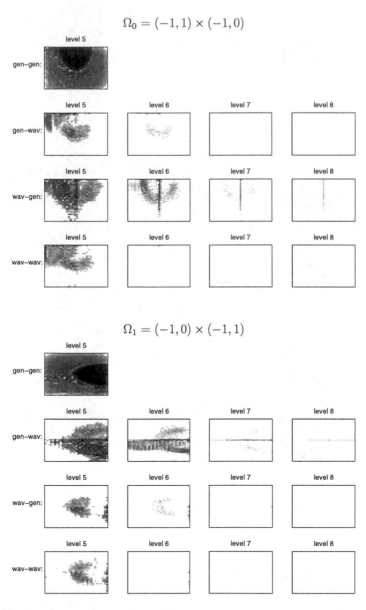

Figure 4.12: Distribution of wavelet coefficients of the final approximation produced by **SD_SOLVE** on Ω_0 (upper part) and Ω_1 (lower part) for $d = 4$.

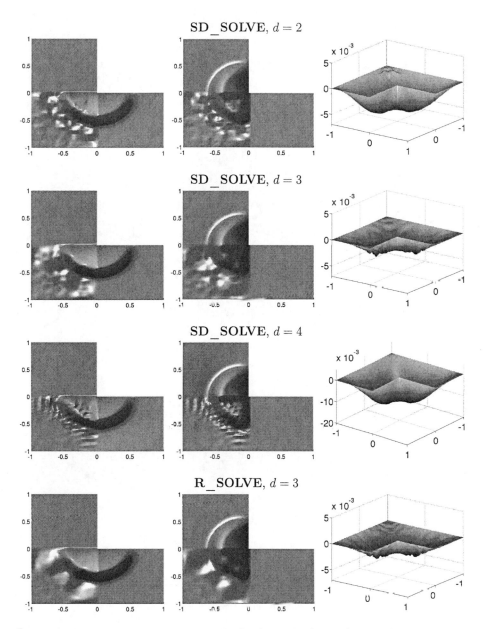

Figure 4.13: Local approximations on Ω_0 (left), on Ω_1 (middle), and the corresponding global pointwise errors (right).

117

SD_SOLVE usually outperforms the adaptive Richardson method in case the relaxation parameter is chosen too small. Clearly, an overestimation of α would even destroy the convergence of the scheme. The above tests with a (near) optimal relaxation parameter are based on a preceding run of the steepest descent method. With regard to the fact that in the context of frame discretizations there does not seem to be another way to get a reliable estimate of α, we may conclude that **SD_SOLVE** really represents an improvement.

Unfortunately, the quantitative performance of the methods so far is not fully satisfactory, which, as we emphasize, is not primarily caused by the properties of the wavelets used for the discretization. On the one hand, the methods suffer from the rather slow convergence of the underlying iterative schemes, which thus represents the first starting point for further advancement.

On the other hand, Figures 4.5, 4.6 for the one-dimensional case, and Figures 4.10–4.12 and 4.13 for the two-dimensional case, have revealed that the amount of redundancy in the overlapping region can be relatively high. Here, the involved local bases more or less equally contribute to the approximation. Thus, it would be desirable to modify the algorithm in such a way that if multiple patches overlap at a certain location, the degrees of freedom are predominantly chosen from one patch only. One is seeking for an algorithm with which one has more control on the special frame representation of the solution that is generated. These ideas and guidelines are indeed going to be realized in Chapter 6.

Chapter 5

Computation of Differential Operators in Frame Coordinates

The intention of this chapter is twofold. The first aim is to show that the stiffness matrix $\mathbf{A}$ from (2.5.5), arising in the discretization of problem (1.3.4) with respect to the aggregated wavelet Gelfand frame Ψ_{L_2} from (2.3.13), is s^*-compressible (recall Definition 4.2) for some $s^* > 0$. To this end, essential use will be made of the specific properties of the scaled wavelet collection $\Psi = \mathbf{D}^{-1}\Psi_{L_2} =: \{\psi_\lambda\}_{\lambda\in\Lambda}$, being a frame for $H_0^t(\Omega)$, $t \in \mathbb{N}_0$, that have been all collected in §2.4. We shall also reuse the notation introduced there. For convenience, we recall that for the order of polynomial reproduction $d > 0$ we assume $d \geq t + 1$ and $\frac{d-t}{n} \geq \frac{1}{2}$.

The second task is to verify the s^*-computability of $\mathbf{A}$ for some $s^* > 0$ in the sense of Definition 4.3. Both properties are mandatory for the availability of a routine **APPLY** with the properties required at the beginning of §4.3, and thus for the optimality of the adaptive frame methods. Furthermore, in view of Remark 4.6, the aim is to prove that $s^* > \frac{d-t}{n}$ can be ensured.

As in the preceding chapter, for any countable index set $\mathcal{I}$, we use $\|\cdot\|$ to denote $\|\cdot\|_{\ell_2(\mathcal{I})}$ or $\|\cdot\|_{\ell_2(\mathcal{I})\to\ell_2(\mathcal{I})}$. Moreover, we always assume the coefficients $a_{\alpha,\beta}(x)$ from (1.3.1) to be sufficiently smooth functions.

5.1 Compressibility

We start with the compressibility of $\mathbf{A}$ and split $\mathbf{A}$ into $\mathbf{A}^{(\mathrm{r})} + \mathbf{A}^{(\mathrm{s})}$, where $\mathbf{A}^{(\mathrm{r})}$ contains those entries $a(\psi_\lambda, \psi_{\lambda'})$ of $\mathbf{A}$ with

$$\begin{cases} \operatorname{supp} \psi_\lambda \subset \overline{\Xi_{\lambda',i'}}, & \text{for some } 1 \leq i' \leq K, \quad \text{when } |\lambda| > |\lambda'|, \\ \operatorname{supp} \psi_{\lambda'} \subset \overline{\Xi_{\lambda,i}}, & \text{for some } 1 \leq i \leq K, \quad \text{when } |\lambda| < |\lambda'|, \end{cases} \qquad (5.1.1)$$

and zeros at the remaining locations in $\Lambda \times \Lambda$. Thus $\mathbf{A}^{(\mathrm{s})}$ is the matrix containing the remaining entries of $\mathbf{A}$, and zeros otherwise; see Figure 5.1. The indices "r" and "s" refer to *regular* and *singular*, respectively.

In order not to be forced to handle $n = 1$ as an exceptional, although easy case, unless explicitly stated otherwise, in the following we will always assume that

$$n > 1. \qquad (5.1.2)$$

Figure 5.1: Left: $\operatorname{supp}\psi_\lambda \subset \overline{\Xi_{\lambda',i'}}$, i.e., an entry of $\mathbf{A}^{(\mathrm{r})}$. Right: $\operatorname{supp}\psi_\lambda \cap \operatorname{sing\,supp}\psi_{\lambda'} \neq \emptyset$, i.e., an entry of $\mathbf{A}^{(\mathrm{s})}$.

Theorem 5.1. *For $j \in \mathbb{N}_0$, we define the infinite matrices $\mathbf{A}_j^{(\mathrm{r})}$ and $\mathbf{A}_j^{(\mathrm{s})}$ by dropping the entries $\mathbf{A}_{\lambda,\lambda'} = a(\psi_{\lambda'}, \psi_\lambda)$ from $\mathbf{A}^{(\mathrm{r})}$ or $\mathbf{A}^{(\mathrm{s})}$ when*

$$\big||\lambda| - |\lambda'|\big| > \tfrac{j}{n} \quad or \quad \big||\lambda| - |\lambda'|\big| > \tfrac{j}{n-1}, \quad respectively. \tag{5.1.3}$$

Then, the number of non-zero entries in each row and column of $\mathbf{A}_j^{(\mathrm{r})}$ and $\mathbf{A}_j^{(\mathrm{s})}$ is of order 2^j, and

$$\|\mathbf{A}^{(\mathrm{r})} - \mathbf{A}_j^{(\mathrm{r})}\| \lesssim 2^{-j(\frac{t+\tilde{d}}{n})}, \quad \|\mathbf{A}^{(\mathrm{s})} - \mathbf{A}_j^{(\mathrm{s})}\| \lesssim 2^{-j(\frac{d-1/2-t}{n-1})}, \tag{5.1.4}$$

for the latter estimate assuming that $\tilde{d} \geq d - 2t - 1$.

Remark 5.1. Theorem 5.1 shows that $\mathbf{A}^{(\mathrm{r})}$ is s^*-compressible with $s^* \geq \frac{d-t}{n}$ or $s^* > \frac{d-t}{n}$ when $\tilde{d} \geq d-2t$ or $\tilde{d} > d-2t$, and that $\mathbf{A}^{(\mathrm{s})}$ is s^*-compressible with $s^* \geq \frac{d-t}{n}$ or $s^* > \frac{d-t}{n}$ when $\frac{d-t}{n} \geq \frac{1}{2}$ or $\frac{d-t}{n} > \frac{1}{2}$, and $\tilde{d} \geq d - 2t - 1$.

In order to prove the theorem, we start with bounding the individual entries of $\mathbf{A}$.

Lemma 5.1. *We have*

$$|\mathbf{A}_{\lambda,\lambda'}^{(\mathrm{r})}| \lesssim 2^{-\big||\lambda|-|\lambda'|\big|(\frac{n}{2}+t+\tilde{d})}, \quad |\mathbf{A}_{\lambda,\lambda'}^{(\mathrm{s})}| \lesssim 2^{-\big||\lambda|-|\lambda'|\big|(\frac{n}{2}+d-1-t)}, \tag{5.1.5}$$

for the latter estimate assuming that $\tilde{d} \geq d - 2t - 1$.

Proof. Let $|\lambda| \geq |\lambda'|$, $e(\lambda) \neq 0$. By a transformation of coordinates, we can write

$$a(\psi_{\lambda'}, \psi_\lambda) = \sum_{|\alpha|,|\beta|\leq t} \int_{\kappa_{p(\lambda)}^{-1}(\operatorname{supp}\psi_\lambda)} \tilde{a}_{\alpha,\beta} D^\alpha(\psi_{\lambda'} \circ \kappa_{p(\lambda)}) D^\beta(\psi_\lambda \circ \kappa_{p(\lambda)}), \tag{5.1.6}$$

for some smooth $\tilde{a}_{\alpha,\beta}$ depending on the coefficients $a_{\alpha,\beta}$ and $\kappa_{p(\lambda)}$. Since bounding the lower order terms is easier, we consider a term of the right-hand side of (5.1.6) for arbitrary $|\alpha| = |\beta| = t$.

When $d - 1 \leq 2t$, select a $\gamma \leq \beta$ with $|\alpha + \gamma| = d - 1$ and so $|\beta - \gamma| = 2t - (d - 1)$. Using the homogeneous Dirichlet boundary conditions for the case that $\operatorname{supp} \psi_\lambda \cap \operatorname{supp} \psi_{\lambda'} \cap \partial\Omega \neq \emptyset$, integration by parts, $\operatorname{vol}(\operatorname{supp} \psi_\lambda) \lesssim 2^{-|\lambda|n}$, and (2.4.6) show that

$$
\left| \int\limits_{\kappa_{p(\lambda)}^{-1}(\operatorname{supp} \psi_\lambda)} \tilde{a}_{\alpha,\beta} D^\alpha (\psi_{\lambda'} \circ \kappa_{p(\lambda)}) D^\beta (\psi_\lambda \circ \kappa_{p(\lambda)}) \right|
$$

$$
= \left| \int\limits_{\kappa_{p(\lambda)}^{-1}(\operatorname{supp} \psi_\lambda)} (-1)^{|\gamma|} D^\gamma (\tilde{a}_{\alpha,\beta} D^\alpha (\psi_{\lambda'} \circ \kappa_{p(\lambda)})) D^{\beta-\gamma} (\psi_\lambda \circ \kappa_{p(\lambda)}) \right|
$$

$$
\lesssim 2^{-|\lambda|n} 2^{|\lambda'|(\frac{n}{2}+d-1-t)} 2^{|\lambda|(\frac{n}{2}+2t-(d-1)-t)} = 2^{-(|\lambda|-|\lambda'|)(\frac{n}{2}+d-1-t)}.
$$

When $d - 1 > 2t$, by additionally using (2.4.7) with $\tilde{d} \geq d - 2t - 1$, and the abbreviation $b_\lambda := |\det D\kappa_{p(\lambda)}|^{1/2}$, we obtain

$$
\left| \int\limits_{\kappa_{p(\lambda)}^{-1}(\operatorname{supp} \psi_\lambda)} \tilde{a}_{\alpha,\beta} D^\alpha (\psi_{\lambda'} \circ \kappa_{p(\lambda)}) D^\beta (\psi_\lambda \circ \kappa_{p(\lambda)}) \right|
$$

$$
= \left| \int\limits_{\kappa_{p(\lambda)}^{-1}(\operatorname{supp} \psi_\lambda)} (-1)^{|\beta|} D^\beta (\tilde{a}_{\alpha,\beta} D^\alpha (\psi_{\lambda'} \circ \kappa_{p(\lambda)})) (\psi_\lambda \circ \kappa_{p(\lambda)}) \right|
$$

$$
= \left| \int\limits_{\kappa_{p(\lambda)}^{-1}(\operatorname{supp} \psi_\lambda)} (-1)^{|\beta|} D^\beta (\tilde{a}_{\alpha,\beta} D^\alpha (\psi_{\lambda'} \circ \kappa_{p(\lambda)})) b_\lambda^{-1} (\psi_\lambda \circ \kappa_{p(\lambda)}) b_\lambda \right|
$$

$$
\lesssim 2^{-|\lambda|n} \inf_{P \in \Pi_{d-2t-2}} \| b_\lambda^{-1} (-1)^{|\beta|} D^\beta (\tilde{a}_{\alpha,\beta} D^\alpha (\psi_{\lambda'} \circ \kappa_{p(\lambda)})) - P \|_{L_\infty(\kappa_{p(\lambda)}^{-1}(\operatorname{supp} \psi_\lambda))} 2^{|\lambda|(\frac{n}{2}-t)}
$$

$$
\lesssim 2^{-|\lambda|n} \operatorname{diam}(\kappa_{p(\lambda)}^{-1}(\operatorname{supp} \psi_\lambda))^{d-2t-1}
$$

$$
\cdot \| b_\lambda^{-1} (-1)^{|\beta|} D^\beta (\tilde{a}_{\alpha,\beta} D^\alpha (\psi_{\lambda'} \circ \kappa_{p(\lambda)})) \|_{W_\infty^{d-2t-1}} 2^{|\lambda|(\frac{n}{2}-t)}
$$

$$
\lesssim 2^{-|\lambda|n} 2^{-|\lambda|(d-1-2t)} 2^{|\lambda'|(\frac{n}{2}+d-1-t)} 2^{|\lambda|(\frac{n}{2}-t)} = 2^{-(|\lambda|-|\lambda'|)(\frac{n}{2}+d-1-t)},
$$

where the Whitney estimate from (2.2.30) has been utilized in the fourth step. This completes the proof of the second estimate.

Finally, when $\operatorname{supp} \psi_\lambda \subset \overline{\Xi_{\lambda',i'}}$ for some $1 \leq i' \leq K$, i.e., when $a(\psi_{\lambda'}, \psi_\lambda) = \mathbf{A}_{\lambda,\lambda'}^{(\mathrm{r})}$, using (2.4.3) and again the mentioned Whitney estimate, we find

$$
\left| \int\limits_{\kappa_{p(\lambda)}^{-1}(\operatorname{supp} \psi_\lambda)} \tilde{a}_{\alpha,\beta} D^\alpha (\psi_{\lambda'} \circ \kappa_{p(\lambda)}) D^\beta (\psi_\lambda \circ \kappa_{p(\lambda)}) \right|
$$

$$
\lesssim 2^{-|\lambda|n} \inf_{P \in \Pi_{\tilde{d}-1}} \| b_\lambda^{-1} (-1)^{|\beta|} D^\beta (\tilde{a}_{\alpha,\beta} D^\alpha (\psi_{\lambda'} \circ \kappa_{p(\lambda)})) - P \|_{L_\infty(\kappa_{p(\lambda)}^{-1}(\operatorname{supp} \psi_\lambda))} 2^{|\lambda|(\frac{n}{2}-t)}
$$

$$
\lesssim 2^{-|\lambda|n} 2^{-|\lambda|\tilde{d}} 2^{|\lambda'|(\frac{n}{2}+\tilde{d}+2t-t)} 2^{|\lambda|(\frac{n}{2}-t)} = 2^{-(|\lambda|-|\lambda'|)(\frac{n}{2}+t+\tilde{d})}.
$$

$\square$

121

Proof of Theorem 5.1. The locality (2.4.2) of the wavelets shows that for a fixed index λ, the number of indices λ' for which $\mathrm{vol}(\mathrm{supp}\,\psi_\lambda \cap \mathrm{supp}\,\psi_{\lambda'}) > 0$ is of the order $\max\{1, 2^{(|\lambda'|-|\lambda|)n}\}$. Moreover, from the piecewise smoothness of the wavelets, we may infer that the number of indices λ' with fixed $|\lambda'| > |\lambda|$ such that $\mathrm{vol}(\mathrm{supp}\,\psi_\lambda \cap \mathrm{supp}\,\psi_{\lambda'}) > 0$ but $\mathrm{supp}\,\psi_{\lambda'}$ is not contained in some $\overline{\Xi}_{\lambda,i}$ is of order $2^{(|\lambda'|-|\lambda|)(n-1)}$. Therefore, the number of non-zero entries in each row of $\mathbf{A}_{l,l'}^{(\mathrm{r})} := [\mathbf{A}_{\lambda,\lambda'}^{(\mathrm{r})}]_{|\lambda|=l,\,|\lambda'|=l'}$ and column of $\mathbf{A}_{l',l}^{(\mathrm{r})}$ is $\mathcal{O}(\max\{1, 2^{(l'-l)n}\})$, whereas the number of non-zero entries in each row of $\mathbf{A}_{l,l'}^{(\mathrm{s})} := [\mathbf{A}_{\lambda,\lambda'}^{(\mathrm{s})}]_{|\lambda|=l,\,|\lambda'|=l'}$ and column of $\mathbf{A}_{l',l}^{(\mathrm{s})}$ is $\mathcal{O}(\max\{1, 2^{(l'-l)(n-1)}\})$. The definition of $\mathbf{A}_j^{(\mathrm{r})}$ and $\mathbf{A}_j^{(\mathrm{s})}$ shows that in each row and column the number of non-zero entries is of the order

$$\sum_{\||\lambda'|-|\lambda|\| \le \frac{j}{n}} \max\{1, 2^{(|\lambda'|-|\lambda|)n}\} \approx 2^j \tag{5.1.7}$$

or

$$\sum_{\||\lambda'|-|\lambda|\| \le \frac{j}{n-1}} \max\{1, 2^{(|\lambda'|-|\lambda|)(n-1)}\} \approx 2^j, \tag{5.1.8}$$

respectively, so that the proof of the first statement is completed.

Estimating $\|\mathbf{A}_{l,l'}^{(\mathrm{r})}\|^2$ and $\|\mathbf{A}_{l,l'}^{(\mathrm{s})}\|^2$ on the products of their maximal absolute row- and column sums, taking into account Lemma 5.1, we find

$$\|\mathbf{A}_{l,l'}^{(\mathrm{r})}\|^2 \lesssim 2^{|l'-l|n} 2^{-|l'-l|(\frac{n}{2}+t+\bar{d})2}, \quad \|\mathbf{A}_{l,l'}^{(\mathrm{s})}\|^2 \lesssim 2^{|l'-l|(n-1)} 2^{-|l'-l|(\frac{n}{2}+d-1-t)2}. \tag{5.1.9}$$

Now, we may apply the estimate

$$\|\mathbf{A}^{(\mathrm{r})} - \mathbf{A}_j^{(\mathrm{r})}\|^2 \le \max_{l'} \sum_{\{l:|l-l'|>\frac{j}{n}\}} \|\mathbf{A}_{l,l'}^{(\mathrm{r})}\| \times \max_l \sum_{\{l':|l-l'|>\frac{j}{n}\}} \|\mathbf{A}_{l,l'}^{(\mathrm{r})}\|.$$

For the first factor we obtain

$$\max_{l'} \sum_{\{l:|l-l'|>\frac{j}{n}\}} \|\mathbf{A}_{l,l'}^{(\mathrm{r})}\| \lesssim \max_{l'} \sum_{\{l:|l-l'|>\frac{j}{n}\}} 2^{|l'-l|\frac{n}{2}} 2^{-|l'-l|(\frac{n}{2}+t+\bar{d})} \tag{5.1.10}$$

$$= \max_{l'} \sum_{\{l:|l-l'|>\frac{j}{n}\}} 2^{-|l'-l|(t+\bar{d})} \lesssim 2^{-j(\frac{t+\bar{d}}{n})}. \tag{5.1.11}$$

The second factor can be estimated analogously, completing the first part of (5.1.4). Finally, in the same way, using

$$\|\mathbf{A}^{(\mathrm{s})} - \mathbf{A}_j^{(\mathrm{s})}\|^2 \le \max_{l'} \sum_{\{l:|l-l'|>\frac{j}{n-1}\}} \|\mathbf{A}_{l,l'}^{(\mathrm{s})}\| \times \max_l \sum_{\{l':|l-l'|>\frac{j}{n-1}\}} \|\mathbf{A}_{l,l'}^{(\mathrm{s})}\|$$

and (5.1.9), completes the proof of the second estimate in (5.1.4). $\qquad\square$

Remark 5.2. For $n = 1$, the estimate $\|\mathbf{A}^{(\mathrm{r})} - \mathbf{A}_j^{(\mathrm{r})}\| \lesssim 2^{-j(t+\tilde{d})}$ holds with the same definition of $\mathbf{A}^{(\mathrm{r})}$ as in (5.1.3), because the proof of Theorem 5.1 is equally valid for this case. Defining $\mathbf{A}_j^{(s)}$ by dropping the entries $\mathbf{A}_{\lambda,\lambda'}$ if $||\lambda| - |\lambda'|| > k(j)$, where $j \leq k(j) \leq 2^j$ and $k(j) > j\frac{t+\tilde{d}}{d-\frac{1}{2}-t}$, one also gets $\|\mathbf{A}^{(s)} - \mathbf{A}_j^{(s)}\| \lesssim 2^{-j(t+\tilde{d})}$. This can be shown by reconsidering the proof of Theorem 5.1. Doing so, one realizes that in (5.1.8) we get $\sum_{||\lambda'|-|\lambda||\leq k(j)\leq 2^j} \max\{1, 2^{(|\lambda'|-|\lambda|)(n-1)}\} \approx 2^j$ instead, and the right estimate in (5.1.9) becomes $\|\mathbf{A}_{l,l'}^{(s)}\|^2 \lesssim 2^{-|l'-l|(d-\frac{1}{2}-t)2}$. For the latter note that Lemma 5.1 is also valid for $n = 1$. Proceeding as in (5.1.10) then gives

$$
\max_{l'} \sum_{\{l:|l-l'|>k(j)\}} \|\mathbf{A}_{l,l'}^{(s)}\| \lesssim \max_{l'} \sum_{\{l:|l-l'|>k(j)\}} 2^{-|l'-l|(d-\frac{1}{2}-t)}
$$
$$
\lesssim 2^{-k(j)(d-\frac{1}{2}-t)} \lesssim 2^{-j(t+\tilde{d})},
$$

because $k(j) > j\frac{t+\tilde{d}}{d-\frac{1}{2}-t}$.

5.2 Computability

The next step is to verify the s^*-computability of $\mathbf{A}$ for an $s^* > \frac{d-t}{n}$. To this end, in the following, the results from [108] are summarized, where, for this goal, suitable quadrature rules have been developed. For some of the rather technical proofs, the reader will be referred to [108].

5.2.1 The quadrature paradigm

We use the splitting of $\mathbf{A}$ into $\mathbf{A}^{(\mathrm{r})} + \mathbf{A}^{(s)}$ as above. Is is sufficient to prove s^*-computability of these matrices. Thus, one encounters the task of approximating integrals of the form (5.1.6), where in this section, for notational convenience and without loss of generality, we may assume that $\kappa_{p(\lambda)} = \mathrm{id}$, so that

$$
\mathbf{A}_{\lambda,\lambda'} = \sum_{i=1}^{K} \sum_{|\alpha|,|\beta|\leq t} \int_{\Xi_{\lambda,i}} a_{\alpha,\beta} D^\alpha \psi_{\lambda'} D^\beta \psi_\lambda, \tag{5.2.1}
$$

each $\Xi_{\lambda,i}$ being an n-cube aligned with the Cartesian coordinates, and $\psi_\lambda|_{\Xi_{\lambda,i}} \in Q_{d-1}$.

Depending on whether $\mathbf{A}_{\lambda,\lambda'}$ represents an entry of $\mathbf{A}_j^{(\mathrm{r})}$ or $\mathbf{A}_j^{(s)}$, it is $\operatorname{supp}\psi_\lambda \cap \operatorname{sing\,supp}\psi_{\lambda'} = \emptyset$ or not. Obviously, in general, the second case is the more complicated one. However, our strategy is to to exclusively apply composite quadrature rules of *variable rank* $N \in \mathbb{N}$, depending on j and $||\lambda| - |\lambda'||$, but *fixed order* $p \in \mathbb{N}$, on $\Xi_{\lambda,i}$. That means, we subdivide the n-cube $\Xi_{\lambda,i}$ under consideration into N equal subcubes, and, on each of these subcubes $\square$, we apply a quadrature rule that is *exact on* $Q_{p-1}(\square)$. Furthermore, the results presented below are based on the assumption that this rule is internal, i.e., that all abscissae are in the closure of the subcube, and that it is uniformly stable, in the sense that the sum of the absolute values of the weights can be bounded by an absolute multiple of the volume of the subcube. Finally, having a fixed p, we can assume that the total number of abscissae is $\mathcal{O}(N)$.

The key tool for the verification of the s^*-computability is given by the following technical lemma from [108] which is going to be applied for $\mathbf{A}^{(\mathrm{r})}$ with $k = n$, and for $\mathbf{A}^{(\mathrm{s})}$ with $k = n - 1$ below.

Lemma 5.2 (see [108]). *For some fixed $k \in \mathbb{N}$, and all $j \in \mathbb{N}_0$, let $\mathbf{B}_j = ((\mathbf{B}_j)_{\lambda,\lambda'})_{\lambda,\lambda' \in \Lambda}$ be a matrix such that the number of possible non-zero entries in each row of $(\mathbf{B}_j)_{l,l'} :=$ $[(\mathbf{B}_j)_{\lambda,\lambda'}]_{|\lambda|=l,\,|\lambda'|=l'}$ and column of $(\mathbf{B}_j)_{l',l}$ is $\mathcal{O}(\max\{1, 2^{(l'-l)k}\})$, and*

$$(\mathbf{B}_j)_{\lambda,\lambda'} = 0 \text{ when } \big|\,|\lambda| - |\lambda'|\,\big| > \tfrac{j}{k}.$$

Let $\mathbf{B}_j^*$ be an approximation for $\mathbf{B}_j$, zero on positions where $\mathbf{B}_j$ is known to be zero, and for which the computation of $(\mathbf{B}_j^*)_{\lambda,\lambda'}$ takes $\mathcal{O}(N_{j,\lambda,\lambda'})$ operations otherwise, where, for some absolute constants $r, q \geq 0$, $r \neq q$,

$$|(\mathbf{B}_j)_{\lambda,\lambda'} - (\mathbf{B}_j^*)_{\lambda,\lambda'}| \lesssim N_{j,\lambda,\lambda'}^{-q} 2^{-\big|\,|\lambda|-|\lambda'|\,\big|(k/2+rk)}. \tag{5.2.2}$$

For some $\rho \in (1, r/q)$ when $r > q$, and $\rho \in (r/q, 1)$ when $r < q$, and $\theta \leq \min\{1, \rho\}$, select

$$N_{j,\lambda,\lambda'} \approx \max\left\{1, 2^{j\theta - \big|\,|\lambda|-|\lambda'|\,\big|\rho k}\right\}. \tag{5.2.3}$$

Then, the work for computing each column of $\mathbf{B}_j^*$ is $\mathcal{O}(2^j)$, and

$$\|\mathbf{B}_j - \mathbf{B}_j^*\| \lesssim \begin{cases} 2^{-jq\theta} & \text{when } r > q, \\ 2^{-j(r+(\theta-\rho)q)} & \text{when } r < q. \end{cases} \tag{5.2.4}$$

In particular, taking $\theta = \min\{1, \rho\}$, we have $\|\mathbf{B}_j - \mathbf{B}_j^*\| \lesssim 2^{-j\min\{q,r\}}$.

Remark 5.3. For $r = q$, choosing $\theta = \rho = 1$, one can compute a $\mathbf{B}_j^*$ taking $\mathcal{O}(2^j)$ operations per column, with for any $\varepsilon > 0$, $\|\mathbf{B}_j - \mathbf{B}_j^*\| \lesssim 2^{-j(q-\varepsilon)}$; see [108] for details.

By (5.2.3), the granularity $N_{j,\lambda,\lambda'}$ of the composite rule for the approximation of an entry $(\mathbf{B}_j)_{\lambda,\lambda'}$ is specified. Concerning its dependence on the level difference $\big|\,|\lambda|-|\lambda'|\,\big|$, note that in the extremal case $\big|\,|\lambda|-|\lambda'|\,\big| = \tfrac{j}{k}$, $N_{j,\lambda,\lambda'} \approx \max\{1, 2^{j(\theta-\rho)}\} = 1$, while for $\big|\,|\lambda|-|\lambda'|\,\big| = 0$, one gets $N_{j,\lambda,\lambda'} \approx \max\{1, 2^{j\theta}\} = 2^{j\theta}$. That means that the decisive amount of work is spend on the computation of the entries close to the diagonal. In addition, $N_{j,\lambda,\lambda'}$ also depends on j, thus on the total amount of work one is going to afford in a single column. Thinking of Lemma 5.2 being applied for $\mathbf{B}_j = \mathbf{A}_j^{(\mathrm{r})}$ and $k = n$, or $\mathbf{B}_j = \mathbf{A}_j^{(\mathrm{s})}$ and $k = n - 1$, in the singular case a larger $N_{j,\lambda,\lambda'}$ is chosen, and thus some more work invested. In principle, the amount of work to compute the single entries of a column in $\mathbf{B}_j$ is distributed in such a way that *on average* the approximation of one entry takes $\mathcal{O}(1)$ operations.

Clearly, with regard to (5.2.2), at first quadrature error estimates for $\mathbf{A}^{(r)}$ and $\mathbf{A}^{(s)}$ have to be derived.

5.2.2 s^*-computability of $\mathbf{A}^{(\mathrm{r})}$

We first consider the task of approximating the entries of $\mathbf{A}_j^{(\mathrm{r})}$. Evidently, it is expedient to apply the composite quadrature rule on the parameter space of the wavelet that has the highest level of the two involved in an entry. Therefore, without loss of generality, throughout this subsection and in §5.2.3 we assume that

$$|\lambda| \geq |\lambda'|.$$

Proposition 5.1. *Let* $\operatorname{supp} \psi_\lambda \subset \overline{\Xi_{\lambda',i'}}$ *for some* $1 \leq i' \leq K$; *see Figure 5.1 (left) for an illustration. Let* $\mathbf{A}_{\lambda,\lambda'}^{(\mathrm{r}),*}$ *be the result of the application of a composite rule of rank* N *and order* p *applied to each of the integrals from (5.2.1). Then,*

$$|\mathbf{A}_{\lambda,\lambda'}^{(\mathrm{r})} - \mathbf{A}_{\lambda,\lambda'}^{(\mathrm{r}),*}| \lesssim N^{-p/n} 2^{-(|\lambda|-|\lambda'|)(n/2+p-d+1)}.$$

A proof of this statement is given in [108].

Corollary 5.1. *Let* $\tilde{d} > d - 2t$ *and* $p > 2d - t - 1$, *then* $\mathbf{A}^{(\mathrm{r})}$ *is* s^*-*computable with* $s^* > \frac{d-t}{n}$.

Proof. Recall that the number of non-zero entries in each row of $\mathbf{A}_{l,l'}^{(\mathrm{r})}$ and column of $\mathbf{A}_{l',l}^{(\mathrm{r})}$ is $\mathcal{O}(\max\{1, 2^{(l'-l)n}\})$. Using Proposition 5.1, an application of Lemma 5.2 for $k = n$ yields a matrix $\mathbf{A}_j^{(\mathrm{r}),*}$, for which the computation of each column takes $\mathcal{O}(2^j)$ operations, and for which $\|\mathbf{A}_j^{(\mathrm{r})} - \mathbf{A}_j^{(\mathrm{r}),*}\| \lesssim 2^{-j\left(\frac{p-d+1}{n}\right)}$ when $d > 1$, and $\|\mathbf{A}_j^{(\mathrm{r})} - \mathbf{A}_j^{(\mathrm{r}),*}\| \lesssim 2^{-j\left(\frac{p-d+1}{n}-\varepsilon\right)}$ for any $\varepsilon > 0$ otherwise (cf. Remark 5.3). Using that $\|\mathbf{A}^{(\mathrm{r})} - \mathbf{A}_j^{(\mathrm{r})}\| \lesssim 2^{-j\left(\frac{t+\tilde{d}}{n}\right)}$ by Theorem 5.1, the proof is completed. $\qquad\square$

5.2.3 s^*-computability of $\mathbf{A}^{(\mathrm{s})}$

The next step is to consider the approximation of the non-zero entries $\mathbf{A}_{\lambda,\lambda'}$ of $\mathbf{A}_j^{(\mathrm{s})}$. Note that for these entries $\operatorname{supp} \psi_\lambda$ will have a non-empty intersection with the singular support of $\psi_{\lambda'}$. As a consequence, for p not too small, generally the decay of the quadrature error will not be as fast as function of the rank $N \to \infty$ or $||\lambda|-|\lambda'|| \to \infty$ as with the entries of $\mathbf{A}_j^{(\mathrm{r})}$. However, since the number of non-zero entries in $\mathbf{A}_{l,l'}^{(\mathrm{s})}$ increases less fast as function of $|l - l'| \to \infty$ compared to that in $\mathbf{A}_{l,l'}^{(\mathrm{r})}$, as shown in Lemma 5.2, this effect can be compensated by investing some more work in their computation without increasing the overall complexity.

Remark 5.4. Let (λ, λ') correspond to a non-zero entry of $\mathbf{A}^{(\mathrm{s})}$, i.e., $\operatorname{supp} \psi_\lambda \not\subset \overline{\Xi_{\lambda',i'}}$ for any $1 \leq i' \leq K$, but such that for all $1 \leq i \leq K$ there exists an $1 \leq i'(i) \leq K$ with $\Xi_{\lambda,i} \subset \Xi_{\lambda',i'(i)}$, meaning that $\operatorname{sing\,supp} \psi_{\lambda'} \cap \operatorname{supp} \psi_\lambda \subset \operatorname{sing\,supp} \psi_\lambda$; see Figure 5.2. Then, it is obvious that the bound of the quadrature error from Proposition 5.1 is still valid. This situation occurs when $\psi_{\lambda'}$ and ψ_λ are piecewise smooth with respect to partitions that are *nested* as function of the level, e.g., in the case of an aggregated wavelet frame, when $\psi_{\lambda'}$ and ψ_λ are wavelets lifted by the same parametric mapping. In order not to complicate the exposition, in the following we will ignore this fact, and use also for such entries the less favorable bound on the quadrature error from the following proposition.

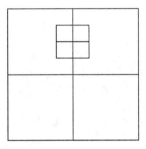

Figure 5.2: $\operatorname{supp}\psi_\lambda \not\subset \overline{\Xi_{\lambda',i'}}$ for any $1 \leq i' \leq K$, i.e., an entry of $\mathbf{A}^{(s)}$, but $\Xi_{\lambda,i} \subset \Xi_{\lambda',i'(i)}$ for any i.

Proposition 5.2. *Let $\mathbf{A}^*_{\lambda,\lambda'}$ be the result of the application of a composite rule of rank N and order p applied to each of the integrals from (5.2.1). Then,*

$$|\mathbf{A}_{\lambda,\lambda'} - \mathbf{A}^*_{\lambda,\lambda'}| \lesssim N^{-p/n} 2^{-(|\lambda|-|\lambda'|)(n/2+p-d+1)} + N^{-(d-t)/n} 2^{-(|\lambda|-|\lambda'|)(n/2+d-1-t)},$$

which is valid without any assumption on the location of $\operatorname{sing\,supp}\psi_\lambda$.

We again refer to [108] for a proof. Note that the first term in the upper bound is equal to the bound given in Proposition 5.1.

Corollary 5.2. *Let $p \geq \max\{d-t, 2d-2-t\}$, $\frac{d-t}{n} > \frac{1}{2}$, and $\tilde{d} \geq d-2t-1$. Then $\mathbf{A}^{(s)}$ is s^*-computable with $s^* = \frac{d-t}{n}$.*

Proof. Recall that the number of non-zero entries in each row of $\mathbf{A}^{(s)}_{l,l'}$ and column of $\mathbf{A}^{(s)}_{l',l}$ is $\mathcal{O}(\max\{1, 2^{(l'-l)(n-1)}\})$. The condition $p \geq \max\{d-t, 2d-2-t\}$ shows that in the bound from Proposition 5.2 the first term is never larger than the second one which can be written as $N^{-(d-t)/n} 2^{-(|\lambda|-|\lambda'|)((n-1)/2+\frac{d-1/2-t}{n-1}(n-1))}$. The condition $\frac{d-t}{n} > \frac{1}{2}$ shows that $r := \frac{d-1/2-t}{n-1} > \frac{d-t}{n} =: q$, and so an application of Lemma 5.2 for $k = n-1$ yields a matrix $\mathbf{A}^{(s),*}_j$, for which the computation of each column takes $\mathcal{O}(2^j)$ operations, and for which $\|\mathbf{A}^{(s)}_j - \mathbf{A}^{(s),*}_j\| \lesssim 2^{-j(\frac{d-t}{n})}$. Using that by $\frac{d-t}{n} \geq \frac{1}{2}$ and $\tilde{d} \geq d-2t-1$, $\|\mathbf{A}^{(s)} - \mathbf{A}^{(s)}_j\| \lesssim 2^{-j(\frac{d-t}{n})}$ by Theorem 5.1, the proof is completed. $\square$

Above result is not fully satisfactory, since actually we need $s^* > \frac{d-t}{n}$ (cf. Remark 4.6, [104]). Therefore, we reconsider the task of approximately computing the entries of $\mathbf{A}_j^{(s)}$. For $\eta \leq \alpha$, $|\beta| + |\eta| \leq d - 1$, *integration by parts* shows that

$$\int_\Omega a_{\alpha\beta} D^\alpha \psi_{\lambda'} D^\beta \psi_\lambda = (-1)^{|\eta|} \sum_{i=1}^K \int_{\Xi_{\lambda,i}} D^{\alpha-\eta} \psi_{\lambda'} D^\eta (a_{\alpha,\beta} D^\beta \psi_\lambda), \qquad (5.2.5)$$

so that alternatively one can apply the composite quadrature rule to each term on the right-hand side, with the advantage that $D^{\alpha-\eta} \psi_{\lambda'}$ is smoother than $D^\alpha \psi_{\lambda'}$. Considering this approach for the *largest possible η*, i.e., $|\eta| = \min\{|\alpha|, d - 1 - |\beta|\}$, one can prove the following result.

Proposition 5.3 (see [108]). *Let $\mathbf{A}_{\lambda,\lambda'}^*$ be the result of the application of a composite rule of rank N and order p applied to each of the integrals from (5.2.5) for each $|\alpha|, |\beta| \leq t$, where for each of these integrals the largest possible η is taken. Then,*

$$|\mathbf{A}_{\lambda,\lambda'} - \mathbf{A}_{\lambda,\lambda'}^*| \lesssim N^{-p/n} 2^{-(|\lambda|-|\lambda'|)(n/2+p-d+1-\boxed{\min\{t,d-1-t\}})}$$
$$+ N^{-(d-t+\boxed{\min\{t,d-1-t\}})/n} 2^{-(|\lambda|-|\lambda'|)(n/2+d-1-t)},$$

which is valid without any assumption on the location of sing supp ψ_λ. *(The terms in the exponents within frames indicate the differences to the bound from Proposition 5.2.)*

Corollary 5.3. *Let $p \geq 2d - 2 - t + \min\{t, d - 1 - t\}$, $\frac{d-t}{n} > \frac{1}{2}$, $d - 1 > t > 0$ and $\tilde{d} \geq d - 2t - 1$. Then $\mathbf{A}^{(s)}$ is s^*-computable for some $s^* > \frac{d-t}{n}$.*

Proof. The number of non-zero entries in each row of $\mathbf{A}_{l,l'}^{(s)}$ and column of $\mathbf{A}_{l',l}^{(s)}$ is $\mathcal{O}(\max\{1, 2^{(l'-l)(n-1)}\})$. The conditions $p \geq 2d - 2 - t + \min\{t, d - 1 - t\}$, and $d - 1 > t > 0$, the latter implying $d \geq 2$, show that in the bound from Proposition 5.3 the first term is never larger than the second one which can be written as $N^{-(d-t+\min\{t,d-1-t\})/n} 2^{-(|\lambda|-|\lambda'|)((n-1)/2+\frac{d-1/2-t}{n-1}(n-1))}$. The condition $\frac{d-t}{n} > \frac{1}{2}$ shows that $r := \frac{d-1/2-t}{n-1} > \frac{d-t}{n}$, and the conditions $t > 0$ and $d - 1 > t$ show that $\min\{t, d - 1 - t\} > 0$ and thus that $q := \frac{d-t+\min\{t,d-1-t\}}{n} > \frac{d-t}{n}$. So an application of Lemma 5.2 for $k = n - 1$ yields a matrix $\mathbf{A}_j^{(s),*}$, for which the computation of each column takes $\mathcal{O}(2^j)$ operations, and for which $\|\mathbf{A}_j^{(s)} - \mathbf{A}_j^{(s),*}\| \lesssim 2^{-j \min\{q,r\}}$, or $\|\mathbf{A}_j^{(s)} - \mathbf{A}_j^{(s),*}\| \lesssim 2^{-j(q-\varepsilon)}$ for any $\varepsilon > 0$ in case $q = r$ (see Remark 5.3). Using that by $\tilde{d} \geq d - 2t - 1$, $\|\mathbf{A}^{(s)} - \mathbf{A}_j^{(s)}\| \lesssim 2^{-j(\frac{d-1/2-t}{n-1})}$ by Theorem 5.1, and $\frac{d-1/2-t}{n-1} > \frac{d-t}{n}$ by the assumption that $\frac{d-t}{n} > \frac{1}{2}$, the proof is completed. $\square$

Summarizing, we can say that under the conditions of Corollary 5.1 and 5.2, $\mathbf{A}$ is s^*-computable with $s^* \geq \frac{d-t}{n}$, and that under the conditions of Corollary 5.1 and 5.3,

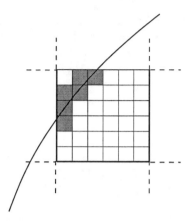

Figure 5.3: The quadrature mesh on $\Xi_{i,\lambda}$. The highlighted squares represent those subcubes on which $\psi_{\lambda'}$ is not arbitrarily smooth.

it is is s^*-computable with $s^* > \frac{d-t}{n}$. Corollary 5.3 requires that $t > 0$ and $d - 1 > t$, the latter meaning that lowest possible order spline wavelets are not covered.

Thus, fully satisfactory results are only obtained, when for the case of an entry of $\mathbf{A}^{(s)}$, quadrature is applied after first applying an integration by parts. For that case, even more favorable bounds, possibly leading to better quantitative behavior of the adaptive frame scheme, could be obtained by applying adaptive quadrature, in the sense that subcubes that intersect the singular support could be more refined than those that do not (cf. Figure 5.3). On the latter subcubes, instead of h-refinement also p-refinement could be considered. These modifications would require more programming efforts though.

5.2.4 A second look at the regular case

We apply fixed order composite quadrature rules mainly because of the approximation of the singular entries of $\mathbf{A}$, i.e, the entries of $\mathbf{A}^{(s)}$, and because of the ease with which they can be adjusted to give approximations within any prescribed tolerance. We have learned from numerical experiments that, when applied to regular entries, their error decreases much faster as function of the difference in levels of the wavelets involved than predicted by the bound of Proposition 5.1. Although, for the goal of proving optimal computational complexity of adaptive frame algorithms, this bound is satisfactory, in view of the quantitative behavior it is interesting and possible to derive a more accurate bound.

In the proofs of the above bounds on the quadrature errors given in [108] the fact that ψ_λ is a wavelet, and, in particular, that it has vanishing moments, does not play a role. Using also this property, and assuming that ψ_λ satisfies the maximal order $d - 2$ of homogeneous boundary conditions, it is possible to show the following improved statement. For the rather technical proof, the reader is referred to [108].

Proposition 5.4. *Let $|\lambda| \geq |\lambda'|$ with supp $\psi_\lambda \subset \overline{\Xi_{\lambda',i'}}$ for some $1 \leq i' \leq K$, and such that ψ_λ satisfies homogeneous Dirichlet boundary conditions of order $d - 2$. Then,*

$$|\mathbf{A}_{\lambda,\lambda'}^{(\mathrm{r})} - \mathbf{A}_{\lambda,\lambda'}^{(\mathrm{r}),*}| \lesssim N^{-p/n} 2^{-(|\lambda|-|\lambda'|)(\frac{n}{2}+t+\tilde{d}+\max(d-t,p+1-d,\lceil\frac{p+1}{2}\rceil))}.$$

To compare with the upper bound from Proposition 5.1, note that $\frac{n}{2}+t+\tilde{d}+\max(d-t,p+1-d,\lceil\frac{p+1}{2}\rceil) - (\frac{n}{2}+p-d+1) = \max(2d+\tilde{d}-p-1, t+\tilde{d}, t+d+\tilde{d}-\lfloor\frac{(p+1)}{2}\rfloor)$.

5.3 Numerical confirmation of the derived estimates

The numerical experiments in this section intend to confirm the sharpness of the different estimates given in Lemma 5.1 and Propositions 5.1–5.4. On domains $\Omega \subset \mathbb{R}^2$, we consider an operator $\mathcal{L} : H_0^1(\Omega) \to H^{-1}(\Omega)$ of order $2t = 2$ defined by

$$(\mathcal{L}u)(v) := \sum_{k=1}^{2} \int_\Omega \frac{\partial u}{\partial x_k} \frac{\partial v}{\partial x_k}, \qquad (5.3.1)$$

which results from the variational formulation of the Poisson equation with homogeneous Dirichlet boundary conditions. We are concerned with the size and the approximation of the entries in the stiffness matrix representing $\mathcal{L}$ with respect to the aggregated wavelet frame Ψ. As spline orders of the reference bases $\Psi_{i,L_2}^\square$ on $(0,1)^2$ used in (2.3.9), we choose $d = 2$ or $d = 3$, and $\tilde{d} = 2$ or $\tilde{d} = 3$ vanishing moments, respectively. Except for affine κ_i, the denominator in (2.3.9) effectively yields smooth, non-polynomial coefficients in the differential operator that we therefore have omitted in (5.3.1).

We consider parametrizations of the type

$$\kappa_i(r,s) = (1-r)(1-s)\mathbf{b}^{(0,0)} + (1-r)s\mathbf{b}^{(0,1)} + r(1-s)\mathbf{b}^{(1,0)} + rs\mathbf{b}^{(1,1)},$$

where $\mathbf{b}^{(k,l)} \in \mathbb{R}^2$, $(k,l) \in \{0,1\}^2$. Thus, provided that the vertices $\mathbf{b}^{(k,l)}$ are ordered appropriately, κ_i maps the unit square to an arbitrary quadrangle in $\mathbb{R}^2$. In case the vertices describe a parallelogram, κ_i is affine and so the denominator in (2.3.9) is a constant.

We consider two different types of overlapping decompositions of Ω. The first type refers to the situation of overlapping rectangular patches with non-matching dyadic grids both being aligned with the Cartesian coordinates, as shown in Figure 5.4 (left).

First of all, we address the decay estimates in Lemma 5.1, which are the essential ingredients for the proof of Theorem 5.1, stating a sufficient compressibility of $\mathbf{A}$.

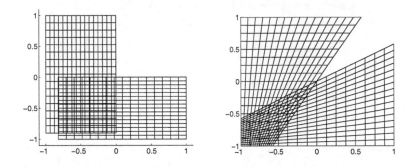

Figure 5.4: Two simple domains made up of two overlapping patches with non-matching dyadic grids.

For the grids on the left in Figure 5.4, one can compute the entries of $\mathbf{A}$ exactly, whereas for the grids on the right in Figure 5.4 we apply our composite quadrature scheme with, for this goal, $N \gg 1$ and a high order p such that the quadrature error is neglectable.

For fixed columns of $\mathbf{A}^{(\mathrm{r})}$ and $\mathbf{A}^{(\mathrm{s})}$, we have computed the largest entry in modulus as function of level difference of row and column indices. The decay of the modulus of this entry is illustrated by the results given in Figure 5.5. Lemma 5.1 predicts the exponential decay rate $n/2 + \tilde{d} + t$ or $n/2 + d - 1 - t$ in base 2 for $\mathbf{A}^{(\mathrm{r})}$ or $\mathbf{A}^{(\mathrm{s})}$, respectively. For $\mathbf{A}^{(\mathrm{r})}$, we observed the rate 4 or 5 for $d = \tilde{d} = 2$ or $d = \tilde{d} = 3$, and for $\mathbf{A}^{(\mathrm{s})}$, we observed the rate 1 or 2 for $d = \tilde{d} = 2$ or $d = \tilde{d} = 3$, all in accordance with the predicted rates.

For investigating the *quadrature errors* of our composite scheme we use product Gaussian quadrature formulas of fixed order p as building block. Figure 5.6 addresses the rate of convergence of the composite quadrature scheme for a fixed entry from $\mathbf{A}^{(\mathrm{r})}$ as function of the granularity or rank N. We have used a quadrature rule of order $p = 4$ for the case $d = \tilde{d} = 2$ and $p = 2$ for $d = \tilde{d} = 3$. We observe the polynomial rates $2 = 4/2 = p/n$ and $1 = 2/2$, respectively, as predicted by both Proposition 5.1 and Proposition 5.4.

For fixed N and $p = d = \tilde{d} = 2$, the decay of the quadrature error in a fixed column of $\mathbf{A}^{(\mathrm{r})}$ as function of the level difference $||\lambda| - |\lambda'||$ of the involved wavelets is examined in Figure 5.7. We observe the exponential rate $n/2 + t + \tilde{d} + \max(d - t, p+1-d, \lceil(p+1)/2\rceil) = 6$ in base 2, as predicted by Proposition 5.4, which is much better than the rate $n/2 + p - d + 1 = 2$ predicted by Proposition 5.1.

Unlike that from Proposition 5.4, as stated in Remark 5.4, the bound from Proposition 5.1 also applies to entries from $\mathbf{A}^{(\mathrm{s})}$ when the singular supports of the corresponding wavelets are nested as function of the level; cf. Figure 5.2. The results given in Figure 5.8 for $p = d = \tilde{d} = 2$ indicate that for those entries the bound from this proposition as function of the level difference is sharp.

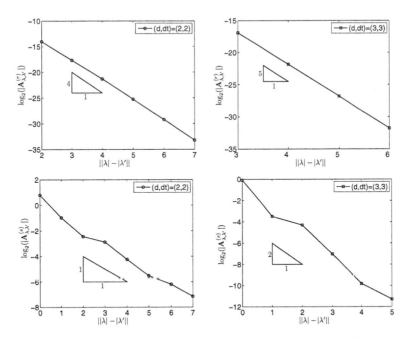

Figure 5.5: Decay of the entries in a column of $\mathbf{A}^{(\mathrm{r})}$ (upper part) and $\mathbf{A}^{(\mathrm{s})}$ (lower part) for $d = \tilde{d} = 2$ and $d = \tilde{d} = 3$.

Figure 5.9 addresses the rate of convergence of the composite quadrature scheme for entries in $\mathbf{A}^{(\mathrm{s})}$ as function of the rank N. We have used a quadrature rule of order $p = 2$ for the case $d = \tilde{d} = 2$ and $p = 4$ for $d = \tilde{d} = 3$. We observe the polynomial rates $(d - t)/n = 1/2$ and 1, respectively, in accordance with the second term from the bound of Proposition 5.2. Since in these cases $p \geq \max\{d-t, 2d-2-t\}$, as stated in the proof of Corollary 5.2, the second term in this bound is always dominating. Recall that for the entries of $\mathbf{A}^{(\mathrm{s})}$ the integrand of any integral in (5.2.1) may be discontinuous. Consequently, if N is successively increased, the ratio of the number of quadrature knots on either side of the singularity may differ, even for uniform dyadic N-refinement, as we have used it in our experiments. This causes the oscillatory behavior of the error that can be observed in Figure 5.9.

For fixed N, the decay of the quadrature error in a fixed column of $\mathbf{A}^{(\mathrm{s})}$ as function of the level difference $||\lambda| - |\lambda'||$ of the involved wavelets is examined in Figure 5.10. The results confirm the exponential rate $n/2 + d - 1 - t$ in base 2 given by the second term from the bound of Proposition 5.2.

Since for $d = 3$, $d - 1 > t = 1$, for this case alternatively we can apply the composite quadrature to the right-hand side in (5.2.5), i.e., *after integration by parts*. The results shown in the lower error diagram of Figure 5.9, obtained with $p = 4$,

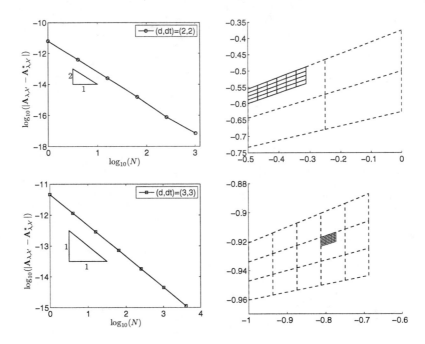

Figure 5.6: Quadrature errors for single entries of $\mathbf{A}^{(r)}$ for different granularities N with $p = 4$ (and $d = \tilde{d} = 2$) and $p = 2$ ($d = \tilde{d} = 3$), respectively. The right pictures show the the singular supports of the wavelets involved.

confirm the improved polynomial rate $3/2 = (d - t + \min\{t, d - 1 - t)\})/n$ predicted by Proposition 5.3, and illustrate an improved quantitative performance. Indeed, in the lower error diagram of Figure 5.9 the initial error for $N = 1$ is more than ten times smaller than without the integration by parts trick.

To conclude we can say that in our tests all estimates have shown to be sharp. More-over, also the quantitative performances of our quadrature scheme for the approxima-tion of the matrices $\mathbf{A}^{(r)}$ have turned out to be quite promising; cf. Figure 5.6 and 5.7. Naturally, the computation of entries in $\mathbf{A}^{(s)}$ is much harder. Nevertheless, applying an additional integration by parts as suggested in (5.2.5), both higher convergence rates as function of the rank N and an improvement of constants can be achieved.

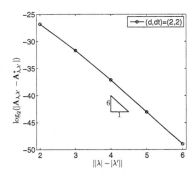

Figure 5.7: Decay of the quadrature error in a column of $\mathbf{A}^{(\mathrm{r})}$ as function of $||\lambda| - |\lambda'||$ for fixed N, and $p = d = \tilde{d} = 2$.

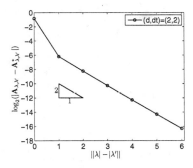

Figure 5.8: Decay of the quadrature error for entries $\mathbf{A}^{(\mathrm{s})}_{\lambda,\lambda'}$ as function of $||\lambda| - |\lambda'||$ for fixed λ' and N, $p = d = \tilde{d} = 2$, where the singular supports of ψ_λ and $\psi_{\lambda'}$ are nested like in Figure 5.2, because $\lambda = (i, \mu)$ and $\lambda' = (i, \mu')$, i.e., they are lifted by the same κ_i.

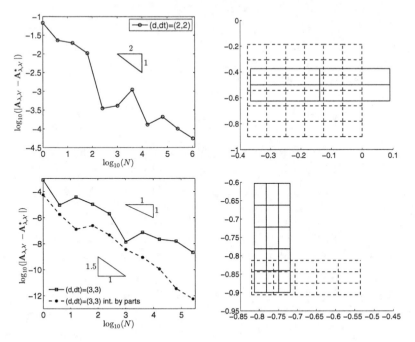

Figure 5.9: Quadrature errors for single entries of $\mathbf{A}^{(s)}$ for different granularities N with $p = d = \tilde{d} = 2$ (top left) and $p = 4$, $d = \tilde{d} = 3$ (bottom left), respectively. The right pictures show the singular supports of the wavelets involved. The dashed line in the lower left picture refers to the case where the composite rule was applied to the right-hand side of (5.2.5).

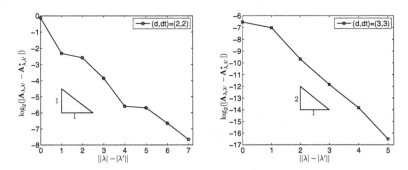

Figure 5.10: Decay of the quadrature error in a column of $\mathbf{A}^{(s)}$ as function of $||\lambda| - |\lambda'||$ for fixed N and $p = 2$, $d = \tilde{d} = 2$, and $p = 4$, $d = \tilde{d} = 3$.

Chapter 6

Adaptive Schwarz Domain Decomposition Solvers

The main chapter of this thesis now centers around another class of adaptive frame methods based on *overlapping domain decomposition algorithms*. The preceding chapters have shown that aggregated wavelet frames are indeed well-suited for the adaptive solution of elliptic operator equations, in the sense that optimal adaptive schemes can be derived and all the essential numerical routines can actually be realized in practice. However, there are still some bottlenecks to overcome.

In Chapter 4 we have learned that an essential task in the development of adaptive frame schemes is to control the redundant parts in the iterates. In the context of the verification of optimality of **SD_SOLVE**, Assumption 4.2 was needed which, so far, could not be rigorously verified for most cases of interest. Up to now, the only alternative is to include the application of the projector $\mathbf{P}$ from (4.3.36) to resolve this theoretical problem. From a practical point of view, however, this strategy seems less attractive, due to the additional amount of work arising. One of the main achievements in this chapter is that, for the algorithms being developed, in most cases optimality can be shown without Assumption 4.2. The sufficient sparsity of the discrete iterates with uniformly bounded ℓ_τ^w-norms will be guaranteed automatically, without taking measures that introduce additional work such as the application of $\mathbf{P}$. Moreover, in view of the numerical results presented in §4.4, one has to state that the quantitative performance of the steepest descent method is not fully satisfactory. This performance essentially depends on three factors.

The first one is clearly the amount of redundancy in the discrete iterates. In fact, it is desirable to keep the number of coefficients from the different patches Ω_i, corresponding to wavelets in the overlapping region, rather small. In this respect, looking at the results in Figures 4.5–4.6, and Figures 4.10–4.13, the approximations produced by Algorithm 1 though seem to be quite uneconomic, leaving the space for considerable improvement.

The second factor is the condition number $\kappa(\mathbf{A}) = \|\mathbf{A}\| \|\mathbf{A}^\dagger\|$ of the stiffness matrix $\mathbf{A}$ which essentially depends on the reference wavelet bases on $(0,1)^n$. Once these bases are fixed, the only way to improve condition numbers is preconditioning. With regard to these observations, the methods being developed in the sequel are a convenient choice, because preconditioning and controlling the amount of redundancy in the overlapping region will be inherent properties already of the exact iterative scheme considered on the full space $\ell_2(\Lambda)$.

The third factor is the quadrature. The results from Chapter 5 show that the approximation of the matrix-vector products using the routine **APPLY**, up to a given tolerance, can be performed with an amount of work staying proportional to the degrees of freedom of the output vector of **APPLY** (smoothness of the coefficients provided). However, in case non-matching grids cannot be avoided, the quantitative performance of the numerical integration can be poor, if not more involved techniques such as adaptive quadrature is applied; cf. §5.3. Fortunately, the algorithms developed later will also come along with a strategy that considerably reduces the number of coefficients from $\mathbf{A}_j^{(s)}$ (cf. §5.1) that have to be computed.

Finally, we shall discover that the schemes fit very well into the framework of aggregated frames, so that an implementation of the methods is quite simple. As another welcome property, the domain decomposition algorithms carry the potential of being parallelized in a rather straightforward fashion.

6.1 A multiplicative Schwarz adaptive wavelet frame method

In this section, an extended presentation of the results from [109] is given.

6.1.1 The exact multiplicative Schwarz method

As the first instance of an overlapping domain decomposition scheme, we introduce the well-known *multiplicative Schwarz method*. We consider again the solution of an elliptic equation

$$a(u,v) = f(v), \quad v \in H_0^t(\Omega), \tag{6.1.1}$$

where $a(\cdot,\cdot)$ represents a bounded, symmetric and elliptic bilinear form on $H_0^t(\Omega)$, $t \in \mathbb{N}$. In this chapter, we shall denote by $\|\cdot\| := a(\cdot,\cdot)^{\frac{1}{2}}$ the energy norm on $H_0^t(\Omega)$ which is equivalent to $\|\cdot\|_{H^t(\Omega)}$. In addition, the corresponding operator norm on $L(H_0^t(\Omega), H_0^t(\Omega))$ will also be denoted with $\|\cdot\|$.

The algorithm will be presented and discussed on the continuous level as an iterative scheme in $H_0^t(\Omega)$. Let us again consider a covering of the domain $\Omega \subset \mathbb{R}^n$ with m overlapping subdomains $\Omega = \bigcup_{i=0}^{m-1} \Omega_i$. Then, the exact multiplicative Schwarz method for the solution of (6.1.1) reads as follows.

Algorithm 4. *Multiplicative Schwarz method.*

$u_0 := 0$

`for` $k = 1, 2, \ldots,$

    $i := (k-1) \bmod m$

    Determine $e_{k-1} \in H_0^t(\Omega_i)$ *as the solution of the problem*

      $a(e_{k-1}, v) = f(v) - a(u_{k-1}, v), \quad v \in H_0^t(\Omega_i).$

    $u_k := u_{k-1} + e_{k-1}$

`endfor`

For $i = 0, \ldots, m-1$, let us denote with $P_i : H_0^t(\Omega) \to H_0^t(\Omega_i) \hookrightarrow H_0^t(\Omega)$ the $a(\cdot, \cdot)$-orthogonal projector onto $H_0^t(\Omega_i)$. Clearly, for a $v \in H_0^t(\Omega)$, $P_i v$ is defined by the relation

$$a(P_i v, w) = a(v, w), \quad \text{for all } w \in H_0^t(\Omega_i). \tag{6.1.2}$$

Then, by $u_k = u_{k-1} + e_{k-1}$, it is $u - u_k = u - u_{k-1} - e_{k-1}$, and by definition of e_{k-1}, using (6.1.1), one also gets $e_{k-1} = P_i(u - u_{k-1})$, so that we arrive at

$$u - u_k = (I - P_i)(u - u_{k-1}), \quad i = (k-1) \bmod m, \ k \geq 1. \tag{6.1.3}$$

As a consequence, for $k \in m\mathbb{N}$, it is

$$\|u - u_k\| \leq \|(I - P_{m-1})(I - P_{m-2}) \cdots (I - P_0)\| \|u - u_{k-m}\|.$$

Hence, convergence of the method is ensured by the following theorem.

Theorem 6.1. *Let us assume that for any $v \in H_0^t(\Omega)$ there exists a decomposition $v = \sum_{i=0}^{m-1} v_i$ with $v_i \in H_0^t(\Omega_i)$ such that for some absolute constant C_0 it is*

$$\sum_{i=0}^{m-1} \|v_i\|_{H^t(\Omega)}^2 \leq C_0 \|v\|_{H^t(\Omega)}^2. \tag{6.1.4}$$

Then,

$$\rho := \|(I - P_{m-1})(I - P_{m-2}) \cdots (I - P_0)\| < 1, \tag{6.1.5}$$

where ρ only depends on the constant C_0 and the maximal number of subdomains that intersect with one subdomain.

This theorem is a direct consequence of [118, Theorem 4.4, Lemma 4.5, and Lemma 4.6]. The theory developed there is formulated for finite dimensional Hilbert spaces, but it directly carries over to the infinite dimensional case.

Remark 6.1. It is important to note that the condition (6.1.4) is equivalent to requiring that $\{H_0^t(\Omega_i); \|\cdot\|_{H^t(\Omega_i)}\}_{i=0}^{m-1}$ forms a stable splitting of $\{H_0^t(\Omega); \|\cdot\|_{H^t(\Omega)}\}$ in the sense of Definition 2.6. (To see this, note that one of the estimates in (2.3.1) can be directly inferred using the triangle inequality.) By Proposition 2.7, the latter is equivalent to the frame property of the aggregated system of local frames for $H_0^t(\Omega_i)$. In case a partition of unity $\{\sigma_i\}_{i=0}^{m-1}$ subject to Definition 2.7 exists, which we always require, we may simply take the decomposition $v = \sum_{i=0}^{m-1} \sigma_i v$. That means that when we are dealing with aggregated (wavelet) frames, Algorithm 4 converges under the same assumption that is anyway needed to guarantee the frame property.

The basic idea of overlapping domain decomposition methods originates from the work of H. A. Schwarz (see [103]), which we shall shortly describe in the following.

Equation (6.1.1) represents, e.g., the weak formulation of the boundary value problem (1.1.4), (1.1.5). Now, if the solution u was known on $\partial\Omega_i \cap \Omega$, then the only remaining task would be to solve the *local* boundary value problem

$$Lu_i = g \qquad \text{in } \Omega_i,$$

$$\frac{\partial^k u_i}{\partial n^k} = 0 \qquad \text{on } \partial\Omega_i \cap \partial\Omega, \quad k = 0, \ldots, t-1,$$

$$\frac{\partial^k u_i}{\partial n^k} = \frac{\partial^k(u|_{\Omega_i})}{\partial n^k} \quad \text{on } \partial\Omega_i \cap \Omega, \qquad k = 0, \ldots, t-1,$$

to obtain $u|_{\Omega_i} = u_i$. Algorithm 4 can be seen as a method to approximate the interface data, i.e., the function values of u on $\partial\Omega_i \cap \Omega$, $i = 0, \ldots, m-1$. In fact, recalling §1.4, we learn that the local problem $a(e_{k-1}, v) = f(v) - a(u_{k-1}, v)$, $v \in H_0^t(\Omega_i)$, in combination with $u_i := u_{k-1}|_{\Omega_i} + e_{k-1}$ represents the weak formulation of the boundary value problem

$$Lu_i = g \qquad \text{in } \Omega_i,$$

$$\frac{\partial^k u_i}{\partial n^k} = 0 \qquad \text{on } \partial\Omega_i \cap \partial\Omega, \quad k = 0, \ldots, t-1, \qquad (6.1.6)$$

$$\frac{\partial^k u_i}{\partial n^k} = \frac{\partial^k(u_{k-1}|_{\Omega_i})}{\partial n^k} \quad \text{on } \partial\Omega_i \cap \Omega, \qquad k = 0, \ldots, t-1.$$

Thus, on each subdomain the original problem is solved with inhomogeneous boundary data, which, on $\partial\Omega_i \cap \Omega$, is given by the trace values of the current global iterate. In this context, it is also important to state that in general there is some freedom in the choice of the function u_0 in (1.4.4). Indeed, in (6.1.6) we may also replace u_{k-1} by some other function $\tilde{u}_{k-1} \in H_0^t(\Omega)$, having the same trace values on $\partial\Omega_i \cap \Omega$. That means, Algorithm 4 has the following important property.

Lemma 6.1. *If in Algorithm 4 a function $w_k \in H_0^t(\Omega_i)$, $i = (k-1) \bmod m$, is subtracted from the global iterate u_{k-1} before the determination of e_{k-1}, then this has no effect on the next global iterate u_k.*

Proof. Let $\tilde{e}_{k-1} \in H_0^t(\Omega_i)$ be the solution of the problem $a(\tilde{e}_{k-1}, v) = f(v) - a(u_{k-1} - w_k, v)$, $v \in H_0^t(\Omega_i)$, and $\tilde{u}_k := u_{k-1} - w_k + \tilde{e}_{k-1}$. Then, it is $a(\tilde{e}_{k-1} - w_k, v) = f(v) - a(u_{k-1}, v)$, $v \in H_0^t(\Omega_i)$. Hence, $\tilde{e}_{k-1} - w_k = e_{k-1}$. Thus, it follows, $\tilde{u}_k = u_{k-1} - w_k + \tilde{e}_{k-1} = u_{k-1} - w_k + e_{k-1} + w_k = u_{k-1} + e_{k-1} = u_k$, completing the proof. $\qquad\square$

6.1.2 Preparations for the design of the adaptive method

As in Chapter 4, an adaptive method subject to a discretization of (6.1.1) with respect to the aggregated wavelet Gelfand frame Ψ_{L_2} for $(H_0^t(\Omega), L_2(\Omega), H^{-t}(\Omega))$ shall be developed. We set again $\Psi := \mathbf{D}^{-1}\Psi_{L_2} =: \{\psi_\lambda\}_{\lambda \in \Lambda}$, where $\Psi = \bigcup_{i=0}^{m-1} \Psi^{(i)}$, and the collections $\Psi^{(i)} := \{\psi_{i,\lambda}\}_{\lambda \in \Lambda_i}$ form biorthogonal wavelet Riesz bases for $H_0^t(\Omega_i)$ with the dual collections $\tilde{\Psi}^{(i)} := \{\tilde{\psi}_{i,\lambda}\}_{\lambda \in \Lambda_i} \subset H^{-t}(\Omega_i)$. Throughout the present chapter we will also assume that the order d of polynomial reproduction of the bases $\Psi^{(i)}$ satisfies (2.4.1). Note that for the stiffness matrix $\mathbf{A} = \{a(\psi_\mu, \psi_\lambda)\}_{\lambda,\mu \in \Lambda}$, one has the relation

$$\|v\| \lesssim \|\mathbf{A}\|_{\ell_2(\Lambda) \to \ell_2(\Lambda)}^{\frac{1}{2}} \inf_{\{\mathbf{v} \in \ell_2(\Lambda) : \mathbf{v}^T \Psi = v\}} \|\mathbf{v}\|_{\ell_2(\Lambda)}, \quad v \subset H_0^t(\Omega), \tag{6.1.7}$$

because for $v = \mathbf{v}^\top \Psi$, with (4.2.1), it is $\|v\|^2 = a(\mathbf{v}^\top \Psi, \mathbf{v}^\top \Psi) = \langle \mathbf{A}\mathbf{v}, \mathbf{v} \rangle_{\ell_2(\Lambda)} = \|\mathbf{v}\|^2 \leq \|\mathbf{A}\|_{\ell_2(\Lambda) \to \ell_2(\Lambda)} \|\mathbf{Q}\mathbf{v}\|_{\ell_2(\Lambda)}^2 \leq \|\mathbf{A}\|_{\ell_2(\Lambda) \to \ell_2(\Lambda)} \|\mathbf{v}\|_{\ell_2(\Lambda)}^2$. The aim is to develop an implementable adaptive method being optimal as in Definition 4.1. It is clear that in view of this aim the auxiliary problems

$$a(e_{k-1}, v) = f(v) - a(u_{k-1}, v), \quad v \in H_0^t(\Omega_i), \tag{6.1.8}$$

should be solved only inexactly. It can be easily seen (recall the discussion on page 5) that in case the functions e_{k-1} are always approximated within some geometrically decreasing tolerances, one still gets majorized linear convergence of the iterates towards the solution u in the sense that

$$\|u - u_k\|_{H^t(\Omega)} \lesssim \varepsilon_k \tag{6.1.9}$$

with $\varepsilon_k = \eta \varepsilon_{k-1}$, for some $\eta < 1$, and thus also $\varepsilon_k \approx \eta^k$, for all $k \geq 1$. In particular, we are going to use an adaptive method for the solution of (6.1.8) being discretized with respect to the wavelet basis $\Psi^{(i)}$.

A necessary condition for optimality

Let us write the kth iterand obtained by this procedure $u_k = \sum_{i=0}^{m-1} u_k^{(i)}$ with $u_k^{(i)} = (\mathbf{u}_k^{(i)})^T \Psi^{(i)}$. In each iteration, the global iterate u_{k-1} is only updated on Ω_i by adding the function $e_{k-1} \in H_0^t(\Omega_i)$. Thus, on the discrete level such a step amounts to an update of the expansion coefficients of u_{k-1} with respect to $\Psi^{(i)}$ only. Consequently, for $k \in i + 1 + m\mathbb{N}_0$ we have

$$u_k^{(i)} - u_{k-m}^{(i)} = u_k - u_{k-1},$$

from which, by the majorized linear convergence of the sequence $\{u_k\}_{k\geq 0}$, follows that $\{u_k^{(i)}\}_{k\in i+1+mN_0}$ is a Cauchy sequence in $H_0^t(\Omega_i)$, and therefore convergent to some limit function $u^{(i)} \in H_0^t(\Omega_i)$ with representation $u^{(i)} = (\mathbf{u}^{(i)})^\top \Psi^{(i)}$. With ε_k from (6.1.9), we may infer

$$\|\mathbf{u}^{(i)} - \mathbf{u}_k^{(i)}\|_{\ell_2(\Lambda_i)} \eqsim \|u^{(i)} - u_k^{(i)}\|_{H^t(\Omega)} \lesssim \varepsilon_k.$$

The last step in this estimate can be seen as follows. From Algorithm 4 we get $u^{(i)} = \sum_{\{j\in\mathbb{N},\ (j-1)\bmod m=i\}} e_{j-1}$, and by $e_{j-1} = P_i(u - u_{j-1})$ clearly $\|e_{j-1}\|_{H^t(\Omega_i)} \lesssim \|u - u_{j-1}\|_{H^t(\Omega)} \lesssim \varepsilon_{j-1} \eqsim \eta^{j-1}$, so that for $(k-1)\bmod m = i$

$$\|u^{(i)} - u_k^{(i)}\|_{H^t(\Omega_i)} = \Big\| \sum_{\substack{j=1 \\ (j-1)\bmod m=i}}^{\infty} e_{j-1} - \sum_{\substack{j=1 \\ j-1\bmod m=i}}^{k} e_{j-1} \Big\|_{H^t(\Omega_i)}$$

$$\leq \sum_{\substack{j=k+1 \\ (j-1)\bmod m=i}}^{\infty} \|e_{j-1}\|_{H^t(\Omega_i)} \lesssim \eta^k \eqsim \varepsilon_k.$$

In Definition 4.1 it is required that for any $\varepsilon > 0$, $\#\operatorname{supp}\mathbf{u}_k \lesssim \varepsilon^{-1/s}|\tilde{\mathbf{u}}|_{\ell_\tau^w(\Lambda)}^{1/s}$, whenever the algorithm is stopped as soon as $\varepsilon_k \leq \varepsilon$. This property, however, can only be fulfilled when

$$\mathbf{u}^{(i)} \in \ell_\tau^w(\Lambda_i), \quad |\mathbf{u}^{(i)}|_{\ell_\tau^w(\Lambda_i)} \lesssim |\tilde{\mathbf{u}}|_{\ell_\tau^w(\Lambda)}, \quad i = 0,\ldots,m-1. \tag{6.1.10}$$

For convenience, we give a formal justification of the latter. For arbitrary $N \geq 1$, we set $\varepsilon^{(N)} := \|\mathbf{u}^{(i)} - \mathbf{u}^{(i),N}\|_{\ell_2(\Lambda_i)}$, in which $\mathbf{u}^{(i),N}$ is a best N-term approximation to $\mathbf{u}^{(i)}$. Let now k be minimal such that $\|\mathbf{u}^{(i)} - \mathbf{u}_k^{(i)}\|_{\ell_2(\Lambda_i)} \leq \varepsilon^{(N)}$. Then, in case the algorithm is optimal, it follows $\#\operatorname{supp}\mathbf{u}_k^{(i)} \lesssim (\varepsilon^{(N)})^{-1/s}|\tilde{\mathbf{u}}|_{\ell_\tau^w(\Lambda)}^{1/s}$, and clearly $N \leq \#\operatorname{supp}\mathbf{u}_k^{(i)}$. We may thus infer, $\varepsilon^{(N)} \lesssim N^{-s}|\tilde{\mathbf{u}}|_{\ell_\tau^w(\Lambda)}$. Hence, $\|\mathbf{u}^{(i)} - \mathbf{u}^{(i),N}\|_{\ell_2(\Lambda_i)} N^s = \varepsilon^{(N)} N^s \lesssim N^{-s}|\tilde{\mathbf{u}}|_{\ell_\tau^w(\Lambda)} N^s = |\tilde{\mathbf{u}}|_{\ell_\tau^w(\Lambda)}$. The assertion (6.1.10) then follows from (3.3.15).

A modified multiplicative Schwarz method

Unfortunately, with the standard approach in Algorithm 4, one does not know why the necessary property (6.1.10) should be satisfied, and that is where Lemma 6.1 comes into play. In fact, we shall consider a *modified approach*, where in each iteration not only the expansion coefficients corresponding to the current subdomain are updated, but also those corresponding to neighboring patches. In particular, before the local problem (6.1.8) is solved, we will subtract from u_{k-1} the function $\sum_{j\neq i}(\check{\mathbf{u}}_k^{(j)})^\top \Psi^{(j)}$, where $\check{\mathbf{u}}_k^{(j)}$, $j \neq i$, consists of those coefficients of $\mathbf{u}_{k-1}^{(j)}$ that correspond to wavelets the supports of which are fully contained in Ω_i. In addition, we may also subtract $u_{k-1}^{(i)}$ itself from u_{k-1}. The modified algorithm *with exact subdomain solves* then reads as follows.

Algorithm 5. *Modified multiplicative Schwarz method.*

$u_0 := 0, \; k := 0$

for $l = 1, 2, \ldots,$

   for $i = 0 \ldots, m - 1$

     $k := (l - 1)m + i + 1$

     Determine $u_k = \sum_{j=0}^{m-1} u_k^{(j)}$, $u_k^{(j)} = (\mathbf{u}_k^{(j)})^T \Psi^{(j)}$ *(i.e.,* $u_k = \mathbf{u}_k^T \Psi$

     where $\mathbf{u}_k = (\mathbf{u}_k^{(0)}, \ldots, \mathbf{u}_k^{(m-1)}))$ *as follows:*

     For $j \neq i$, $\mathbf{u}_{k,\lambda}^{(j)} := \begin{cases} 0 & \text{when } \mathrm{supp}\,\psi_{j,\lambda} \cap \Omega \subset \Omega_i, \\ \mathbf{u}_{k-1,\lambda}^{(j)} & otherwise. \end{cases}$

     Determine $u_k^{(i)}$ *as the solution of the problem*

$$a(u_k^{(i)}, v^{(i)}) = f(v^{(i)}) - a(\textstyle\sum_{j \neq i} u_k^{(j)}, v^{(i)}), \quad v^{(i)} \in H_0^t(\Omega_i).$$

   endfor

endfor

We stress that in Algorithm 5 the function $u_k^{(i)}$ is directly given as the solution of the local problem, because the contribution $u_{k-1}^{(i)}$ is also (implicitly) subtracted from u_{k-1} before the local solve. Hence, by Lemma 6.1, the iterates u_k in Algorithm 5 coincide with those generated by Algorithm 4. We shall see that with these modifications the adaptive algorithm developed below meets the necessary requirement (6.1.10). Moreover, solving the local problems up to suitable geometrically decreasing tolerances, and adding a coarsening step, we will show that property (6.1.10) is also *sufficient* to guarantee optimality of the resulting adaptive method.

6.1.3 The adaptive method and its convergence

We are now ready to formulate our multiplicative Schwarz method *with inexact subdomain solves*. The algorithm consists of three nested loops, where in the innermost loop a complete multiplicative Schwarz iteration is performed, i.e., the subsequent (inexact) solving of the local problems on all m subdomains. After K multiplicative Schwarz iterations, being performed in the middle loop, coarsening is applied. In the outermost loop the whole process is repeated L times to guarantee a desired reduction of the initial error.

Algorithm 6. MultSchw$[\varepsilon, \mu] \to \mathbf{u}_\varepsilon$

% *The input should satisfy* $\varepsilon > 0$ *and* $\mu \geq \|u\|$.

% *With* M *being an upper bound for* $\|\mathbf{A}\|^{\frac{1}{2}}_{\ell_2(\Lambda) \to \ell_2(\Lambda)}$, *let the parameters*

% $\sigma > 0$ *and* $K \in \mathbb{N}$ *be such that* $2\rho^K M \sigma < 1$, *where* ρ *is from* (6.1.5).

% *Let* L *be the smallest integer with* $\mu(2\rho^K M \sigma)^L \leq \varepsilon$.

$u_0 = u_0^{(0)} = \cdots = u_0^{(m-1)} := 0,\ k := 0$

for $l = 1, 2, \ldots, L$

  for $p = 1, \ldots, K$

    for $i = 0, \ldots, m-1$

      $k := (l-1)mK + (p-1)m + i + 1$

      Determine $u_k = \sum_{j=0}^{m-1} u_k^{(j)}$, $u_k^{(j)} = (\mathbf{u}_k^{(j)})^T \Psi^{(j)}$ (*i.e.,* $u_k = \mathbf{u}_k^T \Psi$

      where $\mathbf{u}_k = (\mathbf{u}_k^{(0)}, \ldots, \mathbf{u}_k^{(m-1)})$) *as follows:*

      For $j \neq i$, $\mathbf{u}_{k,\lambda}^{(j)} := \begin{cases} 0 & \text{when } \operatorname{supp} \psi_{j,\lambda} \cap \Omega \subset \Omega_i, \\ \mathbf{u}_{k-1,\lambda}^{(j)} & \text{otherwise.} \end{cases}$

      Determine $u_k^{(i)}$ *as an approximation to the solution* $\bar{u}_k^{(i)}$ *of*

        $a(\bar{u}_k^{(i)}, v^{(i)}) = f(v^{(i)}) - a(\sum_{j \neq i} u_k^{(j)}, v^{(i)}), \quad v^{(i)} \in H_0^t(\Omega_i),$

      such that $\|\bar{u}_k^{(i)} - u_k^{(i)}\| \leq \mu(2\rho^K M \sigma)^{l-1} \frac{\rho^p}{mK}$.

    endfor

  endfor

  $\mathbf{u}_k := \mathbf{COARSE}[\mathbf{u}_k, (\sigma - 1/M) 2\rho^K \mu (2\rho^K M \sigma)^{l-1}]$

endfor

$\mathbf{u}_\varepsilon := \mathbf{u}_k$

Note that for $k \in mK\mathbb{N}$, $\mathbf{u}_k$ (and thus also u_k) got redefined by the application of **COARSE**, i.e., for those k, u_k or $\mathbf{u}_k$ has two meanings that will be distinguished carefully in the sequel.

Proposition 6.1. *Let* $\mu \geq \|u\|$. *The iterand* u_{lmK} *inside* $\mathbf{u}_\varepsilon = \mathbf{MultSchw}[\varepsilon, \mu]$, *for* $l > 0$, *after the application of* **COARSE**, *satisfies*

$$\|u - u_{lmK}\| \leq \mu(2\rho^K M \sigma)^l, \tag{6.1.11}$$

and so $\|u - \mathbf{u}_\varepsilon^T \Psi\| \leq \varepsilon$. *Furthermore, for* $1 \leq q \leq mK$,

$$\|u - u_{(l-1)mK+q}\| \leq 2\mu(2\rho^K M \sigma)^{l-1} \rho^{\frac{q}{m}-1}, \tag{6.1.12}$$

where for $q = mK$, u_{lmK} *should read here as the iterand* before *the application of* **COARSE**.

Proof. The first statement is valid for $l = 0$ by $\|u\| \leq \mu$. Suppose it is valid for some $l - 1 \geq 0$. Then for $p = 0, \ldots, K$,

$$\|u - u_{(l-1)mK+pm}\| \leq \mu(1 + \tfrac{p}{K})\rho^p(2\rho^K M\sigma)^{l-1}, \tag{6.1.13}$$

where, when $p = K$, u_{lmK} should read here as the iterand *before* the application of **COARSE**. Indeed, this estimate is valid for $p = 0$ by the above induction hypothesis. Suppose (6.1.13) is valid for some $p - 1 \geq 0$. Let $u^\dagger_{(l-1)mK+pm}$ be the result of an *exact* last step in the innermost loop, starting from $u_{(l-1)mK+pm-1}$, and let $u^\dagger_{(l-1)mK+pm-1}$ be the result of the *exact* second last step in the innermost loop, starting from $u_{(l-1)mK+pm-2}$. Then, using (6.1.3) and $\|I - P_i\| \leq 1$, we get

$$
\begin{aligned}
\|u - u_{(l-1)mK+pm}\| &\leq \|u - u^\dagger_{(l-1)mK+pm}\| + \|u^\dagger_{(l-1)mK+pm} - u_{(l-1)mK+pm}\| \\
&\leq \|(I - P_{m-1})(u - u_{(l-1)mK+pm-1})\| + \mu(2\rho^K M\sigma)^{l-1}\tfrac{\rho^p}{mK} \\
&= \|(I - P_{m-1})(u - u^\dagger_{(l-1)mK+pm-1}) \\
&\quad + (I - P_{m-1})(u^\dagger_{(l-1)mK+pm-1} - u_{(l-1)mK+pm-1})\| + \mu(2\rho^K M\sigma)^{l-1}\tfrac{\rho^p}{mK} \\
&\leq \|(I - P_{m-1})(I - P_{m-2})(u - u_{(l-1)mK+pm-2})\| \\
&\quad + \|u^\dagger_{(l-1)mK+pm-1} - u_{(l-1)mK+pm-1}\| + \mu(2\rho^K M\sigma)^{l-1}\tfrac{\rho^p}{mK} \\
&\leq \|(I - P_{m-1})(I - P_{m-2})(u - u_{(l-1)mK+pm-2})\| + 2\mu(2\rho^K M\sigma)^{l-1}\tfrac{\rho^p}{mK}.
\end{aligned}
$$

An inductive argument, (6.1.5), and (6.1.13) for $p - 1 \geq 0$ thus yields

$$\|u - u_{(l-1)mK+pm}\| \leq \rho\big[\mu(1 + \tfrac{p-1}{K})\rho^{p-1}(2\rho^K M\sigma)^{l-1}\big] + m\mu(2\rho^K M\sigma)^{l-1}\tfrac{\rho^p}{mK},$$

which is (6.1.13).

By (6.1.13) for $p = K$, and the fact that $\|\mathbf{v}^T \Psi\| \leq M\|\mathbf{v}\|_{\ell_2(\Lambda)}$, $\mathbf{v} \in \ell_2(\Lambda)$, by definition of M in Algorithm 6 (cf. (6.1.7)), the subsequent application of **COARSE** introduces an additional error that in $\|\cdot\|$ is not larger than $M(\sigma - 1/M)2\rho^K \mu(2\rho^K M\sigma)^{l-1}$, the proof of the first statement is completed.

Analogous to the proceeding above, using an inductive argument, (6.1.13), with p reading as $p - 1$, $\|I - P_i\| \leq 1$, and the precisions with which the local problems are solved in the algorithm, we find that

$$
\begin{aligned}
\|u &- u_{(l-1)mK+(p-1)m+(i+1)}\| \\
&\leq \|(I - P_i)(I - P_{i-1})\ldots(I - P_0)(u - u_{(l-1)mK+(p-1)m})\| \\
&\quad + (i+1)\mu(2\rho^K M\sigma)^{l-1}\tfrac{\mu^r}{mK} \\
&\leq \|u - u_{(l-1)mK+(p-1)m}\| + (i+1)\mu(2\rho^K M\sigma)^{l-1}\tfrac{\rho^p}{mK} \\
&\leq [1 + \tfrac{p-1}{K} + \tfrac{(i+1)\rho}{mK}]\rho^{p-1}\mu(2\rho^K M\sigma)^{l-1}, \tag{6.1.14}
\end{aligned}
$$

where here when $p = K$ and $i = m - 1$, u_{lmK} should read as the iterand *before* the application of **COARSE**. Note now that $1 + \tfrac{p-1}{K} + \tfrac{(i+1)\rho}{mK} \leq 1 + \tfrac{K-1}{K} + \tfrac{m\rho}{mK} =$

$1 + \frac{K-1+\rho}{K} \leq 2$. For $q = (p-1)m + (i+1)$, from (6.1.14), we get

$$\|u - u_{(l-1)mK+q}\| \leq 2\rho^{(q-(i+1))/m}\mu(2\rho^K M\sigma)^{l-1} \leq 2\mu(2\rho^K M\sigma)^{l-1}\rho^{\frac{q}{m}-1},$$

completing the proof of the second statement. $\qquad\qquad\qquad\qquad\qquad$ □

6.1.4 Optimality

Having established convergence of **MultSchw**, we will now investigate its optimality. The main result will be formulated in Theorem 6.3. The intention of this subsection is to prove Theorem 6.3 using the key result Theorem 6.2 which essentially states that for **MultSchw** the property (6.1.10) actually holds. The verification of Theorem 6.2 will be addressed afterwards in §6.1.5 and §6.1.6.

The key result for the proof of optimality

For a set $A \subset \mathbb{R}^n$ and $\delta \geq 0$, let

$$B(A;\delta) := \{y \in \mathbb{R}^n : \operatorname{dist}(y, A) \leq \delta\}, \tag{6.1.15}$$

and

$$\Omega_i(-\delta) := \{x \in \Omega_i : B(x;\delta) \cap \Omega \subset \Omega_i\}.$$

With regard to the discussion in §6.1.2, the following pivotal result will be the key for the verification of optimality.

Theorem 6.2. *Let the decomposition $\Omega = \bigcup_{i=0}^{m-1} \Omega_i$ be such that*

$$\Omega \subset \bigcup_{i=0}^{m-1} \Omega_i(-\omega), \quad \text{for some } \omega > 0, \tag{6.1.16}$$

and let

$$\eta := \max_{0 \leq i \leq m-1, \lambda \in \Lambda_i} \big(\operatorname{diam}(\operatorname{supp} \psi_{i,\lambda}), \operatorname{diam}(\operatorname{supp} \tilde{\psi}_{i,\lambda})\big) \leq \frac{\omega}{3m-1}. \tag{6.1.17}$$

(a) *Then, for any $0 \leq i \leq m-1$, there exists a $u^{(i)} = (\mathbf{u}^{(i)})^T\Psi^{(i)} \in H_0^t(\Omega_i)$ such that for $l \in \mathbb{N}$, $1 \leq q \leq mK$, with $(q-1) \bmod m = i$,*

$$\|u^{(i)} - u^{(i)}_{(l-1)mK+q}\|_{H^t(\Omega_i)} \lesssim \mu(2\rho^K M\sigma)^{l-1}\rho^{\frac{q}{m}-1}, \tag{6.1.18}$$

*only dependent on m, ω and the Riesz constants of all $\Psi^{(i)}$, and where for $q = mK$, u_{lmK} should read as the iterand before the application of **COARSE**.*

(b) *If, for some $s < \frac{d-t}{n}$, u has some representation $u = \tilde{\mathbf{u}}^T\Psi$ with $\tilde{\mathbf{u}} \in \ell_\tau^w(\Lambda)$, $\frac{1}{\tau} = s + \frac{1}{2}$, then, $|\mathbf{u}^{(i)}|_{\ell_\tau^w(\Lambda_i)} \lesssim |\tilde{\mathbf{u}}|_{\ell_\tau^w(\Lambda)}$.*

The proof of this result shall be postponed to §6.1.5 and §6.1.6 below.

Remark 6.2. (a) Roughly speaking, condition (6.1.16) means that when for each subdomain Ω_i a small strip near the internal boundaries $\partial\Omega_i \cap \Omega$ is removed, the "shrinked" patches still make up an overlapping covering of Ω. A simple example is given in Figure 6.1. It is obvious that in case the Assumption 2.2 (on page 54) is satisfied, condition (6.1.16) holds for a sufficiently small ω, whereas otherwise (take the decomposition of the L-shaped domain as in Figure 2.4) this does not have to be the case. The condition (6.1.16) permits the development of an arbitrary smooth partition of unity, and thus implies (6.1.4); cf. Remark 6.1. The converse is not true, as §2.7 shows.

(b) The condition (6.1.17) means that the minimal overlap between subdomains is sufficiently large compared to the largest diameter of the support of any primal or dual wavelet. If (6.1.16) holds for a fixed $\omega > 0$, then, by the locality of the primal and dual wavelets, (6.1.17) can always be guaranteed by choosing the minimal level j_0 of the primal and dual multiresolution sequences of the bases $\Psi^{(t)}$ sufficiently large.

Some consequences of Theorem 6.2

In the following, we collect some consequences of Theorem 6.2 that will be essential for the verification of optimality.

Proposition 6.2. *Let, for $i = 0, \ldots, m-1$, $\mathbf{u}^{(i)} \in \ell_2(\Lambda_i)$ be the discrete local limits from Theorem 6.2. We define a special sequence $\mathbf{u} = \{\mathbf{u}_\lambda\}_{\lambda \in \Lambda} \in \ell_2(\Lambda)$ by setting*

$$\mathbf{u}_\lambda := \begin{cases} \mathbf{u}_\lambda^{(i)} & when \ \lambda \in \Lambda_i \ and \ \operatorname{supp} \psi_{i,\lambda} \cap \Omega \not\subset \Omega_j \ for \ some \ j > i, \\ 0 & otherwise. \end{cases}$$

Then, for the iterate $u_{lmK} = (\mathbf{u}_{lmK})^\top \Psi$ in **MultSchw**, *before the application of* **COARSE**, *it is*

$$\|\mathbf{u} - \mathbf{u}_{lmK}\|_{\ell_2(\Lambda)} \leq \mu C (2\rho^K M\sigma)^{l-1} \rho^{K-2}, \tag{6.1.19}$$

for some constant $C > 0$ only depending on m, ω, and on the Riesz constants of all the $\Psi^{(i)}$. Moreover, from the sequence $\mathbf{u} \in \ell_2(\Lambda)$, the continuous solution can be reconstructed, i.e., $u = \mathbf{u}^\top \Psi$.

Proof. Note first that *before* the application of **COARSE**, the iterate u_{lmk} is given by

$$\mathbf{u}_{lmK,\lambda} = \begin{cases} \mathbf{u}_{lmK-m+i+1,\lambda}^{(i)} & when \ \lambda \in \Lambda_i \ and \ \operatorname{supp} \psi_{i,\lambda} \cap \Omega \not\subset \Omega_j \ for \\ & \hspace{4.5cm} some \ j > i, \\ 0 & otherwise. \end{cases}$$

145

Let us denote $\mathbf{u} =: (\breve{\mathbf{u}}^{(0)}, \ldots, \breve{\mathbf{u}}^{(m-1)})$. Hence, the sequences $\breve{\mathbf{u}}^{(i)}$ are created by dropping coefficients from $\mathbf{u}^{(i)}$. Now, using the bound (6.1.18), we obtain

$$
\|\mathbf{u} - \mathbf{u}_{lmK}\|_{\ell_2(\Lambda)} = \left(\sum_{i=0}^{m-1} \|\breve{\mathbf{u}}^{(i)} - \mathbf{u}_{lmK}^{(i)}\|_{\ell_2(\Lambda_i)}^2 \right)^{1/2}
$$

$$
\leq \left(\sum_{i=0}^{m-1} \|\mathbf{u}^{(i)} - \mathbf{u}_{lmK-m+i+1}^{(i)}\|_{\ell_2(\Lambda_i)}^2 \right)^{1/2} \approx \left(\sum_{i=0}^{m-1} \|u^{(i)} - u_{lmK-m+i+1}^{(i)}\|_{H_0^t(\Omega_i)}^2 \right)^{1/2}
$$

$$
\approx \sum_{i=0}^{m-1} \|u^{(i)} - u_{lmK-m+i+1}^{(i)}\|_{H_0^t(\Omega_i)} \lesssim \mu(2\rho^K M\sigma)^{l-1} \sum_{i=0}^{m-1} \rho^{\frac{mK-m+(i+1)}{m}-1}
$$

$$
= \mu(2\rho^K M\sigma)^{l-1} \rho^{K-2} \sum_{i=0}^{m-1} \rho^{\frac{i+1}{m}} = \mu(2\rho^K M\sigma)^{l-1} \rho^{K-2} \rho^{\frac{1}{m}} \frac{1-\rho}{1-\rho^{\frac{1}{m}}}
$$

$$
\leq m\mu(2\rho^K M\sigma)^{l-1} \rho^{K-2},
$$

so that the first statement is verified. Furthermore, we clearly have $\|u - \mathbf{u}^\top \Psi\| \leq \|u - u_{lmK}\| + \|(\mathbf{u}_{lmK})^\top \Psi - \mathbf{u}^\top \Psi\| \lesssim \|u - u_{lmK}\| + \|\mathbf{u}_{lmK} - \mathbf{u}\|_{\ell_2(\Lambda)}$, so that by (6.1.11) and (6.1.19) follows $u = \mathbf{u}^\top \Psi$. $\square$

It is remarkable that, unlike the results for the adaptive steepest descent method (Algorithm 1), we are now able to proof convergence of the sequence of iterates in $\ell_2(\Lambda)$ to a limit in $\ell_2(\Lambda)$. So far this has only been possible for the projected iterates $\mathbf{Q}\mathbf{u}_k$ (or $\mathbf{P}\mathbf{u}_k$; see [104, Proposition 2.3]), whereas convergence of their kernel components was not guaranteed.

Moreover, the next proposition provides some information on the support sizes of the vectors $\mathbf{u}_{lmK}$ in **MultSchw** and their ℓ_τ^w-norms after the application of **COARSE**.

Proposition 6.3. *If in* **MultSchw** *the parameter σ is chosen sufficiently large to ensure that with $C > 0$ from (6.1.19) it is*

$$
(\sigma - 1/M)2 > C\rho^{-2}, \tag{6.1.20}
$$

*and K correspondingly such that $2\rho^K M\sigma < 1$, then, after the evaluation of $\mathbf{u}_{lmK} := $* **COARSE**$[\mathbf{u}_{lmK}, (\sigma - 1/M)2\rho^K \mu(2\rho^K M\sigma)^{l-1}]$, *we have*

$$
\#\mathrm{supp}\,\mathbf{u}_{lmK} \lesssim \left[\mu(2\rho^K M\sigma)^l \right]^{-1/s} |\tilde{\mathbf{u}}|_{\ell_\tau^w(\Lambda)}^{1/s}, \quad and \tag{6.1.21}
$$

$$
|\mathbf{u}_{lmK}|_{\ell_\tau^w(\Lambda)} \lesssim |\tilde{\mathbf{u}}|_{\ell_\tau^w(\Lambda)}, \tag{6.1.22}
$$

uniformly in l, for a $\tilde{\mathbf{u}}$ as in Theorem 6.2 (b).

Proof. For the sequence $\mathbf{u} \in \ell_2(\Lambda)$ from Proposition 6.2, using Theorem 6.2 (b), we infer that $|\mathbf{u}|_{\ell_\tau^w(\Lambda)} \lesssim |\tilde{\mathbf{u}}|_{\ell_\tau^w(\Lambda)}$. Defining $\zeta := ((\sigma - 1/M)2\rho^2)/C > 1$ and $\bar\varepsilon := \mu C(2\rho^K M\sigma)^{l-1}\rho^{K-2}$, we get

$$
\zeta\bar\varepsilon = \frac{(\sigma - 1/M)2\rho^2}{C}\mu C(2\rho^K M\sigma)^{l-1}\rho^{K-2} = (\sigma - 1/M)2\rho^K \mu(2\rho^K M\sigma)^{l-1}.
$$

Then, from Proposition 4.4 one gets

$$\# \operatorname{supp} \mathbf{u}_{lmK} \lesssim \bar{\varepsilon}^{-1/s} |\tilde{\mathbf{u}}|_{\ell_\tau^w(\Lambda)}^{1/s} \quad \text{and} \quad |\mathbf{u}_{lmK}|_{\ell_\tau^w(\Lambda)} \lesssim |\tilde{\mathbf{u}}|_{\ell_\tau^w(\Lambda)}.$$

Moreover, $\bar{\varepsilon} = \mu C(2\rho^K M\sigma)^{l-1}\rho^{K-2} > \mu C(2\rho^K M\sigma)^l \rho^{K-2}$, because $2\rho^K M\sigma < 1$. Thus, $\bar{\varepsilon} \gtrsim \mu(2\rho^K M\sigma)^l$, finishing the proof. $\qquad\square$

Remark 6.3. Note that if the parameter σ is chosen large enough to satisfy (6.1.20), as required in Proposition 6.3, the parameter K has to be chosen such that $2\rho^K M\sigma < 1$. For the proof of convergence, this additional requirement on σ (possibly resulting in a larger K) was not needed.

With a suitable implementation of the inexact subdomain solves, the estimates (6.1.21) and (6.1.22) will be the key to a proof of the optimality of **MultSchw**.

Another application of Theorem 6.2 yields the following.

Lemma 6.2. *If in the algorithm* **MultSchw***, for the approximation of the solution $\bar{u}_k^{(i)}$ on subdomain Ω_i, we use the previous iterate $u_{k-m}^{(i)}$ as the starting value of an iterative solver, then, the error has to be reduced by a constant factor only, independent of l.*

Proof. Let $p > 1$. Using the precisions with which the local problems on the subdomains are solved, we get

$$\|\bar{u}_k^{(i)} - u_{k-m}^{(i)}\| \leq \|\bar{u}_k^{(i)} - \bar{u}_{k-m}^{(i)}\| + \|\bar{u}_{k-m}^{(i)} - u_{k-m}^{(i)}\|$$

$$\leq \|\bar{u}_k^{(i)} - \bar{u}_{k-m}^{(i)}\| + \mu(2\rho^K M\sigma)^{l-1}\frac{\rho^{p-1}}{mK}$$

$$\lesssim \|\bar{u}_k^{(i)} - u_k^{(i)}\| + \|u_k^{(i)} - u^{(i)}\| + \|u^{(i)} - u_{k-m}^{(i)}\| + \|u_{k-m}^{(i)} - \bar{u}_{k-m}^{(i)}\|$$
$$+ \mu(2\rho^K M\sigma)^{l-1}\rho^{p-1}$$

$$\lesssim \mu(2\rho^K M\sigma)^{l-1}\rho^p + \|u_k^{(i)} - u^{(i)}\| + \|u^{(i)} - u_{k-m}^{(i)}\| + \mu(2\rho^K M\sigma)^{l-1}\rho^{p-1}.$$

It is $k = (l-1)mK + (p-1)m + i + 1$, thus $k - m = (l-1)mK + (p-2)m + i + 1$. Theorem 6.2 (a) yields

$$\|u_k^{(i)} - u^{(i)}\| \lesssim \mu(2\rho^K M\sigma)^{l-1}\rho^{((p-1)m+i+1)/m-1}$$
$$= \mu(2\rho^K M\sigma)^{l-1}\rho^{p-2+(i+1)/m} < \mu(2\rho^K M\sigma)^{l-1}\rho^{p-2},$$

and similarly $\|u^{(i)} - u_{k-m}^{(i)}\| \lesssim \mu(2\rho^K M\sigma)^{l-1}\rho^{p-3}$, so that when we insert this into the above estimate, we obtain $\|\bar{u}_k^{(i)} - u_{k-m}^{(i)}\| \lesssim \mu(2\rho^K M\sigma)^{l-1}\rho^{p-3}$. Consequently, the initial error has to be reduced about a fixed constant times ρ^3 (uniformly in l) to attain the target accuracy $\mu(2\rho^K M\sigma)^{l-1}\frac{\rho^p}{mK}$. The case $p = 1$ can be treated in a similar fashion. $\qquad\square$

Adaptive solution of the local problems on Ω_i

The next step is to develop a suitable adaptive solver for the local elliptic problems. With $\mathbf{A}^{(i,j)} := \{a(\psi_{j,\mu}, \psi_{i,\lambda})\}_{\lambda \in \Lambda_i, \mu \in \Lambda_j}$, the matrix $\mathbf{A}$ has a natural block partition as $\mathbf{A} = \{\mathbf{A}^{(i,j)}\}_{0 \leq i,j \leq m-1}$. For $k := (l-1)mK + (p-1)m + i + 1$, the determination of $u_k^{(i)}$ inside **MultSchw**, on the discrete level, amounts to approximating the solution $\bar{\mathbf{u}}_k^{(i)}$ of

$$\mathbf{A}^{(i,i)} \bar{\mathbf{u}}_k^{(i)} = \mathbf{f}|_{\Lambda_i} - \sum_{j \neq i} \mathbf{A}^{(i,j)} \mathbf{u}_k^{(j)}, \tag{6.1.23}$$

within tolerance $\delta := \mu(2\rho^K M\sigma)^{l-1} \frac{\rho^p}{mK}$ in $\| \cdot \|_{\mathbf{A}^{(i,i)}} := \langle \mathbf{A}^{(i,i)} \cdot, \cdot \rangle_{\ell_2(\Lambda_i)}^{\frac{1}{2}} \approx \| \cdot \|_{\ell_2(\Lambda_i)}$. In Chapter 5 we have learned that $\mathbf{A}$ is s^*-compressible and s^*-computable, for some $s^* \geq \frac{d-t}{n}$, using the wavelet properties collected in §2.4. As subblocks of $\mathbf{A}$, this holds as well for both the diagonal blocks $\mathbf{A}^{(i,i)}$ and the non-diagonal blocks $\mathbf{A}^{(i,j)}$, $j \neq i$. Thus, a suitable routine **APPLY**$[\mathbf{A}^{(i,j)}, \mathbf{w}, \varepsilon]$ with the properties described in §4.3 for the approximate application of the blocks $\mathbf{A}^{(i,j)}$ exists. In case a quasi-optimal routine **RHS** for the approximation of $\mathbf{f}$ is available, one may clearly apply it to the restricted sequences $\mathbf{f}|_{\Lambda_i}$.

Remark 6.4. The respective sequences $\mathbf{u}_k^{(j)}$ in (6.1.23) are created from $\mathbf{u}_{k-1}^{(j)}$ by deleting all coefficients $\mathbf{u}_{k-1,\lambda}^{(j)}$, $\lambda \in \Lambda_j$, for which supp $\psi_{j,\lambda} \cap \Omega \subset \Omega_i$. The results in §5.2 and §5.3 have shown that in a quantitative sense the (approximate) computation of the non-diagonal blocks is more expensive than that of the diagonal blocks. This means that removing coefficients from $\mathbf{u}_{k-1}^{(j)}$, $j \neq i$, before a solve on Ω_i, which significantly reduces the number of coefficients from $\mathbf{A}^{(i,j)}$ that have to computed, can be expected to give quantitative advantages in the numerical performance of the method. In fact, for the approximation of the right-hand side of the local problem on Ω_i, only integrals $a(\psi_{j,\mu}, \psi_{i,\lambda})$ have to be computed where (supp $\psi_{j,\mu} \setminus \partial$ supp $\psi_{j,\mu}) \cap (\partial\Omega_i \cap \Omega) \neq \emptyset$; see also Figure 6.1.

As a consequence of Lemma 6.2, we know that, taking the sequence $\mathbf{u}_{k-m}^{(i)}$ as initial guess, there exists a constant $D > 0$ such that for the initial error holds

$$\|\bar{\mathbf{u}}_k^{(i)} - \mathbf{u}_{k-m}^{(i)}\|_{\mathbf{A}^{(i,i)}} \leq D\delta. \tag{6.1.24}$$

We discuss in the following a simple adaptive damped Richardson iteration for the approximation of $\bar{\mathbf{u}}_k^{(i)}$. In principle, this approach will correspond to the second adaptive wavelet algorithm proposed by Cohen, Dahmen, and DeVore in [28]. Alternatively, one can also apply the adaptive wavelet Galerkin method introduced by these authors in [27]; cf. [65] as well. The latter method is somewhat more difficult to describe, but in practice it leads to better numerical performances, for which reason it is going to be used in the numerical experiments discussed in Chapter 7.

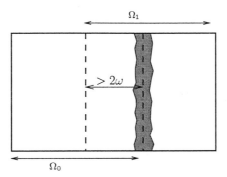

Figure 6.1: Location of supports of wavelets $\psi_{1,\lambda}$ being relevant for the computation of the right-hand side of a local problem in Ω_0.

The matrices $\mathbf{A}^{(i,i)}$ are boundedly invertible, symmetric positive definite operators from $\ell_2(\Lambda_i)$ to $\ell_2(\Lambda_i)$. It is known (for convenience cf. [104] or [38]) that for a sufficiently small $\alpha > 0$, it is

$$\xi := \|\mathbf{I} - \alpha\mathbf{A}^{(i,i)}\|_{\mathbf{A}^{(i,i)}} < 1, \tag{6.1.25}$$

where in the latter equation the norm has to be understood as the operator norm on $L(\ell_2(\Lambda_i), \ell_2(\Lambda_i))$ induced by the norm $\|\cdot\|_{\mathbf{A}^{(i,i)}}$ on $\ell_2(\Lambda_i)$ from above. That means the exact damped Richardson iteration with parameter α for the approximation of $(\mathbf{A}^{(i,i)})^{-1}\mathbf{g}$, for any $\mathbf{g} \in \ell_2(\Lambda_i)$, converges in $\|\cdot\|_{\mathbf{A}^{(i,i)}}$ with the rate $\xi < 1$. Based on this preliminary observation, we propose the following algorithm.

Algorithm 7. LocSolve$[i, \delta] \to \mathbf{u}_k^{(i)}$

% Let ξ be defined as in (6.1.25), and D as in (6.1.24).

% Let P be the smallest integer for which $2\xi^P(D + \frac{1}{2}) \le \frac{1}{2}$.

$\mathbf{v}_0 := \mathbf{u}_{k-m}^{(i)}$

$\eta := \|(\mathbf{A}^{(i,i)})^{-1}\|_{\ell_2(\Lambda_i) \to \ell_2(\Lambda_i)}^{-\frac{1}{2}} \delta/4$

$\mathbf{r} := \mathbf{RHS}[\mathbf{f}|_{\Lambda_i}, \eta] - \sum_{j \neq i} \mathbf{APPLY}[\mathbf{A}^{(i,j)}, \mathbf{u}_k^{(j)}, \frac{\eta}{m-1}]$

for $l = 1, 2, \ldots, P$

$\qquad \mathbf{v}_l := \mathbf{v}_{l-1} + \alpha(\mathbf{r} - \mathbf{APPLY}[\mathbf{A}^{(i,i)}, \mathbf{v}_{l-1}, \frac{\xi^P(D+\frac{1}{2})\delta}{\alpha P\|\mathbf{A}^{(i,i)}\|_{\ell_2(\Lambda_i) \to \ell_2(\Lambda_i)}^{1/2}}])$

endfor

$\mathbf{u}_k^{(i)} := \mathbf{v}_P$

Lemma 6.3. *For the sequence* $\mathbf{u}_k^{(i)} = \mathbf{LocSolve}[i, \delta]$, *one has the estimate*

$$\|\bar{\mathbf{u}}_k^{(i)} - \mathbf{u}_k^{(i)}\|_{\mathbf{A}^{(i,i)}} \leq \delta. \tag{6.1.26}$$

Proof. From the definition of η and the properties of **APPLY** and **RHS**, one easily infers

$$\|\bar{\mathbf{u}}_k^{(i)} - (\mathbf{A}^{(i,i)})^{-1}\mathbf{r}\|_{\mathbf{A}^{(i,i)}} \leq \frac{\delta}{2}. \tag{6.1.27}$$

Combining this with (6.1.24) yields $\|\mathbf{v}_0 - (\mathbf{A}^{(i,i)})^{-1}\mathbf{r}\|_{\mathbf{A}^{(i,i)}} = \|\mathbf{u}_{k-m}^{(i)} - (\mathbf{A}^{(i,i)})^{-1}\mathbf{r}\|_{\mathbf{A}^{(i,i)}} \leq (D + \frac{1}{2})\delta$. Now, let us abbreviate $\gamma := \frac{\xi^P(D+\frac{1}{2})\delta}{\alpha P\|\mathbf{A}^{(i,i)}\|_{\ell_2(\Lambda_i)\to\ell_2(\Lambda_i)}^{1/2}}$. Let also $\mathbf{v}_1^\dagger$ be the result of an exact step in the `for`-loop (with $\gamma = 0$) starting from $\mathbf{v}_0$. Using the relation $\|\cdot\|_{\mathbf{A}^{(i,i)}} \leq \|\mathbf{A}^{(i,i)}\|_{\ell_2(\Lambda_i)\to\ell_2(\Lambda_i)}^{1/2}\|\cdot\|_{\ell_2(\Lambda_i)}$, we get

$$\|\mathbf{v}_1 - (\mathbf{A}^{(i,i)})^{-1}\mathbf{r}\|_{\mathbf{A}^{(i,i)}} \leq \|\mathbf{v}_1 - \mathbf{v}_1^\dagger\|_{\mathbf{A}^{(i,i)}} + \|\mathbf{v}_1^\dagger - (\mathbf{A}^{(i,i)})^{-1}\mathbf{r}\|_{\mathbf{A}^{(i,i)}}$$

$$\leq \alpha\|\mathbf{A}^{(i,i)}\|_{\ell_2(\Lambda_i)\to\ell_2(\Lambda_i)}^{1/2}\gamma + \xi(D + \frac{1}{2})\delta.$$

Hence, an easy inductive argument using the same trick shows

$$\|\mathbf{v}_P - (\mathbf{A}^{(i,i)})^{-1}\mathbf{r}\|_{\mathbf{A}^{(i,i)}} \leq \alpha P\|\mathbf{A}^{(i,i)}\|_{\ell_2(\Lambda_i)\to\ell_2(\Lambda_i)}^{1/2}\gamma + \xi^P(D + \frac{1}{2})\delta$$

$$= 2\xi^P(D + \frac{1}{2})\delta \leq \frac{1}{2}\delta.$$

Thus, using (6.1.27) again, we get $\|\bar{\mathbf{u}}_k^{(i)} - \mathbf{u}_k^{(i)}\|_{\mathbf{A}^{(i,i)}} = \|\bar{\mathbf{u}}_k^{(i)} - \mathbf{v}_P\|_{\mathbf{A}^{(i,i)}} \leq \|\bar{\mathbf{u}}_k^{(i)} - (\mathbf{A}^{(i,i)})^{-1}\mathbf{r}\|_{\mathbf{A}^{(i,i)}} + \|(\mathbf{A}^{(i,i)})^{-1}\mathbf{r} - \mathbf{v}_P\|_{\mathbf{A}^{(i,i)}} \leq \delta$. □

The optimality result

We are now in a position to formulate and prove one of the main results of this thesis, stating optimality of **MultSchw**.

Theorem 6.3. *Assume* (6.1.16) *and* (6.1.17). *Let* σ *be sufficiently large such that* (6.1.20) *is valid, let* $K \in \mathbb{N}$ *be such that* $2\rho^K M\sigma < 1$, *and* $\mu \lesssim \|u\|$. *For some* $s^* \geq \frac{d-t}{n}$, *let* $\mathbf{A}$ *be* s^*-*compressible,* s^*-*computable, and assume* **RHS** *to be quasi-optimal. Let* u *have some representation* $u = \tilde{\mathbf{u}}^T\Psi$ *with* $\tilde{\mathbf{u}} \in \ell_\tau^w(\Lambda)$, *for some* $s \in (0, \frac{d-t}{n})$, $\frac{1}{\tau} = s + \frac{1}{2}$. *Then, for* $\varepsilon \in (0, \mu)$, $\mathbf{u}_\varepsilon := \mathbf{MultSchw}[\varepsilon, \mu]$ *satisfies* $\|u - \mathbf{u}_\varepsilon^T\Psi\| \leq \varepsilon$ *and* $\#\text{supp } \mathbf{u}_\varepsilon \lesssim \varepsilon^{-1/s}|\tilde{\mathbf{u}}|_{\ell_\tau^w(\Lambda)}^{1/s}$, *where the number of operations required by the call can be bounded on some absolute multiple of* $\varepsilon^{-1/s}|\tilde{\mathbf{u}}|_{\ell_\tau^w(\Lambda)}^{1/s}$.

Proof. Using the routine **LocSolve**, it is ensured that in order to get from $\mathbf{u}_{(l-1)mK}$ to $\mathbf{u}_{lmK}$ in **MultSchw**, a fixed number of approximate applications of $\mathbf{A}^{(i,j)}$, $i, j \in \{0, \ldots, m-1\}$, by calls of **APPLY** and a fixed number of calls of **RHS** for the approximation of $\mathbf{f}|_{\Lambda_i}$ are needed. Therefore, from (6.1.22), (4.3.11), and (4.3.12), we get

$$|\mathbf{u}_k|_{\ell_\tau^w(\Lambda)} \lesssim |\tilde{\mathbf{u}}|_{\ell_\tau^w(\Lambda)}, \qquad (6.1.28)$$

uniformly in k. Keeping this in mind, noting that the tolerances with which these calls are made coincide with $\mu(2\rho^K M\sigma)^{l-1}\rho^p$ up to a constant, using (6.1.21), (4.3.2) and (4.3.4), we infer that

$$\# \operatorname{supp} \mathbf{u}_k \lesssim [\mu(2\rho^K M\sigma)^{l-1}\rho^p]^{-1/s}|\tilde{\mathbf{u}}|_{\ell_\tau^w(\Lambda)}^{1/s}, \qquad (6.1.29)$$

uniformly in l. Note that by $k = (l-1)mK + (p-1)m + (i+1)$, it is $(l-1) + \frac{p}{K} = \frac{k}{mK} + \frac{1}{K} - \frac{i+1}{mK}$. We may also write

$$\mu(2\rho^K M\sigma)^{l-1}\rho^p \approx \mu(2\rho^K M\sigma)^{l-1}(2M\sigma)^{p/K}(\rho^K)^{p/K}$$
$$= \mu(2\rho^K M\sigma)^{(l-1)+p/K} \approx \mu(2\rho^K M\sigma)^{k/mK},$$

thus, (6.1.29) can be written as

$$\# \operatorname{supp} \mathbf{u}_k \lesssim [\mu(2\rho^K M\sigma)^{k/mK}]^{-1/s}|\tilde{\mathbf{u}}|_{\ell_\tau^w(\Lambda)}^{1/s}, \qquad (6.1.30)$$

which also holds uniformly in k. Then, using the assumption $\mu \lesssim \|u\|$, (4.3.3), (4.3.5), (4.3.7), the fact that in **LocSolve** only a fixed number of calls of **APPLY** and **RHS** are made, and a geometric series argument (similar to (4.3.19) in the proof of Theorem 4.1), shows that the number of operations needed to compute $\mathbf{u}_k$ can be bounded by an absolute multiple of $[\mu(2\rho^K M\sigma)^{k/mK}]^{-1/s}|\tilde{\mathbf{u}}|_{\ell_\tau^w(\Lambda)}^{1/s}$. Note that $\mu \lesssim \|u\|$ implies $1 \lesssim \mu^{-1/s}\|u\|^{1/s} \approx \mu^{-1/s}\|\tilde{\mathbf{u}}^\top \Psi\|_{H^t(\Omega)}^{1/s} \lesssim \mu^{-1/s}\|\tilde{\mathbf{u}}\|_{\ell_2(\Lambda)}^{1/s} \lesssim \mu^{-1/s}|\tilde{\mathbf{u}}|_{\ell_\tau^w(\Lambda)}^{1/s} < \varepsilon^{-1/s}|\tilde{\mathbf{u}}|_{\ell_\tau^w(\Lambda)}^{1/s}$, so that the constant terms in the bounds for the cost of **APPLY**, **RHS**, and **COARSE** never dominate the other terms. That the log-term in (4.3.7) is harmless has already been shown in the proof of Theorem 4.2. Since $\|u - u_{LmK}\| \lesssim \mu(2\rho^K M\sigma)^L \approx \varepsilon$, by (6.1.11) and the choice of L in **MultSchw**, the proof is completed. $\qquad\square$

Remark 6.5. In Theorem 6.3, s^*-compressibility (and computability) was assumed only for $s^* \geq \frac{d-t}{n}$, while for Theorem 4.2, $s^* > \frac{d-t}{n}$ was needed (recall Remark 4.6).

6.1.5 Construction of the limits on the subdomains: Proof of Theorem 6.2 (a)

As announced, we prove now part (a) of Theorem 6.2, i.e., the convergence of the sequence $\{u_k^{(i)}\}_{k \in i+1+m\mathbb{N}_0}$ in $H_0^t(\Omega_i)$, for each fixed $i \in \{0, \ldots, m-1\}$. The aim is to give explicit expressions for the limit functions, which, in §6.1.6, will be shown to be sufficiently smooth to ensure Theorem 6.2 (b). Before giving a formal proof for the general case of having m subdomains, the basic idea for the construction of the limit functions for the case of two subdomains shall be sketched.

Construction of the local limits for the case of two subdomains, $m = 2$

In the remainder of this chapter, we will sometimes abbreviate the dual form $\langle \cdot, \cdot \rangle_{H^{-t}(\Omega_i) \times H_0^t(\Omega_i)}$ by just writing $\langle \cdot, \cdot \rangle$. The addressed subdomain Ω_i will always be clear from the context.

For *odd* k, $\mathbf{u}_{k,\mu}^{(1)}$ is defined to be zero, when $\operatorname{supp} \psi_{1,\mu} \cap \Omega \subset \Omega_0$. Now, from $\operatorname{diam}(\operatorname{supp} \psi_{1,\mu})$, $\operatorname{diam}(\operatorname{supp} \psi_{0,\lambda})$, $\operatorname{diam}(\operatorname{supp} \tilde{\psi}_{0,\lambda}) \leq \eta$, for $\mu \in \Lambda_1, \lambda \in \Lambda_0$, and $\operatorname{supp} \psi_{0,\lambda} \cap \operatorname{supp} \tilde{\psi}_{0,\lambda} \neq \emptyset$, $\lambda \in \Lambda_0$, and the assumption (6.1.17) that $5\eta \leq \omega$, for any $\lambda \in \Lambda_0$ with $\operatorname{supp} \psi_{0,\lambda} \cap \Omega \not\subset \Omega_1$, we get

$$\mathbf{u}_{k,\lambda}^{(0)} = \langle \tilde{\psi}_{0,\lambda}, u_k^{(0)} \rangle = \langle \tilde{\psi}_{0,\lambda}, u_k - u_k^{(1)} \rangle = \langle \tilde{\psi}_{0,\lambda}, u_k \rangle \to \langle \tilde{\psi}_{0,\lambda}, u \rangle, \quad k \text{ odd} \to \infty;$$

see Figure 6.2. Obviously, already the requirement $3\eta \leq 2\omega$ would be sufficient. For k *even*, $\mathbf{u}_{k,\lambda}^{(0)}$ is defined to be zero when $\operatorname{supp} \psi_{0,\lambda} \cap \Omega \subset \Omega_1$, and it is equal to $\mathbf{u}_{k-1,\lambda}^{(0)}$ otherwise. From $u_k = u_k^{(0)} + u_k^{(1)} \to u$, for $k \to \infty$, we conclude that

$$u_k^{(1)} \to u - \sum_{\{\lambda \in \Lambda_0 : \operatorname{supp} \psi_{0,\lambda} \cap \Omega \not\subset \Omega_1\}} \langle \tilde{\psi}_{0,\lambda}, u \rangle \psi_{0,\lambda} \in H_0^t(\Omega_1), \quad k \text{ even} \to \infty. \quad (6.1.31)$$

To relate this result with the analysis for the general case given below, let us consider smooth functions ϕ_i on Ω that vanish outside Ω_i and that are identically one on Ω_i, except on a sufficiently small strip near $\Omega \setminus \Omega_i$. Let $u^{(0,0)} := \phi_0 u \in H_0^t(\Omega_0)$ with wavelet representation $u^{(0,0)} = (\mathbf{u}^{(0,0)})^{\top} \Psi^{(0)}$. Note that if $\operatorname{supp} \psi_{0,\lambda} \cap \Omega \not\subset \Omega_1$, it is $\mathbf{u}_{\lambda}^{(0,0)} = \langle \tilde{\psi}_{0,\lambda}, u^{(0,0)} \rangle = \langle \tilde{\psi}_{0,\lambda}, \phi_0 u \rangle = \langle \tilde{\psi}_{0,\lambda}, u \rangle$, because ϕ_0 is identically one on $\operatorname{supp} \tilde{\psi}_{0,\lambda}$. Furthermore, ϕ_1 can be chosen to be identically one on $\operatorname{supp}(u - u^{(0,0)}) \cap \Omega$ such that $\phi_1(u - u^{(0,0)}) = u - u^{(0,0)}$. Therefore, one finds

$$u = u - u^{(0,0)} + u^{(0,0)} = \phi_1(u - u^{(0,0)}) + u^{(0,0)} = \phi_1(u - u^{(0,0)}) + \sum_{\lambda \in \Lambda_0} \mathbf{u}_{\lambda}^{(0,0)} \psi_{0,\lambda}$$

$$= \phi_1(u - u^{(0,0)}) + \sum_{\{\lambda \in \Lambda_0 : \operatorname{supp} \psi_{0,\lambda} \cap \Omega \subset \Omega_1\}} \mathbf{u}_{\lambda}^{(0,0)} \psi_{0,\lambda} + \sum_{\{\lambda \in \Lambda_0 : \operatorname{supp} \psi_{0,\lambda} \cap \Omega \not\subset \Omega_1\}} \mathbf{u}_{\lambda}^{(0,0)} \psi_{0,\lambda}.$$

The limit function in $H_0^t(\Omega_1)$ given in (6.1.31) can thus be written as

$$u^{(1)} := \phi_1(u - u^{(0,0)}) + \sum_{\{\lambda \in \Lambda_0 : \operatorname{supp} \psi_{0,\lambda} \cap \Omega \subset \Omega_1\}} \mathbf{u}_{\lambda}^{(0,0)} \psi_{0,\lambda} \in H_0^t(\Omega_1).$$

For this limit function the respective estimate (6.1.18) has to be proved. Moreover, one has to show that it possesses expansion coefficients $\mathbf{u}^{(1)}$ with respect to $\Psi^{(1)}$, satisfying $|\mathbf{u}^{(1)}|_{\ell_\tau^w(\Lambda_1)} \lesssim |\tilde{\mathbf{u}}|_{\ell_\tau^w(\Lambda)}$; see Theorem 6.2 (b). Note that we have here identified the local limit function on Ω_1 for k even $\to \infty$. With an analogous argument, one can identify the local limit $u^{(0)} \in H_0^t(\Omega_0)$ for k odd $\to \infty$.

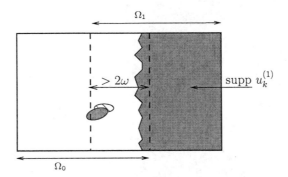

Figure 6.2: $\operatorname{supp}\psi_{0,\lambda}$ and $\operatorname{supp}\tilde{\psi}_{0,\lambda}$ for $\lambda \in \Lambda_0$ with $\operatorname{supp}\psi_{0,\lambda} \cap \Omega \not\subset \Omega_1$, and $\operatorname{supp}u_k^{(1)}$ for some odd k assuming (6.1.17).

The general case of m subdomains

As a preliminary remark, let us state that **MultSchw** produces a sequence $\{u_k\}_{k\in\mathbb{N}_0}$ of approximations to u such that $u_k = \sum_{i=0}^{m-1} u_k^{(i)}$, $u_k^{(i)} = (\mathbf{u}_k^{(i)})^T \Psi^{(i)}$, where for $k \geq 1$ and $j \neq (k-1)\bmod m$,

$$\mathbf{u}_{k,\lambda}^{(j)} := \begin{cases} 0 & \text{when } \operatorname{supp}\psi_{j,\lambda} \cap \Omega \subset \Omega_{(k-1)\bmod m}, \\ \mathbf{u}_{k-1,\lambda}^{(j)} \text{ or } 0 & \text{otherwise.} \end{cases} \tag{6.1.32}$$

Note that the possibility that $\mathbf{u}_{k,\lambda}^{(j)}$ is set to zero also when $\operatorname{supp}\psi_{j\lambda} \cap \Omega \not\subset \Omega_{(k-1)\bmod m}$ allows us to include the effect of the coarsening step. In that case u_k should read as the iterand after coarsening.

Because of the cyclic character of (6.1.32), it is sufficient to prove the convergence and to identify a limit of the sequence $\{u_k^{(m-1)}\}_{k\in m\mathbb{N}_0}$.

The first step is to identify a candidate for the local limit function. With η from (6.1.17), for some $\theta \geq -\frac{1}{3}$, $0 \leq i \leq m-1$, let $\phi_{i,\theta} \in C_0^\infty(\mathbb{R}^n)$ with

$$\operatorname{supp}\phi_{i,\theta} \cap \Omega \subset \Omega_i(-(1+3\theta)\eta), \quad \phi_{i,\theta} \equiv 1 \text{ on } \Omega_i(-(2+3\theta)\eta). \tag{6.1.33}$$

We define $(\chi_{i,\theta})_{0\leq i\leq m-1}$ by $\chi_{0,\theta} = \phi_{0,\theta}$, and for $i > 0$ by

$$\chi_{i,\theta} = \phi_{i,\theta}(1 - \chi_{i-1,\theta}) + \chi_{i-1,\theta}. \tag{6.1.34}$$

These functions have the following properties.

Proposition 6.4. (i) $\operatorname{supp}\chi_{i,\theta} \cap \Omega \subset \bigcup_{j=0}^{i} \Omega_j(-(1+3\theta)\eta)$,

(ii) $\chi_{i,\theta} \equiv 1$ on $\bigcup_{j=0}^{i} \Omega_j(-(2+3\theta)\eta)$.

Proof. First, we realize that

(a) if for a fixed $k \in \{0, \ldots, i\}$, $\phi_{k,\theta}(x) = 1$, then $\chi_{i,\theta}(x) = 1$,

(b) if for all $k \in \{0, \ldots, i\}$, $\phi_{k,\theta}(x) = 0$, then $\chi_{i,\theta}(x) = 0$.

Indeed, if for an $x \in \Omega$, and $k \in \{0, \ldots, i\}$, it is $\phi_{k,\theta}(x) = 1$, then, $\chi_{k,\theta}(x) = \phi_{k,\theta}(x)(1 - \chi_{k-1,\theta}(x)) + \chi_{k-1,\theta}(x) = 1$, and so $\chi_{j,\theta}(x) = 1$, $j = k, \ldots, i$, by (6.1.34). In case $\phi_{j,\theta}(x) = 0$, for all $j = 0, \ldots, i$, then, again by (6.1.34), we get $\chi_{i,\theta}(x) = 0$.

Now, for the proof of (i), let $x \notin \bigcup_{j=0}^{i} \Omega_j(-(1+3\theta)\eta)$. It follows $x \notin \text{supp}\, \phi_{j,\theta} \cap \Omega$, $j = 0, \ldots, i$, thus $\phi_{j,\theta}(x) = 0$, $j = 0, \ldots, i$, from which (i) can be deduced using (b).

Let now $x \in \bigcup_{j=0}^{i} \Omega_j(-(2+3\theta)\eta)$, then there exists a $k \in \{0, \ldots, i\}$ such that $x \in \Omega_k(-(2+3\theta)\eta)$, thus $\phi_{k,\theta}(x) = 1$, meaning that $\chi_{i,\theta}(x) = 1$, proving (ii). $\qquad\square$

Next, we define special functions $u^{(i,j)} \in H_0^t(\Omega_i)$, $j = 0, \ldots, m-1, 0 \leq i \leq j$, where $u^{(m-1)} := u^{(m-1,m-1)}$ represents the candidate for the limit on Ω_{m-1}.

Definition 6.1. For $0 \leq i \leq j \leq m-1$, we define $u^{(i,j)} \in H_0^t(\Omega_i)$ with representation $(\mathbf{u}^{(i,j)})^T \Psi^{(i)}$ as follows:

```
for j = 0, ..., m − 1
```

$$u^{(j,j)} := \phi_{j,-\frac{1}{3}}\Big(u - \sum_{i=0}^{j-1} u^{(i,j-1)}\Big) + \sum_{i=0}^{j-1} \sum_{\{\lambda \in \Lambda_i : \text{supp}\, \psi_{i,\lambda} \cap \Omega \subset \Omega_j\}} \mathbf{u}_\lambda^{(i,j-1)} \psi_{i,\lambda}$$

$$u^{(i,j)} := u^{(i,j-1)} - \sum_{\{\lambda \in \Lambda_i : \text{supp}\, \psi_{i,\lambda} \cap \Omega \subset \Omega_j\}} \mathbf{u}_\lambda^{(i,j-1)} \psi_{i,\lambda}, \quad \text{for } 0 \leq i \leq j-1$$

```
endfor
```

The key result for the proof of Theorem 6.2 (a) is given by the next theorem.

Theorem 6.4. *With* $u^{(m-1)} := u^{(m-1,m-1)}$, *for* $l \in \mathbb{N}_0$, *it holds*

$$\|u^{(m-1)} - u_{(l+1)m}^{(m-1)}\|_{H^t(\Omega_{m-1})} \lesssim \sum_{i=0}^{m-1} \|u - u_{lm+i+1}\|_{H^t(\Omega)}. \tag{6.1.35}$$

In fact, we can conclude Theorem 6.2 (a) in the following way. Let $\tilde{l} \in \mathbb{N}$. If in Theorem 6.2 (a), $(q-1) \bmod m = m-1$, it is $q = mp$, for a $p \in \{1, \ldots, K\}$. That means

$$\|u^{(m-1)} - u_{(\tilde{l}-1)mK+q}^{(m-1)}\|_{H^t(\Omega)} = \|u^{(m-1)} - u_{(\tilde{l}-1)mK+pm}^{(m-1)}\|_{H^t(\Omega)}$$
$$= \|u^{(m-1)} - u_{((\tilde{l}-1)K+p)m}^{(m-1)}\|_{H^t(\Omega)}.$$

From Theorem 6.4, using (6.1.12), one infers

$$\|u^{(m-1)} - u^{(m-1)}_{((\bar{l}-1)K+p)m}\|_{H^t(\Omega_{m-1})} \lesssim \sum_{i=0}^{m-1} \|u - u_{((\bar{l}-1)K+p-1)m+i+1}\|_{H^t(\Omega)}$$

$$= \sum_{i=0}^{m-1} \|u - u_{(\bar{l}-1)mK+(p-1)m+i+1}\|_{H^t(\Omega)} \lesssim 2\mu(2\rho^K M\sigma)^{\bar{l}-1} \sum_{i=0}^{m-1} \rho^{\frac{(p-1)m+i+1}{m}-1}$$

$$\leq 2\mu(2\rho^K M\sigma)^{\bar{l}-1} m\rho^{\frac{1}{m}-1}\rho^{\frac{pm}{m}-1} \approx \mu(2\rho^K M\sigma)^{\bar{l}-1}\rho^{\frac{q}{m}-1}. \tag{6.1.36}$$

The statement for $i \neq m - 1$ can be proven analogously.

Theorem 6.4, in turn, is induced by the following lemma, which also involves the functions $\chi_{j,\theta}$ defined in (6.1.34) with the choice $\theta = j$.

Lemma 6.4. *For $0 \leq j \leq m - 1$, $l \in \mathbb{N}$, it is*

$$\|\chi_{j,j}(u^{(j,j)} - u^{(j)}_{lm+j+1})\|_{H^t(\Omega_j)} \lesssim \sum_{i=0}^{j} \|u - u_{lm+i+1}\|_{H^t(\Omega)}.$$

Before giving the proof of this lemma, we infer the statement of Theorem 6.4.

Proof of Theorem 6.4. By Proposition 6.4 (ii), $\chi_{m-1,m-1} \equiv 1$ on $\bigcup_{i=0}^{m-1} \Omega_i(-(2+3(m-1))\eta) = \bigcup_{i=0}^{m-1} \Omega_i(-(3m-1)\eta)$. By (6.1.17), we have $(3m-1)\eta \leq \omega$, and by the assumption (6.1.16) therefore $\Omega \subset \bigcup_{i=0}^{m-1} \Omega_i(-(3m-1)\eta)$, so that $\chi_{m-1,m-1} \equiv 1$ on Ω. Hence, Lemma 6.4 applied for $j = m - 1$ immediately completes the proof. $\square$

The rest of this subsection is dedicated to the remaining verification of Lemma 6.4. To this end, two further propositions have to be stated and proved first.

Proposition 6.5. (i) *For $0 \leq i < j \leq m - 1$, $u^{(i,j)}$ vanishes on*

$$\Omega_{i+1}(-\eta) \cup \cdots \cup \Omega_j(-\eta).$$

(ii) *For $0 \leq i \leq m - 1$, $u_k^{(i)}$ vanishes on*

$$\Omega_{i+1}(-\eta) \cup \cdots \cup \Omega_{(k-1) \bmod m}(-\eta),$$

when $(k - 1) \bmod m > i$, and on

$$\Omega_{i+1}(-\eta) \cup \cdots \cup \Omega_{m-1}(-\eta) \cup \Omega_0(-\eta) \cup \cdots \cup \Omega_{(k-1) \bmod m}(-\eta),$$

when $(k - 1) \bmod m < i$ (and $k \geq m$).

155

Proof. We prove the claim (i) by an induction on j. For any $0 \leq i < m - 1$, it is $u^{(i,i+1)} = \sum_{\{\lambda \in \Lambda_i : \text{supp}\, \psi_{i,\lambda} \cap \Omega \not\subset \Omega_{i+1}\}} \mathbf{u}_\lambda^{(i,i)} \psi_{i,\lambda}$. The definition of η in (6.1.17) yields that $u^{(i,i+1)}$ vanishes on $\Omega_{i+1}(-\eta)$. Now suppose that for $0 \leq i < j-1 < m-1$, $u^{(i,j-1)}$ vanishes on $\Omega_{i+1}(-\eta) \cup \cdots \cup \Omega_{j-1}(-\eta)$, then, by $u^{(i,j)} = \sum_{\{\lambda \in \Lambda_i : \text{supp}\, \psi_{i,\lambda} \cap \Omega \not\subset \Omega_j\}} \mathbf{u}_\lambda^{(i,j-1)} \psi_{i,\lambda}$, we infer by the definition of η in (6.1.17) that $u^{(i,j)}$ vanishes on $\Omega_j(-\eta)$, and by induction hypothesis that it also vanishes on $\Omega_{i+1}(-\eta) \cup \cdots \cup \Omega_{j-1}(-\eta)$, so that (i) is proved.

The relation $(k-1) \bmod m > i$ means that in the algorithm, in the current cycle of subsequent subdomain solves, the patch Ω_i has already been treated, so that $u_k^{(i)}$ is created by subsequent sparsening of the last update on patch Ω_i, meaning that the first part of (ii) is obvious by the choice of η. The second case $(k-1) \bmod m < i$ means that the local solve on Ω_i, during the current cycle, has not been performed yet. Since the last update on Ω_i, the algorithm has visited $\Omega_{i+1}, \ldots, \Omega_{m-1}$ and then $\Omega_0, \ldots, \Omega_{(k-1) \bmod m}$, every time with the respective sparsening of the former update on Ω_i, so that also the second part of (ii) is clear. $\qquad\square$

Proposition 6.6. *It holds that* $\chi_{j,-\frac{1}{3}} u = \sum_{i=0}^{j} u^{(i,j)}$.

Proof. For $j = 0$, we have $\chi_{0,-\frac{1}{3}} u = \phi_{0,-\frac{1}{3}} u = u^{(0,0)}$. Let us suppose that the statement is valid for $j - 1 > 0$. Then, we have

$$
\chi_{j,-\frac{1}{3}} u = \phi_{j,-\frac{1}{3}}(u - \chi_{j-1,-\frac{1}{3}} u) + \chi_{j-1,-\frac{1}{3}} u
$$

$$
= \phi_{j,-\frac{1}{3}}\left(u - \sum_{i=0}^{j-1} u^{(i,j-1)}\right) + \sum_{i=0}^{j-1} u^{(i,j-1)}
$$

$$
= u^{(j,j)} - \sum_{i=0}^{j-1} \sum_{\{\lambda \in \Lambda_i : \text{supp}\, \psi_{i,\lambda} \cap \Omega \subset \Omega_j\}} \mathbf{u}_\lambda^{(i,j-1)} \psi_{i,\lambda} + \sum_{i=0}^{j-1} u^{(i,j-1)}
$$

$$
= u^{(j,j)} + \sum_{i=0}^{j-1} \left(u^{(i,j-1)} - \sum_{\{\lambda \in \Lambda_i : \text{supp}\, \psi_{i,\lambda} \cap \Omega \subset \Omega_j\}} \mathbf{u}_\lambda^{(i,j-1)} \psi_{i,\lambda} \right)
$$

$$
= u^{(j,j)} + \sum_{i=0}^{j-1} u^{(i,j)} = \sum_{i=0}^{j} u^{(i,j)}.
$$

$\qquad\square$

We are finally ready to prove Lemma 6.4.

Proof of Lemma 6.4. It holds $\chi_{0,0} = \phi_{0,0}$ and $\text{supp}\, \phi_{0,0} \cap \Omega \subset \Omega_0(-\eta)$. Moreover, from Proposition 6.5 (ii) we get that for $i \neq 0$ the function $u_{lm+1}^{(i)}$ vanishes on $\Omega_0(-\eta)$. And we also have $\phi_{0,-\frac{1}{3}} \equiv 1$ on $\Omega_0(-\eta)$. From these observations, one obtains

$$
\chi_{0,0}(u^{(0,0)} - u_{lm+1}^{(0)}) = \phi_{0,0}(\phi_{0,-\frac{1}{3}} u - u_{lm+1}^{(0)}) = \phi_{0,0}(u - u_{lm+1}).
$$

The statement for $j = 0$ now follows by the smoothness of $\phi_{0,0}$.

Let us now consider the case $j > 0$. By Proposition 6.4, it is

$$\operatorname{supp} \chi_{j,j} \subset \bigcup_{i=0}^{j} \Omega_i(-(1+3j)\eta) \subset \bigcup_{i=0}^{j} \Omega_i(-\eta), \tag{6.1.37}$$

as well as $\chi_{j,-\frac{1}{3}} \equiv 1$ on $\bigcup_{i=0}^{j} \Omega_i(-(2+3(-\frac{1}{3}))\eta) = \bigcup_{i=0}^{j} \Omega_i(-\eta)$, hence,

$$\chi_{j,j} = \chi_{j,j} \chi_{j,-\frac{1}{3}}. \tag{6.1.38}$$

Again Proposition 6.5 (ii) (second part) tells us that for $i > j$, $u_{lm+j+1}^{(i)}$ vanishes on $\bigcup_{i=0}^{j} \Omega_i(-\eta)$. Together with (6.1.37), (6.1.38), and Proposition 6.6, we get

$$\chi_{j,j}(u - u_{lm+j+1}) = \chi_{j,j}\chi_{j,-\frac{1}{3}}u - \chi_{j,j}u_{lm+j+1}$$

$$= \chi_{j,j}\sum_{i=0}^{j} u^{(i,j)} - \chi_{j,j}u_{lm+j+1} = \chi_{j,j}\sum_{i=0}^{j} u^{(i,j)} - \chi_{j,j}\sum_{i=0}^{m-1} u_{lm+j+1}^{(i)}$$

$$= \chi_{j,j}\sum_{i=0}^{j} u^{(i,j)} - \chi_{j,j}\sum_{i=0}^{j} u_{lm+j+1}^{(i)} = \chi_{j,j}\sum_{i=0}^{j}(u^{(i,j)} - u_{lm+j+1}^{(i)})$$

$$= \chi_{j,j}(u^{(j,j)} - u_{lm+j+1}^{(j)}) + \chi_{j,j}\sum_{i=0}^{j-1}(u^{(i,j)} - u_{lm+j+1}^{(i)}). \tag{6.1.39}$$

Note now that for $i < j$, $u^{(i,j)}$ and $u_{lm+j+1}^{(i)}$ vanish on $\Omega_j(-\eta)\cup\Omega_{j-1}(-\eta)\cup\cdots\cup\Omega_{i+1}(-\eta)$ by Proposition 6.5. We want to show next that

$$\chi_{j,j}(u_{lm+j+1}^{(i)} - u^{(i,j)}) = \chi_{i,j}(u_{lm+j+1}^{(i)} - u^{(i,j)}) \tag{6.1.40}$$

for those i. And, indeed, from the definition of $\chi_{j,j}$, we get $\chi_{j,j}(u_{lm+j+1}^{(i)} - u^{(i,j)}) = \phi_{j,j}(1 - \chi_{j-1,j})(u_{lm+j+1}^{(i)} - u^{(i,j)}) + \chi_{j-1,j}(u_{lm+j+1}^{(i)} - u^{(i,j)})$. Clearly, $\operatorname{supp}(\phi_{j,j}(1 - \chi_{j-1,j})) \subset \operatorname{supp} \phi_{j,j} \subset \Omega_j(-(1+3j)\eta) \subset \Omega_j(-\eta)$, on which $u_{lm+j+1}^{(i)} - u^{(i,j)}$ vanishes. Repeating this argument, we may infer (6.1.40). Using (6.1.40) and (6.1.39) yields

$$\chi_{j,j}(u^{(j,j)} - u_{lm+j+1}^{(j)}) = \chi_{j,j}(u - u_{lm+j+1}) + \sum_{i=0}^{j-1} \chi_{i,j}(u_{lm+j+1}^{(i)} - u^{(i,j)})$$

$$= \chi_{j,j}(u - u_{lm+j+1}) + \sum_{i=0}^{j-1} \chi_{i,j} \sum_{\lambda \in \Lambda_{i,j}} (\mathbf{u}_{lm+i+1,\lambda}^{(i)} - \mathbf{u}_\lambda^{(i,i)})\psi_{i,\lambda},$$

where $\Lambda_{i,j} := \{\lambda \in \Lambda_i : \operatorname{supp} \psi_{i,\lambda} \cap \Omega \not\subset \Omega_q, \ i+1 \le q \le j\}$ or a subset of that when in (6.1.32) additional coefficients are set to zero due to coarsening. We have

$\|\chi_{j,j}(u-u_{lm+j+1})\|_{H^t(\Omega_j)} \lesssim \|u-u_{lm+j+1}\|_{H^t(\Omega)}$ by the smoothness of $\chi_{j,j}$. Furthermore, using Proposition 6.4, and recalling (6.1.15), we detect that for $i < j$, we have

$$B(\operatorname{supp}\chi_{i,j};2\eta) \cap \Omega \subset \bigcup_{k=0}^{i} \Omega_k(-(1+3j)\eta+2\eta) = \bigcup_{k=0}^{i} \Omega_k(-(3j-1)\eta)$$

$$\subset \bigcup_{k=0}^{i} \Omega_k(-(2+3i)\eta), \tag{6.1.41}$$

where for the last step the fact that by $j > i$ it is $3j - 1 \geq 2 + 3i$ was used. By Proposition 6.4 (ii), $\chi_{i,i} \equiv 1$ on the last set in the chain in (6.1.41), thus also on $B(\operatorname{supp}\chi_{i,j};2\eta) \cap \Omega$. From this finding and because of $\operatorname{diam}(\operatorname{supp}\psi_{i,\lambda}) \leq \eta$, $\operatorname{diam}(\operatorname{supp}\tilde{\psi}_{i,\lambda}) \leq \eta$, we get

$$\chi_{i,j}(\mathbf{u}_{lm+i+1,\lambda}^{(i)} - \mathbf{u}_{\lambda}^{(i,i)})\psi_{i,\lambda} = \chi_{i,j}\langle\tilde{\psi}_{i,\lambda}, u_{lm+i+1}^{(i)} - u^{(i,i)}\rangle\psi_{i,\lambda}$$

$$= \chi_{i,j}\langle\tilde{\psi}_{i,\lambda}, \chi_{i,i}(u_{lm+i+1}^{(i)} - u^{(i,i)})\rangle\psi_{i,\lambda}.$$

Since $\chi_{i,j}$ is smooth, and $\Psi^{(i)}$ is a Riesz basis for $H_0^t(\Omega_i)$, we arrive at

$$\left\|\chi_{i,j}\sum_{\lambda\in\Lambda_{i,j}}(\mathbf{u}_{lm+i+1,\lambda}^{(i)} - \mathbf{u}_{\lambda}^{(i,i)})\psi_{i,\lambda}\right\|_{H^t(\Omega_i)}$$

$$\lesssim \left\|\sum_{\lambda\in\Lambda_{i,j}}\langle\tilde{\psi}_{i,\lambda}, \chi_{i,i}(u_{lm+i+1}^{(i)} - u^{(i,i)})\rangle\psi_{i,\lambda}\right\|_{H^t(\Omega_i)}$$

$$\approx \left(\sum_{\lambda\in\Lambda_{i,j}}|\langle\tilde{\psi}_{i,\lambda}, \chi_{i,i}(u_{lm+i+1}^{(i)} - u^{(i,i)})\rangle|^2\right)^{\frac{1}{2}}$$

$$\leq \left(\sum_{\lambda\in\Lambda_i}|\langle\tilde{\psi}_{i,\lambda}, \chi_{i,i}(u_{lm+i+1}^{(i)} - u^{(i,i)})\rangle|^2\right)^{\frac{1}{2}}$$

$$\approx \|\chi_{i,i}(u_{lm+i+1}^{(i)} - u^{(i,i)})\|_{H^t(\Omega_i)} \lesssim \sum_{q=0}^{i}\|u - u_{lm+q+1}\|_{H^t(\Omega)},$$

by the induction hypothesis. Since this holds for any $0 \leq i \leq j - 1$, the proof is completed. $\qquad\square$

6.1.6 Smoothness of the limits on the subdomains: Proof of Theorem 6.2 (b)

The intention of this section is to prove that if the solution u has a representation $u = \tilde{\mathbf{u}}^\top\Psi$ with $\tilde{\mathbf{u}} \in \ell_\tau^w(\Lambda)$, for some $s \in (0, \frac{d-t}{n})$, $\frac{1}{\tau} = s + \frac{1}{2}$, then, for each of the functions $u^{(i,j)} = (\mathbf{u}^{(i,j)})^T\Psi^{(i)} \in H_0^t(\Omega_i)$ from Definition 6.1, we have

$$|\mathbf{u}^{(i,j)}|_{\ell_\tau^w(\Lambda_i)} \lesssim |\tilde{\mathbf{u}}|_{\ell_\tau^w(\Lambda)}. \tag{6.1.42}$$

Since, in particular, this holds true for $u^{(m-1)} = u^{(m-1,m-1)}$, and analogously, as well for the limit functions on the other subdomains, we may then conclude part (b) of Theorem 6.2.

The key for a proof of the above result is a characterization of those functions $v = \mathbf{v}^\top \Psi^{(i)} \in H_0^t(\Omega_i)$ for which $\mathbf{v} \in \ell_\tau^w(\Lambda_i)$ in terms of membership of a non-standard smoothness space. This issue shall be examined first now.

Recall from (3.3.21) that for $s \in (0, \frac{d-t}{n})$, $sn + t - \frac{1}{\tau} \notin \{0, \ldots, d-2\}$, we have the characterization

$$\mathbf{v} \in \ell_\tau(\Lambda_i) \text{ if and only if } v = \mathbf{v}^T \Psi^{(i)} \in \mathring{B}_{\tau,\tau}^{sn+t}(\Omega_i) \tag{6.1.43}$$

with equivalent (quasi-)norms. For given $s \in (0, \frac{d-t}{n})$, let us select $s_0 < s < s_1$ with for $r = 0, 1$, $s_r \in (0, \frac{d-t}{n})$ and, with $\tau_r := (\frac{1}{2} + s_r)^{-1}$, $s_r n + t - \frac{1}{\tau_r} \notin \{0, \ldots, d-2\}$. Then from (3.3.13) and (6.1.43), using a classical interpolation argument (cf. [11, Chapter 2]), it follows that

$$\mathbf{v} \in \ell_\tau^w(\Lambda_i) \text{ if and only if } v = \mathbf{v}^\top \Psi^{(i)} \in \mathcal{B}^s(\Omega_i) := \left(\mathring{B}_{\tau_0,\tau_0}^{s_0 n + t}(\Omega_i), \mathring{B}_{\tau_1,\tau_1}^{s_1 n + t}(\Omega_i) \right)_{\theta,\infty}, \tag{6.1.44}$$

where $\frac{1}{\tau} - \frac{1}{\tau_0} = \theta(\frac{1}{\tau_1} - \frac{1}{\tau_0})$, with equivalent (quasi-)norms. Note that the interpolation space at the right-hand side is not a Besov space.

Since $s < \frac{d-t}{n}$ and $\frac{d-t}{n} \geq \frac{1}{2}$, it is $sn + t - \frac{1}{\tau} = sn + t - s - \frac{1}{2} = s(n-1) + t - \frac{1}{2} < \frac{d-t}{n}(n-1) + t - \frac{1}{2} = d - (\frac{d-t}{n} + \frac{1}{2}) \leq d - 1$, from Definition 3.3 (ii), we infer $\mathring{B}_{\tau,\tau}^{sn+t}(\Omega_i) \hookrightarrow \mathring{B}_{\tau,\tau}^{sn+t}(\Omega)$ by means of the zero extension. Again by an interpolation argument, we thus get the continuous embedding

$$\mathcal{B}^s(\Omega_i) \hookrightarrow \mathcal{B}^s(\Omega) := \left(\mathring{B}_{\tau_0,\tau_0}^{s_0 n + t}(\Omega), \mathring{B}_{\tau_1,\tau_1}^{s_1 n + t}(\Omega) \right)_{\theta,\infty}. \tag{6.1.45}$$

The verification of (6.1.42) will be based on two auxiliary results. The first building block is the following.

Lemma 6.5. *If the solution $u \in H_0^t(\Omega)$ has a representation $u = \tilde{\mathbf{u}}^\top \Psi$, with $\tilde{\mathbf{u}} \in \ell_\tau^w(\Lambda)$, for some $s \in (0, \frac{d-t}{n})$, $\frac{1}{\tau} = s + \frac{1}{2}$, then, for a function $\phi_i \in C_0^\infty(\mathbb{R}^n)$, with $\text{supp }\phi_i \cap \Omega \subset \Omega_i$, it is $\phi_i u = \mathbf{v}^\top \Psi^{(i)} \in \mathcal{B}^s(\Omega_i)$, with $\mathbf{v} \in \ell_\tau^w(\Lambda_i)$ and $|\mathbf{v}|_{\ell_\tau^w(\Lambda_i)} \lesssim |\tilde{\mathbf{u}}|_{\ell_\tau^w(\Lambda)}$.*

Proof. If $\tilde{\mathbf{u}} = (\tilde{\mathbf{u}}^{(0)}, \ldots, \tilde{\mathbf{u}}^{(m-1)}) \in \ell_\tau^w(\Lambda)$, then, clearly, $\tilde{\mathbf{u}}^{(i)} \in \ell_\tau^w(\Lambda_i)$, $i = 0, \ldots, m-1$, and $(\tilde{\mathbf{u}}^{(i)})^\top \Psi^{(i)} \in \mathcal{B}^s(\Omega_i)$ by (6.1.44), and $u \in \mathcal{B}^s(\Omega)$ by (6.1.45). Because of the smoothness of ϕ_i, the mapping

$$v \mapsto \phi_i v : \mathcal{B}^s(\Omega) \to \mathcal{B}^s(\Omega_i) \text{ is bounded.} \tag{6.1.46}$$

Therefore, it is $\phi_i u \in \mathcal{B}^s(\Omega_i)$, and thus, again using (6.1.44) and (6.1.45), we obtain

$$|\mathbf{v}|_{\ell^w_\tau(\Lambda_i)} \approx \|\phi_i u\|_{\mathcal{B}^s(\Omega_i)} \lesssim \|u\|_{\mathcal{B}^s(\Omega)} \lesssim \sum_{i=0}^{m-1} \|(\tilde{\mathbf{u}}^{(i)})^\top \Psi\|_{\mathcal{B}^s(\Omega_i)}$$

$$\approx \sum_{i=0}^{m-1} |\tilde{\mathbf{u}}^{(i)}|_{\ell^w_\tau(\Lambda_i)} \lesssim |\tilde{\mathbf{u}}|_{\ell^w_\tau(\Lambda)}.$$

$\square$

Remark 6.6. From (6.1.44), (6.1.45), (6.1.46), and the existence of a smooth partition of unity with respect to the covering $\{\Omega_i : 0 \le i \le m-1\}$ of Ω because of condition (6.1.16), it follows that for $s \in (0, \frac{d-t}{n})$, $u = \tilde{\mathbf{u}}^T \Psi$ with $\tilde{\mathbf{u}} \in \ell^w_\tau(\Lambda)$ if and only if $u \in \mathcal{B}^s(\Omega)$. We note that $B^{sn+t}_{\tau,\tau}(\Omega) \cap H^t_0(\Omega) = \mathring{B}^{sn+t}_{\tau,\tau}(\Omega) = \left(\mathring{B}^{s_0 n+t}_{\tau_0,\tau_0}(\Omega), \mathring{B}^{s_1 n+t}_{\tau_1,\tau_1}(\Omega) \right)_{\theta,\tau} \hookrightarrow \mathcal{B}^s(\Omega)$.

The second building block for the proof of (6.1.42) is presented in the following lemma.

Lemma 6.6. *If, for some* $s \in (0, \frac{d-t}{n})$, $\mathbf{v} \in \ell^w_\tau(\Lambda_j)$, *then*

$$v_{i,j} := \sum_{\{\lambda \in \Lambda_j : \operatorname{supp} \psi_{j,\lambda} \cap \Omega \subset \Omega_i\}} \mathbf{v}_\lambda \psi_{j,\lambda} \in \mathcal{B}^s(\Omega_i) \tag{6.1.47}$$

with its (quasi-)norm bounded by some multiple of $|\mathbf{v}|_{\ell^w_\tau(\Lambda_j)}$.

In view of Definition 6.1, using Lemma 6.5 and Lemma 6.6, a straightforward inductive argument finally proves (6.1.42), and thus Theorem 6.2 (b).

Proof of Lemma 6.6. We will make use of the third characterization of the space $\mathring{B}^\alpha_{p,q}(\Omega_i)$ given in Definition 3.3 (iii). Let $s \in (0, \frac{d-t}{n})$. Then, $sn + t - \frac{1}{\tau} = t + s(n-1) - \frac{1}{2} > t - 1$. When $sn + t - \frac{1}{\tau} \notin \{0, \dots, d-2\}$, select $s_0 < s < s_1$, with $s_r \in (0, \frac{d-t}{n})$, $s_r n + t - \frac{1}{\tau_r} \notin \{0, \dots, d-2\}$, $r = 0, 1$, and $\lceil s_0 n + t - \frac{1}{\tau_0} \rceil = \lceil s_1 n + t - \frac{1}{\tau_1} \rceil$. The characterization of $\mathring{B}^\nu_{p,q}(\Omega_i)$ in Definition 3.3 (iii) now shows that $\mathring{B}^{s_1 n+t}_{\tau_1,\tau_1}(\Omega_i) = B^{s_1 n+t}_{\tau_1,\tau_1}(\Omega_i) \cap \mathring{B}^{s_0 n+t}_{\tau_0,\tau_0}(\Omega_i)$, and so

$$\mathcal{B}^s(\Omega_i) = \left(B^{s_0 n+t}_{\tau_0,\tau_0}(\Omega_i) \cap \mathring{B}^{s_0 n+t}_{\tau_0,\tau_0}(\Omega_i), B^{s_1 n+t}_{\tau_1,\tau_1}(\Omega_i) \cap \mathring{B}^{s_0 n+t}_{\tau_0,\tau_0}(\Omega_i) \right)_{\theta,\infty}$$

$$= \left(B^{s_0 n+t}_{\tau_0,\tau_0}(\Omega_i), B^{s_1 n+t}_{\tau_1,\tau_1}(\Omega_i) \right)_{\theta,\infty} \cap \mathring{B}^{s_0 n+t}_{\tau_0,\tau_0}(\Omega_i)$$

$$= \Big\{ v \in \left(B^{s_0 n+t}_{\tau_0,\tau_0}(\Omega_i), B^{s_1 n+t}_{\tau_1,\tau_1}(\Omega_i) \right)_{\theta,\infty} :$$

$$\gamma_l v = 0 \text{ on } \partial\Omega \cap \partial\Omega_i \text{ for } l = 0, \dots, t-1, \tag{6.1.48}$$

$$\gamma_l v = 0 \text{ on } \partial\Omega_i \cap \Omega \text{ for } l = 0, \dots, \min(d-2, \lfloor sn + t - \tfrac{1}{\tau} \rfloor) \Big\}. \tag{6.1.49}$$

Now let $\mathbf{v} \in \ell_\tau^w(\Lambda_j)$, then $v_{i,j} \in \mathcal{B}^s(\Omega)$. Since $\mathcal{B}^s(\Omega)$ is continuously embedded in $\left(B_{\tau_0,\tau_0}^{s_0 n+t}(\Omega), B_{\tau_1,\tau_1}^{s_1 n+t}(\Omega)\right)_{\theta,\infty}$, and the restriction of functions to Ω_i is bounded from $B_{\tau_i,\tau_i}^{s_1 n+t}(\Omega)$ to $B_{\tau_i,\tau_i}^{s_1 n+t}(\Omega_i)$, we conclude that $v_{i,j} \in \left(B_{\tau_0,\tau_0}^{s_0 n+t}(\Omega_i), B_{\tau_1,\tau_1}^{s_1 n+t}(\Omega_i)\right)_{\theta,\infty}$ with (quasi-)norm bounded by some multiple of $|\mathbf{v}|_{\ell_\tau^w(\Lambda_j)}$. Note that the trace operators in (6.1.48) and (6.1.49) applied to $v_{i,j}$ indeed vanish. Hence, we have shown the assertion except for some isolated values of $s \in (0, \frac{d-t}{n})$. But since the mapping $\mathbf{v} \mapsto v_{i,j}$ is linear, the statement for any $s \in (0, \frac{d-t}{n})$ follows by an interpolation argument. $\square$

6.2 An additive Schwarz adaptive wavelet frame method

Next, we consider the *additive* Schwarz method. In contrast to the subsequent computation of local updates on Ω_i, within the additive scheme, these can be performed in parallel.

6.2.1 The exact additive Schwarz method

With a relaxation parameter $\alpha > 0$, the exact iterative scheme in $H_0^t(\Omega)$ reads as follows.

Algorithm 8. *Additive Schwarz method.*

$u_0 = 0$

for $k = 1, 2, \ldots,$

  $u_k := u_{k-1} + \alpha \sum_{i=0}^{m-1} e_{k-1}^{(i)}$,

  where, for $i = 0, \ldots, m-1$, $e_{k-1}^{(i)}$ is the solution of the problem

  $a(e_{k-1}^{(i)}, v) = f(v) - a(u_{k-1}, v), \quad v \in H_0^t(\Omega_i)$.

endfor

Obviously, by (6.1.2), it is

$$\|u - u_k\| = \left\| u - u_{k-1} - \alpha \sum_{i=0}^{m-1} e_{k-1}^{(i)} \right\| = \left\| u - u_{k-1} - \alpha \sum_{i=0}^{m-1} P_i(u - u_{k-1}) \right\|$$

$$\leq \left\| I - \alpha \sum_{i=0}^{m-1} P_i \right\| \|u - u_{k-1}\|. \tag{6.2.1}$$

If α is chosen sufficiently small, Algorithm 8 generates a linearly convergent sequence $\{u_k\}_{k \in \mathbb{N}_0} \subset H_0^t(\Omega)$.

Theorem 6.5. *The operator* $T := \sum_{i=0}^{m-1} P_i : H_0^t(\Omega) \to H_0^t(\Omega)$ *is* $a(\cdot,\cdot)$-*symmetric and positive definite. Furthermore, Algorithm 8 converges linearly in the energy norm* $\|\cdot\|$ *with the rate*

$$\rho := \left\| I - \alpha \sum_{i=0}^{m-1} P_i \right\|, \tag{6.2.2}$$

for $\alpha \in (0, 2/\lambda_{\max}(T))$, *where* $\lambda_{\max}(T)$ *denotes the largest eigenvalue of* T. *The optimal relaxation parameter is given by* $\alpha_{opt} = 2/(\lambda_{\min}(T) + \lambda_{\max}(T))$, *where* $\lambda_{\min}(T)$ *represents the smallest eigenvalue of* T.

Proof. Let, for $i = 0, \ldots, m-1$, $\mathcal{L}_i : H_0^t(\Omega_i) \to H^{-t}(\Omega_i)$ be defined by $\mathcal{L}_i := E_i'\mathcal{L}E_i$, E_i being the continuous embedding of $H_0^t(\Omega_i)$ into $H_0^t(\Omega)$ by means of the zero extension and E_i' the associated dual operator. As $\mathcal{L}$ itself, the operators $\mathcal{L}_i$ are boundedly invertible, positive definite and symmetric operators. Using this notation, from (6.1.2) we learn that P_i can be written as $P_i = E_i\mathcal{L}_i^{-1}E_i'\mathcal{L}$. Therefore, for $v, w \in H_0^t(\Omega)$, it holds

$$\begin{aligned}
a(Tv, w) &= \sum_{i=0}^{m-1} a(P_i v, w) = \sum_{i=0}^{m-1} \langle \mathcal{L}E_i\mathcal{L}_i^{-1}E_i'\mathcal{L}v, w\rangle_{H^{-t}(\Omega)\times H_0^t(\Omega)} \\
&= \sum_{i=0}^{m-1} \langle \mathcal{L}v, E_i\mathcal{L}_i^{-1}E_i'\mathcal{L}w\rangle_{H^{-t}(\Omega)\times H_0^t(\Omega)} = \sum_{i=0}^{m-1} a(v, E_i\mathcal{L}_i^{-1}E_i'\mathcal{L}w) \\
&= a(v, Tw).
\end{aligned}$$

Hence, T and thus also $I - \alpha T$ is $a(\cdot,\cdot)$-symmetric. Moreover, we clearly have

$$\begin{aligned}
a(Tv, v) &= \sum_{i=0}^{m-1} \langle \mathcal{L}E_i\mathcal{L}_i^{-1}E_i'\mathcal{L}v, v\rangle_{H^{-t}(\Omega)\times H_0^t(\Omega)} \\
&= \sum_{i=0}^{m-1} \langle \mathcal{L}_i^{-1}E_i'\mathcal{L}v, E_i'\mathcal{L}v\rangle_{H_0^t(\Omega_i)\times H^{-t}(\Omega_i)}, \quad v \in H_0^t(\Omega),
\end{aligned}$$

hence, T is positive definite. We may thus infer $\|I - \alpha T\| = \max\{\alpha\lambda_{\max}(T) - 1, 1 - \alpha\lambda_{\min}(T)\}$, from which the assertion follows easily using (6.2.1). $\square$

The reason why this algorithm is so desirable is the strict parallelism with which the updates on the subdomains can be performed. But, unfortunately, at first glance this seems to prohibit the incorporation of a similar strategy for the reduction of coefficients in the overlapping regions as for the multiplicative method. In fact, in **MultSchw**, before a local solve on Ω_i, the contributions from the other patches Ω_j, $j \neq i$, are potentially modified, and the solution of the local problem on Ω_i then depends on this modification on Ω_j, $j \neq i$. In the following, we introduce an appropriate reformulation of Algorithm 8 with which we shall show afterwards that the above mentioned strategies can indeed be realized in the additive case as well.

Algorithm 9. *Reformulated additive Schwarz method.*

$u_0 := 0$, $\bar{u}_0 := 0$, $k := 0$

for $l = 1, 2, \ldots,$

    for $i = 0, \ldots, m - 1$

      $k := (l-1)m + i + 1$

      Determine $e_{k-1} \in H_0^t(\Omega_i)$ as the solution of the problem

      $a(e_{k-1}, v) = f(v) - a(\bar{u}_{(l-1)m}, v), \quad v \in H_0^t(\Omega_i).$

      $u_k := \bar{u}_{(l-1)m} + m\alpha e_{k-1}$

    endfor

    $\bar{u}_k := \frac{1}{m} \sum_{j=0}^{m-1} u_{k-j}$

endfor

Lemma 6.7. *The iterates $\bar{u}_{lm}$, $l \in \mathbb{N}_0$, produced by Algorithm 9, coincide with the iterates u_l, $l \in \mathbb{N}_0$, produced by Algorithm 8.*

Proof. The case $l = 0$ is clear. Let us suppose that for $l - 1 \in \mathbb{N}_0$ the statement was true. For an iterate $\bar{u}_k$ of Algorithm 9, we have

$$\bar{u}_{lm} = \frac{1}{m} \sum_{j=0}^{m-1} u_{lm-j} = \frac{1}{m} \sum_{j=0}^{m-1} (\bar{u}_{(l-1)m} + m\alpha e_{lm-j-1})$$

$$= \bar{u}_{(l-1)m} + \frac{1}{m} \sum_{j=0}^{m-1} m\alpha e_{(l-1)m+m-(j+1)}$$

$$= \bar{u}_{(l-1)m} + \alpha \sum_{j=0}^{m-1} e_{(l-1)m+(j+1)-1}. \tag{6.2.3}$$

Now, for $j = 0, \ldots, m - 1$, $a(e_{(l-1)m+(j+1)-1}, v) = f(v) - a(\bar{u}_{(l-1)m}, v)$, $v \in H_0^t(\Omega_j)$, thus $e_{(l-1)m+(j+1)-1} = e_{l-1}^{(j)}$, because $\bar{u}_{(l-1)m}$ coincides with u_{l-1} from Algorithm 8 by induction hypothesis. Hence, the assertion follows from (6.2.3). $\qquad\square$

Corollary 6.1. *For each fixed $i \in \{0, \ldots, m - 1\}$, the sequences $\{u_{lm+i+1}\}_{l\in\mathbb{N}_0}$, as well as the sequence $\{\bar{u}_{lm}\}_{l\in\mathbb{N}_0}$, produced by Algorithm 9, converge in the energy norm to the solution $u \in H_0^t(\Omega)$.*

Proof. The convergence of the sequence $\{\bar{u}_{lm}\}_{l\in\mathbb{N}_0}$ is clear by Lemma 6.7 and Theorem 6.5. Let now $i \in \{0, \ldots, m - 1\}$ be fixed. Then, $\|u - u_{lm+i+1}\| = \|u - \bar{u}_{lm} - m\alpha e_{lm+i}\| \leq \|I - m\alpha P_i\| \|u - \bar{u}_{lm}\|$, so that the proof is completed by the convergence of the sequence $\{\bar{u}_{lm}\}_{l\in\mathbb{N}_0}$. $\qquad\square$

A statement similar to Lemma 6.1 can also be shown to hold.

Lemma 6.8. *If we compute in Algorithm 9 for $k \in \mathbb{N}$, $i = (k-1) \bmod m$,*

$$\tilde{u}_k := (\bar{u}_{(l-1)m} - w_k) + \alpha m \tilde{e}_{k-1}, \text{ where } w_k \in H_0^t(\Omega_i) \text{ and}$$

$$a(\tilde{e}_{k-1}, v) = f(v) - a\Big(\bar{u}_{(l-1)m} - \frac{1}{m\alpha} w_k, v\Big), \quad v \in H_0^t(\Omega_i),$$

then, it is $\tilde{u}_k = u_k$.

Proof. The assertion can be easily verified using the same argument as in the proof of Lemma 6.1. □

6.2.2 The adaptive method and its convergence

Having verified the convergence of Algorithm 9, with Lemma 6.8 we are now in a position to formulate an adaptive additive Schwarz method reading as follows.

Algorithm 10. AddSchw$[\varepsilon, \mu] \to \mathbf{u}_\varepsilon$

% *The input should satisfy $\varepsilon > 0$ and $\mu \geq \|u\|$.*

% *With M being an upper bound for $\|\mathbf{A}\|_{\ell_2(\Lambda) \to \ell_2(\Lambda)}^{\frac{1}{2}}$, let the parameters*

% *$\sigma > 0$ and $K \in \mathbb{N}$ be such that $2\rho^K M\sigma < 1$, with ρ from (6.2.2).*

% *Let L be the smallest integer with $\mu(2\rho^K M\sigma)^L \leq \varepsilon$.*

$u_0 = u_0^{(0)} = \cdots = u_0^{(m-1)} := 0, \ k := 0, \ \bar{u}_0 := 0$

for $l = 1, 2, \ldots,$

  for $p = 1, \ldots, K$

    for $i = 0, \ldots, m-1$

      $k := (l-1)mK + (p-1)m + i + 1$

      Determine $u_k = \sum_{j=0}^{m-1} u_k^{(j)}$, $u_k^{(j)} = (\mathbf{u}_k^{(j)})^T \Psi^{(j)}$ *(i.e., $u_k = \mathbf{u}_k^T \Psi$)*

      where $\mathbf{u}_k = (\mathbf{u}_k^{(0)}, \ldots, \mathbf{u}_k^{(m-1)}))$ *as follows:*

      For $j \neq i$, $\mathbf{u}_{k,\lambda}^{(j)} := \begin{cases} 0 & \text{when } \operatorname{supp} \psi_{j,\lambda} \cap \Omega \subset \Omega_i, \\ \bar{\mathbf{u}}_{(l-1)mK+(p-1)m,\lambda}^{(j)} & otherwise. \end{cases}$

      For $j \neq i$, $\mathbf{w}_k^{(j)} := \bar{\mathbf{u}}_{(l-1)mK+(p-1)m}^{(j)} - \mathbf{u}_k^{(j)}$, $\quad w_k^{(j)} := (\mathbf{w}_k^{(j)})^\top \Psi^{(j)}$.

      $u_k^{(i)} := m\alpha \tilde{u}_k^{(i)}$, *where $\tilde{u}_k^{(i)}$ is an approximation to the solution*

      of the problem

        $a(\breve{u}_k^{(i)}, v^{(i)})$

        $= f(v^{(i)}) - a(\bar{u}_{(l-1)mK+(p-1)m} - \frac{1}{m\alpha}(\tilde{u}_{(l-1)mK+(p-1)m}^{(i)} + \sum_{j \neq i} w_k^{(j)}), v^{(i)})$,

        $v^{(i)} \in H_0^t(\Omega_i)$ *such that* $\|\breve{u}_k^{(i)} - \tilde{u}_k^{(i)}\| \leq \mu(2\rho^K M\sigma)^{l-1} \frac{\rho^p}{\alpha mK}$

    endfor

    $\bar{u}_k := \frac{1}{m} \sum_{j=0}^{m-1} u_{k-j} =: \bar{\mathbf{u}}_k^\top \Psi$, *and* $\bar{\mathbf{u}}_k := (\bar{\mathbf{u}}_k^{(0)}, \ldots, \bar{\mathbf{u}}_k^{(m-1)})$, $\bar{u}_k^{(i)} := (\bar{\mathbf{u}}_k^{(i)})^\top \Psi^{(i)}$

  endfor

  $\bar{\mathbf{u}}_k := \mathbf{COARSE}[\bar{\mathbf{u}}_k, (\sigma - 1/M)2\rho^K \mu(2\rho^K M\sigma)^{l-1}]$

endfor

$\mathbf{u}_\varepsilon := \bar{\mathbf{u}}_k$

Note again that for $k \in mK\mathbb{N}$, $\bar{\mathbf{u}}_k$ and $\bar{u}_k$ got redefined by the application of **COARSE**, so that we will have to carefully distinguish between the versions before and after the thresholding.

The principle structure of this method coincides with the one of **MultSchw**. A few comments on the special construction of the intermediate iterates $u_{(l-1)mK+(p-1)m+i+1}$, $i = 0, \ldots, m-1$, and their properties are in order.

Basically, Lemma 6.8 is used with $w_k := \bar{u}_{(l-1)mK+(p-1)m}^{(i)} + \sum_{j \neq i} w_k^{(j)}$. Here $w_k^{(j)} =$

$(\mathbf{w}_k^{(j)})^\top \Psi^{(j)}$, where for $j \neq i$ $\mathbf{w}_k^{(j)}$ consists of all coefficients $\bar{\mathbf{u}}_{(l-1)mK+(p-1)m,\lambda}^{(j)}$ such that $\operatorname{supp} \psi_{j,\lambda} \cap \Omega \subset \Omega_i$. Hence, after solving the local elliptic problem, the frame expansion of $u_{(l-1)mK+(p-1)m+i+1}$ on Ω_i is made up of wavelets $\psi_{i,\lambda}$ and $\psi_{j,\mu}$, $j \neq i$, the latter intersecting $\partial\Omega_i \cap \Omega$. Note also that in case $m\alpha = 1$ in the right-hand side of the local problem we have

$$\bar{u}_{(l-1)mK+(p-1)m} - \left(\bar{u}_{(l-1)mK+(p-1)m}^{(i)} + \sum_{j\neq i} w_k^{(j)}\right) = \sum_{j\neq i}\left(\bar{u}_{(l-1)mK+(p-1)m}^{(j)} - w_k^{(j)}\right)$$

$$= \sum_{j\neq i}(\bar{\mathbf{u}}_{(l-1)mK+(p-1)m}^{(j)} - \mathbf{w}_k^{(j)})^\top \Psi^{(j)} = \sum_{j\neq i}\mathbf{u}_{(l-1)mK+(p-1)m+i+1}^{(j)}$$

$$= \sum_{j\neq i}\mathbf{u}_k^{(j)},$$

where the vectors $\mathbf{u}_{(l-1)mK+(p-1)m+i+1}^{(j)}$ are created by dropping entries from $\bar{\mathbf{u}}_{(l-1)mK+(p-1)m}^{(j)}$. This means that in this case the right-hand side for the local problem is constructed similar to **MultSchw**. Therefore, with $m\alpha = 1$, for the approximation of this right-hand side, again only integrals $a(\psi_{i,\lambda}, \psi_{j,\mu})$, $(\operatorname{supp}\psi_{j,\mu} \setminus \partial\operatorname{supp}\psi_{j,\mu}) \cap (\partial\Omega_i \cap \Omega) \neq \emptyset$, have to be computed; recall Remark 6.4. We shall see in Chapter 7 that $\alpha = \frac{1}{m}$ often is an admissible and also convenient choice, although in general it might not be the optimal one.

The convergence of **AddSchw** is clarified by the following proposition.

Proposition 6.7. *Let $\mu \geq \|u\|$. The iterand $\bar{u}_{lmK}$ inside $\mathbf{u}_\varepsilon = \mathbf{AddSchw}[\varepsilon, \mu]$, for $l > 0$, after the application of* **COARSE**, *satisfies*

$$\|u - \bar{u}_{lmK}\| \leq \mu(2\rho^K M\sigma)^l, \tag{6.2.4}$$

and so $\|u - \mathbf{u}_\varepsilon^T\Psi\| \leq \varepsilon$. Furthermore, for $1 \leq q \leq mK$,

$$\|u - u_{(l-1)mK+q}\| \leq 2\mu(2\rho^K M\sigma)^{l-1}\rho^{\frac{q}{m}-1}. \tag{6.2.5}$$

Proof. The first statement is valid for $l = 0$ by $\|u\| \leq \mu$. Suppose it is valid for some $l - 1 \geq 0$. Then for $p = 0, \ldots, K$,

$$\|u - \bar{u}_{(l-1)mK+pm}\| \leq \mu\left(1 + \frac{p}{K}\right)\rho^p(2\rho^K M\sigma)^{l-1}, \tag{6.2.6}$$

where, when $p = K$, $\bar{u}_{lmK}$ should read here as the iterand *before* the application of **COARSE**. Indeed, this estimate is valid for $p = 0$ by the above induction hypothesis. Suppose (6.2.6) is valid for some $p - 1 \geq 0$. Note that by Lemma 6.8, for the function $u_{(l-1)mK+pm}^\dagger$ resulting from an exact step of the additive Schwarz method starting from $\bar{u}_{(l-1)mK+(p-1)m}$, we have

$$u_{(l-1)mK+pm}^\dagger = \frac{1}{m}\sum_{i=0}^{m-1}\left(\bar{u}_{(l-1)mK+(p-1)m} + m\alpha P_i(u - \bar{u}_{(l-1)mK+(p-1)m})\right)$$

$$= \frac{1}{m}\sum_{i=0}^{m-1}\left(\left(\sum_{j\neq i}u_{(l-1)mK+(p-1)m+i+1}^{(j)}\right) + m\alpha\breve{u}_{(l-1)mK+(p-1)m+i+1}^{(i)}\right).$$

Note that, in addition, we have

$$\bar{u}_{(l-1)mK+pm} = \frac{1}{m}\sum_{i=0}^{m-1}\left(\left(\sum_{j\neq i}u_{(l-1)mK+(p-1)m+i+1}^{(j)}\right) + m\alpha\tilde{u}_{(l-1)mK+(p-1)m+i+1}^{(i)}\right).$$

Consequently,

$$\|u_{(l-1)mK+pm}^{\dagger} - \bar{u}_{(l-1)mK+pm}\|$$

$$\leq \frac{m\alpha}{m}\sum_{i=0}^{m-1}\|\tilde{u}_{(l-1)mK+(p-1)m+i+1}^{(i)} - \breve{u}_{(l-1)mK+(p-1)m+i+1}^{(i)}\|$$

$$\leq \frac{m\alpha}{m}m\mu(2\rho^K M\sigma)^{l-1}\frac{\rho^p}{\alpha mK} = \mu(2\rho^K M\sigma)^{l-1}\frac{\rho^p}{K}.$$

From this we learn with the aid of (6.2.6) for $p-1$ that it is

$$\|u - \bar{u}_{(l-1)mK+pm}\| \leq \|u - u_{(l-1)mK+pm}^{\dagger}\| + \|u_{(l-1)mK+pm}^{\dagger} - \bar{u}_{(l-1)mK+pm}\|$$

$$\leq \rho\|u - \bar{u}_{(l-1)mK+(p-1)m}\| + \|u_{(l-1)mK+pm}^{\dagger} - \bar{u}_{(l-1)mK+pm}\|$$

$$< \rho\left[\mu(1+\tfrac{p-1}{K})\rho^{p-1}(2\rho^K M\sigma)^{l-1}\right] + \mu(2\rho^K M\sigma)^{l-1}\tfrac{\rho^p}{K},$$

which is (6.2.6). By (6.2.6) for $p = K$, and the fact that $\|\mathbf{v}^T\Psi\| \leq M\|\mathbf{v}\|_{\ell_2(\Lambda)}$, $\mathbf{v} \in \ell_2(\Lambda)$, by definition of M in Algorithm 10 (cf. (6.1.7)), the subsequent application of **COARSE** introduces an additional error that in $\|\cdot\|$ is not larger than $M(\sigma - 1/M)2\rho^K\mu(2\rho^K M\sigma)^{l-1}$, the proof of the first statement is completed.

Making again use of the fact that by Lemma 6.8 it is

$$u_{(l-1)mK+(p-1)m+i+1}^{\dagger\dagger} := \bar{u}_{(l-1)mK+(p-1)m} + m\alpha P_i(u - \bar{u}_{(l-1)mK+(p-1)m})$$

$$= \left(\sum_{j\neq i}u_{(l-1)mK+(p-1)m+i+1}^{(j)}\right) + m\alpha\breve{u}_{(l-1)mK+(p-1)m+i+1}^{(i)},$$

from (6.2.6) with p reading as $p-1$ we get

$$\|u - u_{(l-1)mK+(p-1)m+(i+1)}\|$$

$$\leq \|u - u_{(l-1)mK+(p-1)m+i+1}^{\dagger\dagger}\| + \|u_{(l-1)mK+(p-1)m+i+1}^{\dagger\dagger} - u_{(l-1)mK+(p-1)m+(i+1)}\|$$

$$\leq \|(I - m\alpha P_i)(u - \bar{u}_{(l-1)mK+(p-1)m})\| + m\alpha\mu(2\rho^K M\sigma)^{l-1}\frac{\rho^p}{\alpha mK}$$

$$\leq [1 + \tfrac{p-1}{K} + \tfrac{\rho}{K}]\rho^{p-1}\mu(2\rho^K M\sigma)^{l-1}. \tag{6.2.7}$$

For the latter estimate we assume that $\|I - m\alpha P_i\| \leq 1$, which can be achieved by choosing $\alpha \leq 2/m$; recall that P_i is $a(\cdot,\cdot)$-symmetric and an $a(\cdot,\cdot)$-orthogonal projector. Note now that $1 + \tfrac{p-1+\rho}{K} \leq 1 + \tfrac{K-1+\rho}{K} \leq 2$. For $q = (p-1)m + (i+1)$, from (6.2.7), we get

$$\|u - u_{(l-1)mK+q}\| \leq 2\rho^{(q-(i+1))/m}\mu(2\rho^K M\sigma)^{l-1} \leq 2\mu(2\rho^K M\sigma)^{l-1}\rho^{\frac{q}{m}-1},$$

completing the proof of the second statement. $\qquad\square$

6.2.3 Optimality

It would clearly be desirable to prove optimality of **AddSchw** under the same assumptions as in Theorem 6.3. We will see that this aim can be realized for the case of two patches, $m = 2$. For the general case, we propose to resort to the technique used in §4.3.6, i.e., the incorporation of a projection step to control the kernel components. For an explanation, let us consider the decomposition depicted in Figure 6.3. There,

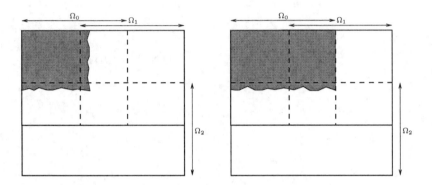

Figure 6.3: A domain covering with three subdomains. Left: The highlighted area represents the support of the contribution from Ω_0 when in **MultSchw** a local problem on Ω_2 has just been solved. Right: The same for **AddSchw**.

a domain covering consisting of three patches is shown. Now, when in **MultSchw** a local problem has been solved on Ω_2, from Ω_0 only wavelets $\psi_{0,\lambda}$ contribute to the global approximation u_k the supports of which are neither fully contained in Ω_1 nor in Ω_2. This is caused by the fact that, after the solve on Ω_0, the contribution from $\Psi^{(0)}$ is subsequently thinned out *twice*. The left picture thus highlights the support of the contribution from Ω_0. The algorithm **AddSchw** is addressed in the right picture. Unlike the multiplicative case, after the solve on Ω_2, also wavelets $\psi_{0,\lambda}$ may appear in $u^{(0)}_{(l-1)3K+(p-1)3+2+1}$ for which supp $\psi_{0,\lambda} \cap \Omega \subset \Omega_1$, because it is obtained by just *once* dropping entries from $\bar{u}^{(0)}_{(l-1)3K+(p-1)3}$ that correspond to wavelets being fully supported in Ω_2. And the wavelets making up $\bar{u}^{(0)}_{(l-1)3K+(p-1)3}$ are, in general, spread over all Ω_0. That means that in the interior of $(\Omega_0 \cap \Omega_1) \setminus \Omega_2$ the frame representation is not guaranteed to be as economic as for **MultSchw**, in the sense that wavelets from Ω_0 *and* Ω_1 contribute, whereas for our multiplicative method predominantly the wavelets from Ω_1 are used to represent the solution. The proof of Theorem 6.2 (a) is essentially based on this special structure.

Consequently, it does not seem to be easy to identify local limit functions for **AddSchw** with $m > 3$.

The case of two subdomains, $m = 2$

The basic ingredient for the proof of optimality of **MultSchw** has been Theorem 6.2. The key question is now whether a similar result can be shown for **AddSchw** as well. And, indeed, in this subsection it will be proved that for the case of two patches, i.e., $\Omega = \Omega_0 \cup \Omega_1$, the intermediate iterates $u_{(l-1)mK+(p-1)m+i+1}$, generated within **AddSchw**, are composed of functions $u^{(i)}_{(l-1)mK+(p-1)m+i+1}$, $i = 0, 1$, which form convergent sequences in $H_0^t(\Omega_i)$. Furthermore, it turns out that the local limit functions coincide with those we have identified for **MultSchw**.

To this end, let us assume (6.1.16) and (6.1.17), and let us consider an *odd* $k = (l-1)2K + (p-1)2 + 1 =: (l-1)2K + q$ and $\lambda \in \Lambda_0$ such that $\operatorname{supp} \psi_{0,\lambda} \cap \Omega \not\subset \Omega_1$. With the same argument as in §6.1.5 (case $m = 2$), it follows that

$$\left| \mathbf{u}^{(0)}_{k,\lambda} - \langle \tilde{\psi}_{0,\lambda}, u \rangle \right| = \left| \langle \tilde{\psi}_{0,\lambda}, u_k^{(0)} - u \rangle \right| = \left| \langle \tilde{\psi}_{0,\lambda}, u_k - u_k^{(1)} - u \rangle \right| = \left| \langle \tilde{\psi}_{0,\lambda}, u_k - u \rangle \right|$$
$$\lesssim \| u_k - u \| \leq 2\mu (2\rho^K M \sigma)^{l-1} \rho^{\frac{q}{2}-1}. \tag{6.2.8}$$

Thus, the sequence $\{\mathbf{u}^{(0)}_{k,\lambda}\}_{k \text{ odd}}$ converges to $\langle \tilde{\psi}_{0,\lambda}, u \rangle$ for k odd $\to \infty$.

For *even* $k = (l-1)2K + (p-1)2 + 2$ and $\lambda \in \Lambda_0$ such that $\operatorname{supp} \psi_{0,\lambda} \cap \Omega \not\subset \Omega_1$, it is

$$\mathbf{u}^{(0)}_{k,\lambda} = \frac{1}{2} \left(\mathbf{u}^{(0)}_{k-2,\lambda} + \mathbf{u}^{(0)}_{k-3,\lambda} \right), \quad \mathbf{u}^{(0)}_{0,\lambda} = \mathbf{u}^{(0)}_{2,\lambda} = 0, \quad k \text{ even} \geq 4. \tag{6.2.9}$$

The latter argument is further explained by Figure 6.4. There, the potential sparsity patterns of the different vectors in **AddSchw** are depicted. The vector $\bar{\mathbf{u}}_k$ is a combination of the vectors $\mathbf{u}_{k-1}$ and $\mathbf{u}_k$ which exhibit the same pattern as the iterates in **MultSchw**. That means, for instance, that $\mathbf{u}_{k-1}$ has zero coefficients at those locations where $\operatorname{supp} \psi_{1,\lambda} \cap \Omega \subset \Omega_0$. It is highlighted in Figure 6.4 that the coefficients $\mathbf{u}^{(0)}_{k,\lambda}$ with $\operatorname{supp} \psi_{0,\lambda} \cap \Omega \not\subset \Omega_1$ are just copied from the vector $\bar{\mathbf{u}}_{k-2}$ which in turn are the average of a convergent component (filled black box, cf. (6.2.8)) and the coefficients $\mathbf{u}_{k-2,\lambda}$, $\operatorname{supp} \psi_{0,\lambda} \cap \Omega \not\subset \Omega_1$.

Now, from (6.2.9), using Lemma 6.9 below, one may also deduce

$$\lim_{k \text{ even} \to \infty} \mathbf{u}^{(0)}_{k,\lambda} = \langle \tilde{\psi}_{0,\lambda}, u \rangle.$$

Hence, we may conclude from $u_k = u_k^{(0)} + u_k^{(1)} \to u$, k even $\to \infty$, that

$$u_k^{(1)} \to u - \sum_{\{\lambda \in \Lambda_0 : \operatorname{supp} \psi_{0,\lambda} \cap \Omega \not\subset \Omega_1\}} \langle \tilde{\psi}_{0,\lambda}, u \rangle \psi_{0,\lambda} \in H_0^t(\Omega_1), \quad k \text{ even} \to \infty. \tag{6.2.10}$$

That means we have verified the following result.

Proposition 6.8. *For $m = 2$, the sequences $\{u_k^{(i)}\}_{k \in i+1+m\mathbb{N}_0}$, generated by the algorithm **AddSchw**, converge in $H_0^t(\Omega_i)$ to exactly the same limit functions $u^{(i)} \in H_0^t(\Omega_i)$, $i = 0, 1$, that have been identified for **MultSchw** in §6.1.5.*

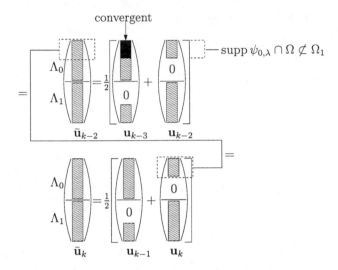

Figure 6.4: Schematic representation of the argument used in (6.2.9) for even k.

Lemma 6.9. *Let $\{b_k\}_{k\in\mathbb{N}_0}$ be a convergent sequence of real numbers with*

$$|b_k - b| \leq C\xi^k, \quad k \geq 0, \tag{6.2.11}$$

for the limit $b \in \mathbb{R}$ and some $\xi \in (0,1)$. Let the sequence $\{a_k\}_{k\in\mathbb{N}_0}$ be recursively defined by

$$a_{k+1} := \frac{1}{2}(a_k + b_k), \quad k \geq 0, \quad a_0 := 0. \tag{6.2.12}$$

Then, the sequence $\{a_k\}_{k\in\mathbb{N}_0}$ is also convergent, and its limit is b.

Proof. By (6.2.12), it is $|a_{k+1} - b| \leq \frac{1}{2}|a_k - b| + \frac{1}{2}|b_k - b|$, so that with a simple induction and (6.2.11), one gets

$$|a_{k+1} - b| \leq 2^{-(k+1)}|a_0 - b| + \sum_{j=0}^{k} 2^{-(k+1-j)}|b_j - b|$$

$$\leq 2^{-(k+1)}|b| + C\sum_{j=0}^{k} 2^{-(k+1-j)}\xi^j$$

$$\leq 2^{-(k+1)}|b| + C(k+1)\max\left\{\frac{1}{2}, \xi\right\}^{k+1} \to 0, \quad k \to \infty.$$

$\square$

Note that if the intermediate iterates $u_{(l-1)mK+(p-1)m+i+1}$, $i = 0, 1$, are composed of locally convergent components, then this holds also true for the averaged functions $\bar{u}_{lmK}$. In the following, it shall be shortly outlined how the optimality of the method can be proven on the basis of the above observations. This should be understood as working plan, where the technical details still have to be elaborated.

Thanks to Proposition 6.8, the smoothness analysis of the limits on Ω_i is already covered by §6.1.6. Thus, Theorem 6.2 (b) carries over. The next step would be to prove a result similar to Theorem 6.2 (a) also for the functions $u_{(l-1)mK+q}^{(i)}$, $1 \leq q \leq mK$, $(q-1) \bmod m = i$ in **AddSchw**. This can, in principle, be realized with the techniques developed in §6.1.5 and the ideas just mentioned, in particular, (6.2.8), (6.2.9), and Lemma 6.9. Following the lines of §6.1.4, we collect the remaining ingredients for the proof of the optimality.

Step 1 The proof of Proposition 6.2 was only based on the special structure of $\mathbf{u}_{lmK}$ in **MultSchw** and the estimates (6.1.18) and (6.1.11). Thus, the first task would be to show an analogous statement for $\bar{\mathbf{u}}_{lmK}$ using an estimate like Theorem 6.2 (a) and (6.2.4).

Step 2 Proposition 6.3 is just a consequence of Proposition 6.2 and the tolerance with which the call of **COARSE** is performed. Hence, its statement can be carried over to $\bar{\mathbf{u}}_{lmK}$ in **AddSchw** once Step 1) has been accomplished.

Step 3 Lemma 6.2 is induced by Theorem 6.2 (a), so that a similar statement should be available for **AddSchw** as well. Consequently, using $\tilde{u}_{k-m}^{(i)}$ as initial guess for the approximation of $\breve{u}_k^{(i)}$, the error within an iterative solver has to be reduced by a constant factor D only. By virtue of this observation, an adaptive Richardson iteration for the solution of the local problems can be constructed similar to the approach outlined in §6.1.4 such that only a fixed number of calls of **APPLY** and **RHS** is needed. In the numerical experiments in Chapter 7, however, we are going to apply the adaptive wavelet Galerkin method from [27].

Some comments on the general case, $m > 2$

As explained at the beginning of §6.2.3, in the general case of $m > 2$ subdomains, it does not always seem to be possible to identify local limit functions. For convenience, in the following we briefly discuss a possible alternative approach for the construction of an adaptive method involving the application of the projector $\mathbf{P}$ from (4.3.36) as in Algorithm 3. In §6.2.4 emphasis will then be placed on the question how the application of $\mathbf{P}$ can be implemented in practice. The numerical results presented in Chapter 7, however, will show that, for the considered non-trivial model problems, **AddSchw** in its present form converges with the optimal rate in linear time also for $m > 2$.

In the first place, the additive Schwarz iteration given in Algorithm 8, on the discrete level, is nothing but a damped Richardson iteration for the preconditioned linear system

$$\mathbf{M}^{-1}\mathbf{A}\mathbf{u} = \mathbf{M}^{-1}\mathbf{f},$$

with the block diagonal matrix $\mathbf{M} = \mathrm{diag}(\mathbf{A}^{(0,0)}, \ldots, \mathbf{A}^{(m-1,m-1)})$. Hence, the easiest way to obtain an adaptive algorithm with a projection step to control the redundant parts in the iterates would be to reuse Algorithm 3 with $\mathbf{A}$ and $\mathbf{f}$ replaced by $\mathbf{M}^{-1}\mathbf{A}$ and $\mathbf{M}^{-1}\mathbf{f}$, respectively. Thus, the decisive step is the approximation of the preconditioned residuals $\mathbf{M}^{-1}(\mathbf{f} - \mathbf{A}\mathbf{u}_k)$. This can be done by first approximating $\mathbf{f} - \mathbf{A}\mathbf{u}_k$ by suitable calls of **APPLY** and **RHS** with some tolerances ε_k, which gives an approximate (unpreconditioned) residual $\bar{\mathbf{r}}$, and by adaptively approximating the solutions of the $m - 1$ independent local elliptic problems

$$\mathbf{A}^{(i,i)}\bar{\mathbf{w}}^{(i)} = (\bar{\mathbf{r}}_\lambda)_{\lambda \in \Lambda_i}, \quad i = 0, \ldots, m - 1, \tag{6.2.13}$$

all with precisions of the order ε_k afterwards. Using Algorithm 3 as point of departure, we may expect that an adaptive method can be constructed for which convergence can be proved analogously to [104, Proposition 2.3], in the sense that after the application of **COARSE** one has

$$\|\mathbf{P}\tilde{\mathbf{u}} - \mathbf{u}_k\|_{\ell_2(\Lambda)} \leq \varepsilon_k, \quad k \geq 0, \tag{6.2.14}$$

where again $\tilde{\mathbf{u}}$ represents some solution of the discrete problem in $\ell_\tau^w(\Lambda)$, and ε_k geometrically decreases for $k \to \infty$.

For the verification of optimality one has to make sure that

$$|\mathbf{u}_k|_{\ell_\tau^w(\Lambda)} \lesssim |\mathbf{P}\tilde{\mathbf{u}}|_{\ell_\tau^w(\Lambda)} \lesssim |\tilde{\mathbf{u}}|_{\ell_\tau^w(\Lambda)}, \tag{6.2.15}$$

which, in combination with (6.2.14), shows that the local parts $\mathbf{u}_k^{(i)} = \{\mathbf{u}_{k,\lambda}\}_{\lambda \in \Lambda_i}$ converge linearly in $\ell_2(\Lambda)$ to a limit with bounded ℓ_τ^w-norm. The latter has to be used to prove that only a fixed number (not depending on the loop index k) of steps of a linearly convergent iterative solver is needed for the approximation of the solution of a problem (6.2.13), if the local solution from the preceding local solve is used as initial guess. Basically, optimality can then be shown by using the properties of the routines **APPLY**, **RHS**, and **COARSE**, as well as (6.2.15), and the geometric decrease of the ε_k.

6.2.4 Implementation of the projector P

We finally add some comments on the practical implementation of $\mathbf{P}$ from (4.3.36). Let in the sequel $\mathbf{v} = (\mathbf{v}^{(0)}, \ldots, \mathbf{v}^{(m-1)}) \in \ell_2(\Lambda)$, $v = \mathbf{v}^\top \Psi \in H_0^t(\Omega)$, as well as $v^{(i)} := (\mathbf{v}^{(i)})^\top \Psi^{(i)}$, $\Psi^{(i)} := \{2^{-t|\lambda|}\psi_{i,\lambda}\}_{\lambda \in \Lambda_i}$ and $\tilde{\Psi}^{(i)} := \{2^{t|\mu|}\tilde{\psi}_{i,\mu}\}_{\mu \in \Lambda_i}$ for $i = 0, \ldots, m - 1$. Thus, we write out the dyadic weights $2^{-t|\lambda|}$ and $2^{t|\mu|}$ explicitly in the following, whereas in the above analysis these were hidden in the wavelet functions (recall the notation introduced at the beginning of §6.1.2). With this notation we get the following statement.

Proposition 6.9. *For the projector* $\mathbf{P} : \ell_2(\Lambda) \to \ell_2(\Lambda)$ *from (4.3.36), it holds that*

$$\mathbf{P}v = \left(\langle \sigma_0 v, \tilde{\Psi}^{(0)} \rangle_{H_0^t(\Omega) \times H^{-t}(\Omega)}, \ldots, \langle \sigma_{m-1} v, \tilde{\Psi}^{(m-1)} \rangle_{H_0^t(\Omega) \times H^{-t}(\Omega)} \right). \qquad (6.2.16)$$

Proof. For $(j, \mu) \in \Lambda$, $\mu \in \Lambda_j$, using (4.3.36), we have

$$
\begin{aligned}
(\mathbf{P}v)_{(j,\mu)} &= \sum_{i=0}^{m-1} \sum_{\lambda \in \Lambda_i} 2^{-t(|\lambda|-|\mu|)} \langle \psi_{i,\lambda}, \sigma_j \tilde{\psi}_{j,\mu} \rangle_{H_0^t(\Omega) \times H^{-t}(\Omega)} \mathbf{v}_\lambda^{(i)} \\
&= \sum_{i=0}^{m-1} 2^{t|\mu|} \left\langle \sigma_j \sum_{\lambda \in \Lambda_i} 2^{-t|\lambda|} \psi_{i,\lambda} \mathbf{v}_\lambda^{(i)}, \tilde{\psi}_{j,\mu} \right\rangle_{H_0^t(\Omega) \times H^{-t}(\Omega)} \\
&= \sum_{i=0}^{m-1} 2^{t|\mu|} \langle \sigma_j (\mathbf{v}^{(i)})^\top \Psi^{(i)}, \tilde{\psi}_{j,\mu} \rangle_{H_0^t(\Omega) \times H^{-t}(\Omega)} \\
&= 2^{t|\mu|} \langle \sigma_j v, \tilde{\psi}_{j,\mu} \rangle_{H_0^t(\Omega) \times H^{-t}(\Omega)}.
\end{aligned}
$$

$\square$

In other words, the coefficients of $\mathbf{P}v$ are obtained by taking the corresponding function $v = \mathbf{v}^\top \Psi$, dividing it into the local parts $\sigma_j v$, and by computing the unique expansion coefficients of $\sigma_j v$ with respect to the basis $\Psi^{(j)}$ for $H_0^t(\Omega_j)$ for $j = 0, \ldots, m-1$. Thus, an alternative way besides the application of $\mathbf{P}$ via the matrix representation (4.3.36) and **APPLY** as proposed in [104] is now the approximation of $\mathbf{P}v$ by solving the zero order problems

$$\{\langle \psi_{i,\lambda}, \psi_{i,\mu} \rangle_{L_2(\Omega)}\}_{\mu,\lambda \in \Lambda_i} \tilde{\mathbf{v}}^{(i)} = \{\langle \sigma_i v, \psi_{i,\mu} \rangle_{L_2(\Omega)}\}_{(i,\mu) \in \Lambda}, \quad i = 0, \ldots, m-1, \tag{6.2.17}$$

because then $\tilde{\mathbf{v}}^{(i)} = \{\langle \sigma_i v, \tilde{\psi}_{i,\mu} \rangle_{L_2(\Omega)}\}_{\lambda \in \Lambda_i}$. This can be realized by using m parallel calls of an adaptive wavelet method. Let us mention that because of $v = \mathbf{v}^\top \Psi$ one coefficient of the right-hand side in (6.2.17) can be written as

$$\langle \sigma_i v, \psi_{i,\mu} \rangle_{L_2(\Omega)} = \sum_{j=0}^{m-1} \sum_{\lambda \in \Lambda_j} \mathbf{v}_\lambda^{(j)} 2^{-t|\lambda|} \langle \sigma_i \psi_{j,\lambda}, \psi_{i,\mu} \rangle_{L_2(\Omega)}.$$

Hence, for fixed $i \in \{0, \ldots, m-1\}$, the approximation of the matrix and the right-hand side in (6.2.17) only requires the computation of inner products between two primal wavelets living on the same subdomain Ω_i and inner products between a wavelet from Ω_i and wavelet from Ω_j, $j \in \{0, \ldots, m-1\}$, multiplied by σ_i, respectively. No dual wavelets are involved anymore. This implicit application of $\mathbf{P}$ can be expected to have better quantitative properties.

6.2.5 Concluding remarks

With **AddSchw** we have constructed an algorithm that is guaranteed to converge for an arbitrary open covering $\{\Omega_i\}_{i=0}^{m-1}$ (Ω_i being a smooth image of the unit cube). For the case $m = 2$, we have outlined that all essential ingredients, in particular the local convergence to suitable limits, needed for a proof of optimality, are available. For $m > 2$, to have a complete theoretical justification of optimality, we have considered the application of the operator $\mathbf{P}$ from §6.2.4. Using such a technique, we expect that an adaptive algorithm can be constructed, for which it is possible to prove optimality for general open coverings of Ω. In addition, we have outlined an implementable strategy for the application of $\mathbf{P}$. However, this approach still causes a considerable, additional, computational effort. On the one hand, the matrices on the left-hand side in (6.2.17) have to be approximated. For some elliptic problems, e.g., a Helmholtz equation, this may not represent an additional cost. On the other hand, also coefficients of the form $\langle \sigma_i \psi_{j,\lambda}, \psi_{i,\mu} \rangle_{L_2(\Omega)}$, $\lambda \in \Lambda_j$, $\mu \in \Lambda_i$, have to be computed. For these reasons, we prefer **AddSchw** in its present form.

In Chapter 7, we shall test **AddSchw** for cases where $m = 2$, $m = 3$, and $m = 4$ to see whether the algorithm may perform optimally also for $m > 2$ in practice. It is commonly known that the convergence rate for the additive case in (6.2.2) is typically somewhat larger than the rate in (6.1.5). However, there is some hope that with a parallel execution of the local problems, **AddSchw** may outperform **MultSchw**. This would indeed justify the consideration of **AddSchw**. Appropriate numerical tests concerning this question are also included in Chapter 7.

Chapter 7

Numerical Tests

The intention of this chapter is to practically confirm the (qualitative) theoretical findings concerning convergence with optimal order and linear complexity of the adaptive Schwarz frame methods worked out in Chapter 6. Moreover, on the one hand, it shall be demonstrated that these algorithms represent a considerable quantitative improvement compared to the steepest descent method from Chapter 4. On the other hand, a comparison of **MultSchw** with a standard adaptive finite element code shall be presented in order to give an evident classification of its numerical performance. As another important aspect, the results obtained with a serial and a first parallel implementation of the additive adaptive Schwarz frame method from §6.2 are discussed. We consider one- and two-dimensional Poisson and biharmonic problems.

7.1 The adaptive multiplicative Schwarz method

The first step is to analyze the algorithm **MultSchw** developed in §6.1. Besides the verification of convergence and optimality, and a comparison with the steepest descent method, the influence of the assumptions (6.1.16) and (6.1.17) shall be investigated in detail.

7.1.1 The Poisson equation in the unit interval

The first model problem is the one-dimensional Poisson equation (4.4.1) in the unit interval $\Omega = (0, 1)$ introduced in §4.4.1. The exact solution is given by the function from (4.4.2) which is depicted in Figure 4.1. The right-hand side is chosen accordingly. For the construction of the frame we proceed as specified in §4.4.1, i.e., we aggregate local wavelet bases from [95] lifted to $\Omega_0 = (0, 0.7)$ and $\Omega_1 = (0.3, 1)$ into a global frame. In particular, we use spline wavelets of order $d = 2, 3, 4$, having $\tilde{d} = 2, 3$, and 6 vanishing moments, respectively. For more details on the setup of this model problem, the reader is referred to the beginning of §4.4.1.

The construction of the primal and dual interval bases is based on a pair of biorthogonal multiresolution sequences $\{V_j\}_{j \geq j_0}$ and $\{\tilde{V}_j\}_{j \geq j_0}$ with a sufficiently large *coarsest scale* j_0. Note that increasing j_0 decreases the maximal size of the supports of the primal and dual scaling functions and wavelets. Thus, assumption (6.1.17) can be fulfilled by choosing j_0 sufficiently large. However, we will not make rigorously sure that condition (6.1.17) holds, and we shall see that a violation of (6.1.17) (at least to some extent) does not spoil the convergence and optimality. For the cases $d = 2, 3$ we use $j_0 = 4$, for $d = 4$ we use $j_0 = 5$, and for $d = 3$ for some experiments also $j_0 = 3$ is chosen. The adaptive solution of the local problems is always performed using the adaptive wavelet Galerkin method introduced in [27]. This approach is usually more efficient than an adaptive Richardson iteration as developed in §6.1.4. The local solver is also always initialized with the preceding approximate solution of the local problem on the respective subdomain. Doing so, the asymptotically optimal complexity of **MultSchw** is retained. We do not give a formal prove of this statement, but the results presented below confirm this claim.

Figure 7.1 shows the decay of the ℓ_2-norm of the residuals of the iterates $\mathbf{u}_k$ after the application of **COARSE** vs. their support length and CPU time, respectively. Recall that the ℓ_2-norm of the residuals is equivalent to $\|u - u_k\| \eqsim \|u - u_k\|_{H^t(\Omega)}$. For the choices $d = 2$, $d = 3$ (the latter with $j_0 = 4$), and $d = 4$, the algorithm appears to converge with the optimal rate $d - 1$. In case $d = 3$, $j_0 = 3$, **MultSchw** performs significantly worse than with $j_0 = 4$ and shows a suboptimal behavior at the end. Let us therefore focus on the case $d = 3$ in the following.

By checking the maximal diameters of the supports of the wavelets, one realizes that in both cases, $j_0 = 4$ and $j_0 = 3$, the sufficient condition (6.1.17) for the existence of a smooth limit on each subdomain is violated, to a different extent though. Despite the violation of (6.1.17), for $d = 3$, $j_0 = 4$, we observe convergence of the iterands on each subdomain. In Figure 7.2 the distribution of the wavelet coefficients of the final approximation u_{2k} on each subdomain is depicted. The corresponding local approximations and global pointwise errors are shown in Figure 7.3. Concerning the case $d = 3$, $j_0 = 4$, we can say that the final local approximation $u_{2k}^{(0)}$ looks very smooth and consists of a rather equally distributed set of wavelets on small scales. Note that by definition of our modified multiplicative Schwarz scheme $u_{2k}^{(0)}$ does not contain contributions from wavelets $\psi_{0,\lambda}$ whose supports are contained in $\Omega_1 = (0.3, 1)$. On Ω_1 the adaptive scheme plainly detects the singularity at 0.5 where the wavelet coefficients show the characteristic tree-like structure. The additional peaks in the coefficient distribution near the left end of Ω_1 can be explained by the fact that the local limit function $u^{(1)} \in H_0^t(\Omega_1)$ equals $u - \sum_{\{\lambda \in \Lambda_0: \text{supp}\, \psi_{0,\lambda} \cap \Omega \not\subset \Omega_1\}} \mathbf{u}_\lambda^{(0)} \psi_{0,\lambda}$, where $u^{(0)} = (\mathbf{u}^{(0)})^T \Psi^{(0)}$ is the local limit function on Ω_0. Since at any fixed point $x \in (0.3, 1)$ at most finitely many wavelets in the sum do not vanish, the local adaptive solver detects the natural singularities of these wavelets.

For $d = 3$, $j_0 = 3$, we observe a completely different behavior. In particular, in view of Figure 7.3 (third row) one realizes that the local contributions produced by **MultSchw** for now look completely different and that their amplitudes show an overshoot of about a factor 3 compared to the exact solution. Moreover, in our tests we have observed that they even seem to grow unboundedly. Thus, contrary to the case $j_0 = 4$, no local convergence of the sequences $(u_{2k}^{(i)})_{k \in \mathbb{N}}$ in $H_0^t(\Omega_i)$ could be

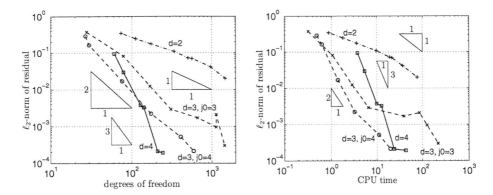

Figure 7.1: ℓ_2-norms of the residuals of $\mathbf{u}_k$ after the application of **COARSE** versus #supp $\mathbf{u}_k$ (left) and versus CPU time (right). Dash-dotted line: $d = 2$, circles: $d = 3$, $j_0 = 4$, crosses: $d = 3$, $j_0 = 3$, solid line: $d = 4$.

observed. In addition, Figure 7.4 (left) reveals the unbounded growth of the ℓ_τ^w-norms of the iterates in this case, which causes the suboptimal convergence we observe in Figure 7.1.

Summarizing, from the different results for $j_0 = 4$ and $j_0 = 3$, we conclude that although condition (6.1.17) is not strictly necessary for optimality, on the other hand it should not be violated to a too large extent. The above results also indicate that the approximations generated in **MultSchw** are much sparser than this was the case for the steepest descent method. This becomes obvious by comparing the first, second, and last row in Figure 7.2 with the results in Figure 4.5. Especially the smoothest case $d = 4$ looks very promising now.

Finally, in this subsection we present a more detailed comparison of **MultSchw** with the steepest descent method (we will refer to this method as "SD"), and with the multiplicative Schwarz method without our modification of removing contributions from other subdomains before solving on Ω_i (referred to as "plain DD"). The three methods are tested for $d = 3$. We choose $j_0 = 3$ for "plain DD" and "SD", because it is the smallest admissible choice in the wavelet construction, and it results in the best performances for "SD" and "plain DD". For **MultSchw**, we choose $j_0 = 4$. The results given in Figure 7.5 show that **MultSchw** outperforms these two methods. To attain the same accuracy, the amount of degrees of freedom within **MultSchw** has been about 4 times smaller than within **SD_SOLVE**. Concerning the computing time, one realizes that the difference is even more than a factor 10. Moreover, **MultSchw** turns out to be more efficient than "plain DD". For the latter method we also observe local convergence as well as bounded ℓ_τ^w-norms of the iterates; cf. Figure 7.4 (left). Note that the local approximations computed with "plain DD" look completely different than the original for **MultSchw**; cf. Figures 7.2 and 7.3 (in each case fourth row). The distribution of active wavelets seems to be as arbitrary as for the steepest descent

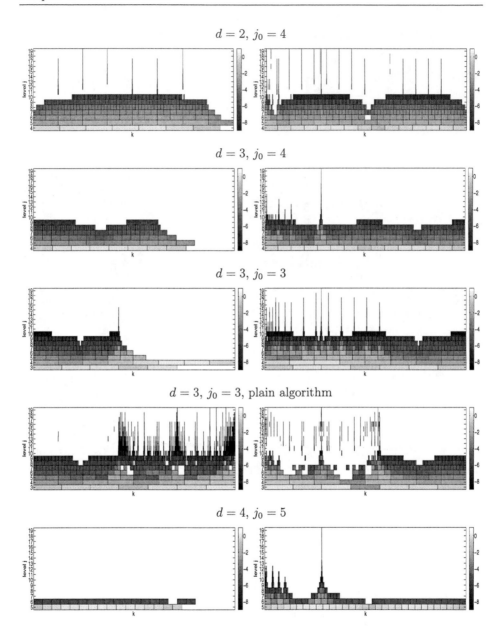

Figure 7.2: Distribution of wavelet coefficients of the final approximation u_{2k} on Ω_0 (left) and on Ω_1 (right). From top to bottom: $d = 2$ with $j_0 = 4$, $d = 3$ with $j_0 = 4$, $d = 3$ with $j_0 = 3$, the multiplicative Schwarz method without the additional removing of coefficients before a local solve, $d = 4$ with $j_0 = 5$.

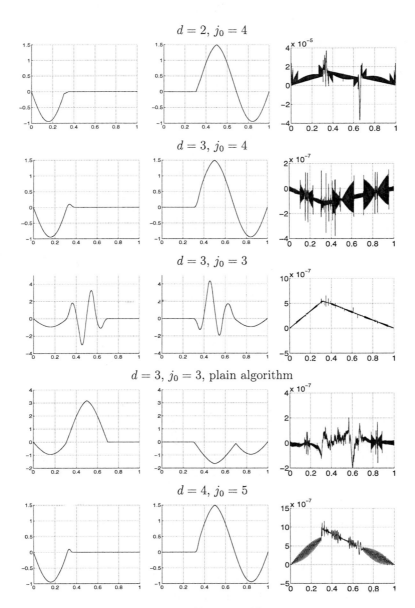

Figure 7.3: Final local approximations $u_{2k}^{(0)}$ (left), $u_{2k}^{(1)}$ (middle) produced by **MultSchw**, and the corresponding global pointwise error (right). From top to bottom: $d = 2$ with $j_0 = 4$, $d = 3$ with $j_0 = 4$, $d = 3$ with $j_0 = 3$, **MultSchw** without the additional removing of coefficients before a local solve, $d = 4$ with $j_0 = 5$.

method.

The moral of these results so far is that convergence with the optimal rate in combination with the best quantitative performance, i.e., very sparse (local) approximations and the shortest computing time, is obtained for **MultSchw** in case the maximal diameter of the supports of the primal and dual wavelets is sufficiently small.

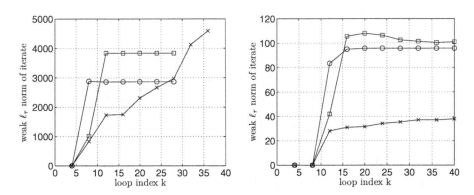

Figure 7.4: Left: ℓ_τ^w-norms of the iterates of **MultSchw** for the 1D Poisson equation, $d = 3$, $j_0 = 3$ (crosses), $j_0 = 4$ (circles), and for "plain DD", $j_0 = 3$ (boxes), $\tau = 2.5^{-1}$. Right: ℓ_τ^w-norms of the iterates of **MultSchw** for the 2D Poisson equation, $d = 3$, $j_0 = 3$ (crosses), $j_0 = 4$ (circles), and for "plain DD", $j_0 = 3$ (boxes), $\tau = 1.5^{-1}$.

7.1.2 The Poisson equation in the L-shaped domain

We now consider the variational form of Poisson's equation (4.4.3) in the L-shaped domain $\Omega = (-1, 1)^2 \backslash [0, 1)^2$. We proceed completely analogous to §4.4.2 and choose the singularity function (4.4.4) as exact solution; see again Figure 4.7 for a plot of this function and the corresponding infinitely smooth right-hand side.

We thus decompose Ω into two overlapping rectangles $\Omega_0 = (-1, 1) \times (-1, 0)$, $\Omega_1 = (-1, 0) \times (-1, 1)$. The wavelet frames used for the discretization of our model problem are constructed by simply lifting wavelet bases on the unit cube to the subdomains and collecting the resulting local bases into a global system of elements. The reference bases are once more chosen to consist of tensor products of the scaling functions and wavelets constructed in [95].

A few comments on the specific choice of the domain decomposition are in order. Firstly, we have to note that assumption (6.1.16) is not satisfied. Indeed, for any $\omega > 0$, $\Omega_0(-\omega) \cup \Omega_1(-\omega)$ does not cover a part of Ω near the re-entrant corner. In principle, this can be overcome by adding another (non-convex) polygonal patch Ω_2 given by the vertices, say, $(0, 0), (1.0, 0), (-1.0, -1.0), (0, 1.0)$ as already mentioned in Figure 2.6. Nevertheless, not to be forced to handle non-matching grids in the

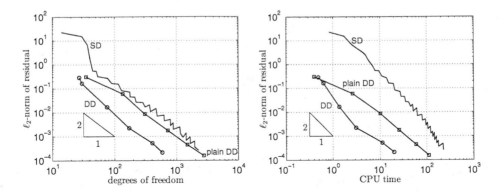

Figure 7.5: Comparison of **MultSchw** (DD), **SD_SOLVE** (SD), and **MultSchw** without additional sparsening (plain DD) for the 1D Poisson equation.

overlapping regions, and to cope with the non-convex supports of the wavelets in the determination of the sparsity pattern of the stiffness matrix, we do not follow this option. Because of the homogeneous Dirichlet boundary conditions at the point where the overlap of the two subdomains gets infinitely small, we have hope that our method is still optimal, although the (sufficient) condition (6.1.17) for optimality is violated. This will be supported by the results given below.

As above, we test **MultSchw** with frames made up of biorthogonal spline wavelet bases of primal order $d = 2, 3, 4$ and the respective number of $\tilde{d} = 2, 3, 6$ vanishing moments.

Figure 7.6 shows the convergence histories with respect to the degrees of freedom and computing time. For all setups the optimal convergence rates $\frac{d-1}{2}$ can be observed. Similar to the one-dimensional example from §7.1.1, in case $d = 3$, the method performs quantitatively better for the choice $j_0 = 4$ compared to $j_0 = 3$, and this time also for $j_0 = 3$ optimal convergence can be observed. Moreover, as Figure 7.7 (third row) reveals, again an exceptional oscillatory behavior of the local contributions $u_{2k}^{(i)}$ on Ω_i to u_{2k} can be found in the latter case; see also Figure 7.10 for a plot of the active wavelet coefficients. This phenomenon vanishes for the choice $j_0 = 4$; cf. Figure 7.7 (second row). The progression of the ℓ_τ^w-norm of $\mathbf{u}_{2k}$ during the iteration is investigated in Figure 7.4 (right), clearly indicating the uniform boundedness for $j_0 = 4$. In case $j_0 = 3$, although generally smaller, the ℓ_τ^w-norms seem to be slowly but constantly increasing.

In Figure 7.13 we once more compare **MultSchw** with "plain DD" and "SD" for $d = 3$. We choose $j_0 = 4$ for **MultSchw** and $j_0 = 3$ for the other algorithms. In this particular example we observe that **MultSchw** performs significantly better than **SD_SOLVE**. On the one hand, 3 times less degrees of freedom are used within **MultSchw**, and the computing time to reach the same precision as **SD_SOLVE**

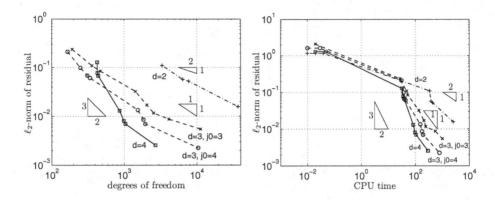

Figure 7.6: ℓ_2-norms of the residuals of $\mathbf{u}_k$ after the application of **COARSE** versus #supp $\mathbf{u}_k$ (left) and versus CPU time (right). Dash-dotted line: $d = 2$, circles: $d = 3$, $j_0 = 4$, crosses: $d = 3$, $j_0 = 3$, solid line: $d = 4$.

is about 13 times smaller. Moreover, one can see that **MultSchw** shows a better quantitative performance than the variant "plain DD" nearly during the whole approximation process. At termination, though, they have reached almost the same accuracy with a similar amount of degrees of freedom and computing time. Note again that "plain DD" has been tested with $j_0 = 3$. This choice gives a quantitative advantage compared to $j_0 = 4$ for this method. It may thus perform almost as well as **MultSchw** for $j_0 = 4$.

Considering once more Figure 7.4 (right), one can say that the ℓ_τ^w-norms for "plain DD" are larger than for **MultSchw** with $j_0 = 3$ and $j_0 = 4$, but they also seem to stay bounded. Looking at Figure 7.7 (fourth row) and Figure 7.11, we realize that "plain DD", as **SD_SOLVE** and **R_SOLVE**, generates local approximations that are nonzero where the global solution actually vanishes. This redundancy mainly occurs on the low levels of resolution (cf. Figure 7.11), and it is caused by a rather moderate number of coefficients so that the performance of the method is not spoiled.

Summarizing, we can say that if the supports of the primal and dual wavelets are chosen sufficiently small, then the local approximations generated within **MultSchw** for all spline orders $d = 2, 3$, and 4 look very smooth, and the sets of wavelets used for the approximation are very sparse; cf. also Figures 7.8, 7.9, and 7.12. Fortunately, to observe this nice behavior, the minimal level of resolution j_0 can still be chosen rather small, i.e., $j_0 = 4$ is sufficient for $d = 3$ and $d = 2$, and $j_0 = 5$ for $d = 4$. The fact that for larger j_0 usually the condition numbers of the stiffness matrices get worse does not do any harm. Concerning the guidelines formulated in §4.4.3 one can state that all aims could be realized. We have discovered a way to significantly reduce the redundancy in the overlapping region in combination with a much better underlying iterative solver also providing some preconditioning. Furthermore, in principle one

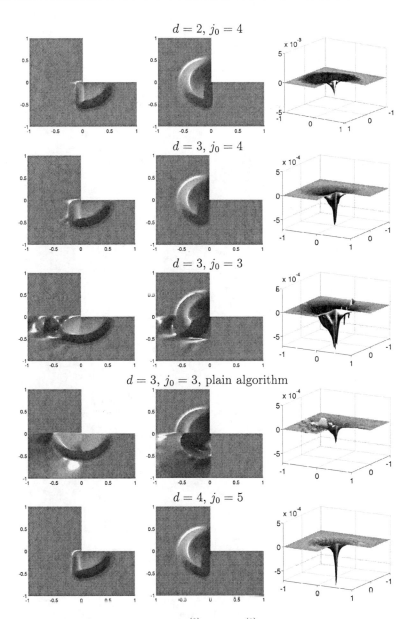

Figure 7.7: Final local approximations $u_{2k}^{(0)}$ (left), $u_{2k}^{(1)}$ (middle) produced by **MultSchw**, and the corresponding global pointwise error (right). From top to bottom: $d = 2$ with $j_0 = 4$, $d = 3$ with $j_0 = 4$, $d = 3$ with $j_0 = 3$, **MultSchw** without the additional removing of coefficients before a local solve, $d = 4$ with $j_0 = 5$.

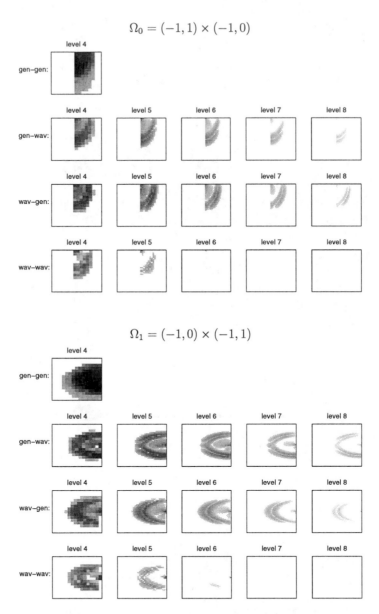

Figure 7.8: Distribution of wavelet coefficients of the final approximation produced by **MultSchw** on Ω_0 (upper part) and Ω_1 (lower part) for $d = 2$, $j_0 = 4$.

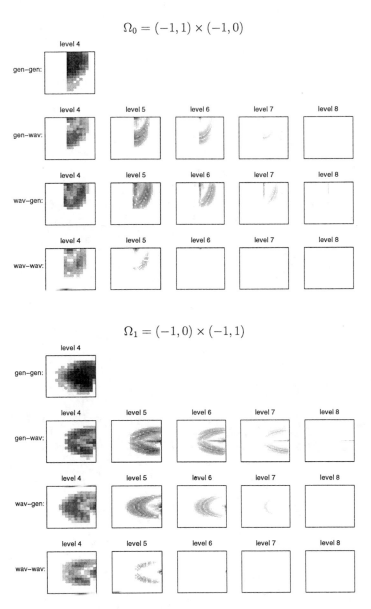

Figure 7.9: Distribution of wavelet coefficients of the final approximation produced by **MultSchw** on Ω_0 (upper part) and Ω_1 (lower part) for $d = 3$, $j_0 = 4$.

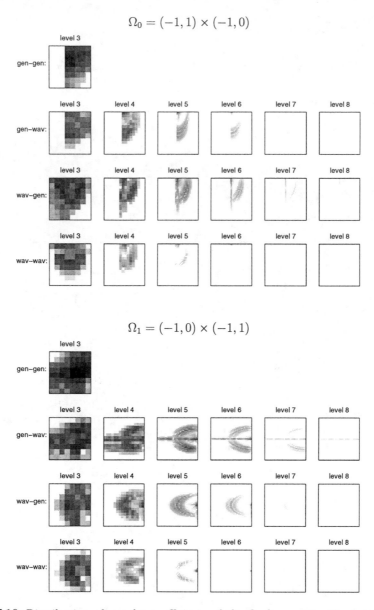

Figure 7.10: Distribution of wavelet coefficients of the final approximation produced by **MultSchw** on Ω_0 (upper part) and Ω_1 (lower part) for $d = 3$, $j_0 = 3$.

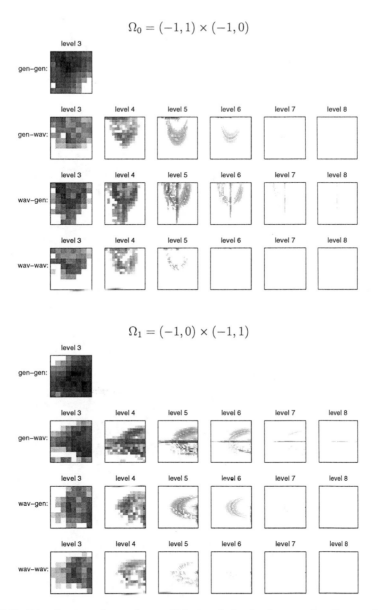

Figure 7.11: Distribution of wavelet coefficients of the final approximation produced by **MultSchw** without the additional removing of coefficients ("plain DD") on Ω_0 (upper part) and Ω_1 (lower part) for $d = 3$, $j_0 = 3$.

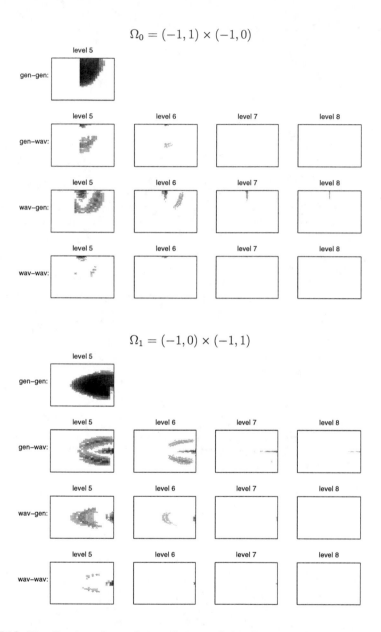

Figure 7.12: Distribution of wavelet coefficients of the final approximation produced by **MultSchw** on Ω_0 (upper part) and Ω_1 (lower part) for $d = 4$, $j_0 = 5$.

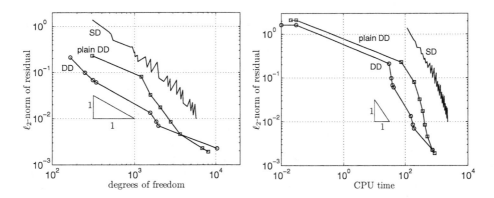

Figure 7.13: Comparison of **MultSchw** (DD), **SD_SOLVE** (SD), and **MultSchw** without additional sparsening (plain DD).

has now control over the local contribution on Ω_i that the algorithm generates.

7.1.3 Comparison with an adaptive finite element code

The next step is to compare the algorithm **MultSchw** with a standard adaptive solver based on a finite element discretization in two space dimensions. We compare with the software package *Differential Equations Analysis Library* (deal.II); see [6]. This software is open source and it realizes a fully adaptive finite element scheme, using a refinement strategy based on an a posteriori error estimator developed by Kelly et al. in [83]. It is based on rectangular partitions of the domain, so that the geometric shapes of the finite elements and the wavelets, that are constructed by forming tensor products, fit together. For these reasons this software seems to be a reasonable choice for a comparison.

As our model problem, we choose the two-dimensional Poisson equation in the L-shaped domain just as in §7.1.2 and §4.4.2. We compare **MultSchw** for $d = 2$ $(j_0 = 4)$, $d = 3$ $(j_0 = 4)$, and $d = 4$ $(j_0 = 5)$ with the adaptive finite element solver for the case of bi-linear, bi-quadratic, and bi-cubic Lagrange elements, respectively. Figure 7.14 shows the decay of the continuous L_2- and H^1-errors (H^1-seminorm) with respect to the number of degrees of freedom. Concerning the H^1-error (right column), one observes that for $d = 2$ both methods perform more or less equally well. For the higher orders, **MultSchw** produces significantly sparser approximations. For $d = 3$ the difference lies between a factor 8 and a factor 3. For $d = 4$ the difference at termination is also still more than a factor 3, whereas at an earlier stage one even observes a factor 7. Hence, the asymptotic convergence rates appear to be the same, while the quantitative performance with respect to the degrees of freedom is considerably better for our adaptive wavelet frame method. Concerning L_2-errors we

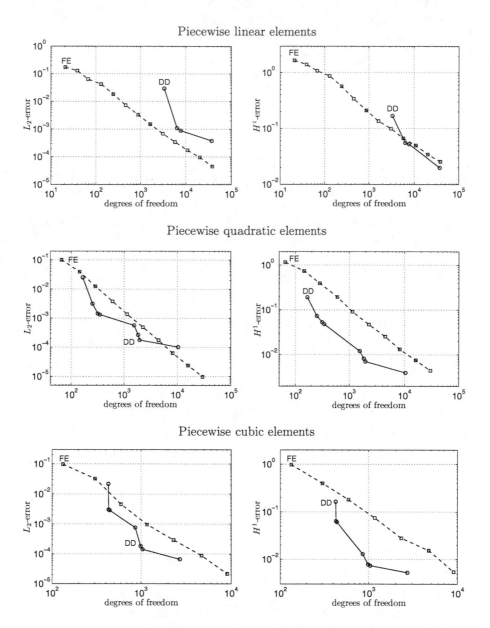

Figure 7.14: Comparison of **MultSchw** (solid line) with an adaptive finite element solver (dashed line). Left: Degrees of freedom vs. L_2-error. Right: Degrees of freedom vs. H^1-error.

can say that only for $d = 4$ **MultSchw** shows a better performance. For convenience, Table 7.1 lists the corresponding L_∞-errors of the algorithms at termination. It becomes visible that at this final stage, in particular for $d = 3$, and $d = 4$, the wavelet scheme uses about 3 times less degrees of freedom, and the L_∞-error is even smaller.

Adaptive solver	#dofs	L_∞-error
FE, piecewise affine	38107	1.291e-03
MultSchw, $d = 2$	36897	2.646e-03
FE, piecewise quadratic	29515	1.487e-03
MultSchw, $d = 3$	10381	6.866e-04
FE, piecewise cubic	9059	1.024e-03
MultSchw, $d = 4$	2690	9.215e-04

Table 7.1: Number of degrees of freedom and L_∞ error for **MultSchw** and the adaptive finite element solver at termination.

7.1.4 The biharmonic equation in the L-shaped domain

As an instance of a fourth order problem, i.e., $t = 2$, let us now consider the variational form of the biharmonic equation with homogeneous Dirichlet boundary conditions in the L-shaped domain $\Omega = (-1, 1)^2 \backslash [0, 1)^2$,

$$\Delta^2 u = f \text{ in } \Omega, \quad u = \frac{\partial u}{\partial n} = 0 \text{ on } \partial\Omega. \tag{7.1.1}$$

The associated $H_0^2(\Omega)$-elliptic bilinear form is given in (1.3.9). We choose the decomposition of Ω into subdomains as in §7.1.2 and also construct the aggregated frame as explained there. The primal wavelets on Ω_i are chosen to satisfy homogeneous boundary conditions of order $t - 1 = 1$ at $\partial\Omega_i$.

Setup of the exact solution and right-hand side

For the setup of a suitable exact solution and right-hand side we state the following result from [70, Theorem 3.4.1] which is in the spirit of Theorem 3.8.

Theorem 7.1. *Let $u \in H_0^2(\Omega)$ be the solution of $\Delta^2 u = f$ with f given in $H^{-1}(\Omega)$. Assume that the two-dimensional polygonal domain Ω has no corners S_j such that for* *the inner angle ω_j at S_j holds $\omega_j = \tan(\omega_j)$. Then, there exists a function $u_R \in H^3(\Omega)$ (called the regular part) and constants $c_{j,m}$, such that u may be written as*

$$u = u_R + \sum_j \sum_{0 < z_{j,m} < 1} c_{j,m} \tilde{S}_{j,m}, \tag{7.1.2}$$

where $z_{j,m}$ are the real roots of the characteristic equation $\sin^2(z\omega_j) = z^2 \sin^2(\omega_j)$ (and $u - u_R$ is called the singular part of u). The functions $\tilde{S}_{j,m}$ have the form

$$\tilde{S}_{j,m}(r_j, \theta_j) = \zeta_j(r_j) r^{1+z_{j,m}} v(z_{j,m}, \theta_j). \tag{7.1.3}$$

Here (r_j, θ_j) denote polar coordinates associated to S_j, and v is defined as

$$\begin{aligned}
v(z_{j,m}, \theta_j) = & \left[(z_{j,m} - 1)^{-1} \sin((z_{j,m} - 1)\omega_j) - (z_{j,m} + 1)^{-1} \sin((z_{j,m} + 1)\omega_j) \right] \\
& \cdot \left[\cos((z_{j,m} - 1)\theta_j) - \cos((z_{j,m} + 1)\theta_j) \right] \\
& - \left[(z_{j,m} - 1)^{-1} \sin((z_{j,m} - 1)\theta_j) - (z_{j,m} + 1)^{-1} \sin((z_{j,m} + 1)\theta_j) \right] \\
& \cdot \left[\cos((z_{j,m} - 1)\omega_j) - \cos((z_{j,m} + 1)\omega_j) \right].
\end{aligned}$$

For Ω being the L-shaped domain, let S_1 denote the re-entrant corner, where we have $\omega_1 = \frac{3}{2}\pi$. In this case at S_1 one has two functions

$$\tilde{S}_{1,1}(r, \theta) = \zeta(r) r^{1+z_{1,1}} v(z_{1,1}, \theta) \text{ and } \tilde{S}_{1,2}(r, \theta) = \zeta(r) r^{1+z_{1,2}} v(z_{1,2}, \theta),$$

with $z_{1,1} \approx 0.5445$ and $z_{1,2} \approx 0.9085$ in (7.1.2). We take $\tilde{S} := \tilde{S}_{1,1} + \tilde{S}_{1,2}$ as the exact solution; see Figure 7.15 for a plot of this function and the corresponding right-hand side. The function ζ represents a cut-off function just as described right after (4.4.4). The right-hand side $\Delta^2 \tilde{S}$ is an infinitely smooth function vanishing in a vicinity of

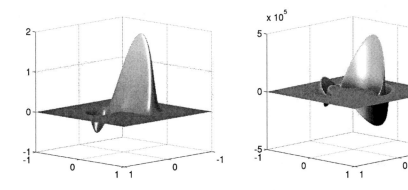

Figure 7.15: Exact solution (left) and right-hand side (right) of the two-dimensional biharmonic test problem.

the origin, so that, as for the above Poisson problem, the singularity of the solution at the re-entrant corner is solely induced by the shape of the domain. Indeed, as addressed after Theorem 3.8, by [70, Theorem 1.2.18], $\tilde{\mathcal{S}}$ is contained in $H^\alpha(\Omega)$ for all $\alpha < 1 + (1 + z_{1,1})$, but $\tilde{\mathcal{S}} \notin H^\alpha(\Omega)$ for $\alpha \geq 1 + (1 + z_{1,1})$. On the other hand, it is contained in all spaces $B^\alpha_{\tau,\tau}(\Omega)$, $\frac{1}{\tau} = \frac{\alpha}{2} + \frac{1}{2}$, $\alpha > 0$ [33]. Hence, by interpolation between these spaces and the Sobolev spaces just mentioned, one obtains $\tilde{\mathcal{S}} \in B^s_{\tau,\tau}(\Omega)$, $\frac{1}{\tau} = \frac{s-(2+\delta)}{2} + \frac{1}{2}$, for any fixed $\delta \in (0, 0.54)$ and all $s > 2 + \delta$ (Proposition 3.1). Consequently, an application of Theorem 3.6 shows that the benchmark for the optimal convergence rate of our adaptive frame scheme is $\frac{d-t}{2} = \frac{d-2}{2}$.

Numerical results

The obligatory convergence diagrams with respect to the number of degrees of freedom and computing time for piecewise quadratic ($d = 3$) and cubic ($d = 4$) wavelets are shown in Figure 7.16, confirming the convergence of **MultSchw** with the optimal rate $\frac{d-2}{2}$ in linear time also for this fourth order problem. For the case $d = 4$, the

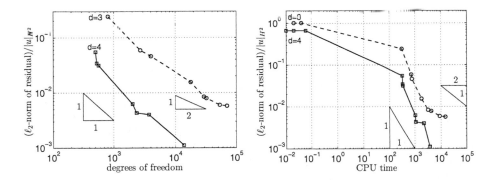

Figure 7.16: ℓ_2-norms of the residuals in **MultSchw** after the application of **COARSE** relative to the H^2-seminorm of the solution of the biharmonic equation versus #supp $\mathbf{u}_k$ (left) and versus CPU time (right). Dashed line: $d = 3$, $j_0 = 4$, solid line: $d = 4$, $j_0 = 5$.

generated local approximations and the pointwise error obtained at termination of the method are depicted in Figure 7.17.

It is remarkable that the higher order wavelets $d = 4$ lead to the better performance. Also for the case of the Poisson equation in one and two dimensions, the choice $d = 4$ has lead to the sparsest approximations and equally well or even faster convergence; consider again Figures 7.1, 7.6. We recall that the results in §4.4 have shown that for **SD_SOLVE** $d = 3$ has been the better choice (Figures 4.2, 4.8). The worse performance for $d = 4$ in these cases was explained by the larger condition numbers of the underlying wavelet bases, and the considerable amount of redundancy. Above

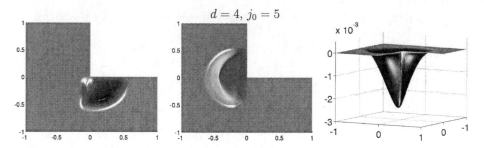

Figure 7.17: Local contributions generated by **MultSchw** for the biharmonic problem in the L-shaped domain and the corresponding pointwise error for $d = 4$, $j_0 = 5$.

results thus indicate that the essential features of **MultSchw**, i.e., preconditioning, and the additional removing of coefficients in the overlapping region, really pay off.

7.2 The adaptive additive Schwarz method

The remaining method to be examined now is the adaptive additive Schwarz method developed in §6.2. We consider two model problems in two space dimensions, i.e., the Poisson equation in the L-shaped domain and in an angular ring-shaped domain. The main objective of this section is to verify that, for these model problems, **AddSchw** also converges with the optimal order in linear time and that its quantitative performance is comparable to the one of **MultSchw**. In addition, the results of some numerical tests of a parallel implementation of **AddSchw** are presented. Firstly, in §7.2.1 below, some comments on the realization of the parallel version of **AddSchw** are collected.

7.2.1 Parallelization strategy

It is obvious that the steps in the inner loop of **AddSchw** can be performed strictly parallel, because for each $k = (l-1)mK + (p-1)m + i + 1$, $i = 0, \ldots, m-1$, the right-hand sides of the local problems on Ω_i all solely depend on the iterate $\bar{u}_{(l-1)mK+(p-1)m}$. We associate with each of the subdomains Ω_i one processor $\mathcal{P}_i$, each of which is assigned to be responsible for the solution of the local problem on Ω_i. After processor $\mathcal{P}_i$ has completed its local solution step and has set up the vector $\mathbf{u}_k$, it sends $\mathbf{u}_k$ to the *master* processor $\mathcal{P}_0$. As soon as $\mathcal{P}_0$ has received all vectors $\mathbf{u}_k$, $k = (l-1)mK + (p-1)m + i + 1$, $i = 0, \ldots, m-1$, $i \neq 0$, $\bar{\mathbf{u}}_{(l-1)mK+pm}$ is computed on $\mathcal{P}_0$ and broadcasted to all other processors, so that the next turn of local solves can be performed, and so on.

For the implementation of this procedure, the *Message Passing Interface (MPI)* has been used. The parallel tests are performed on the *MARC Linux Cluster*, consisting of 142 *Opteron* nodes, placed at the Marburg University Computing Center. For the results of the parallel tests presented below the respective global problem is treated by an homogeneous set of *Opteron* 270, 2.0GHz processors.

Clearly, in the context of parallel adaptive methods one challenging task is to dynamically assign the degrees of freedom to the available processors in such a way that the work load is as equally distributed as possible to get reasonable speedups. With our intuitive way to parallelize **AddSchw**, of course, this may not be the case, for instance, when the solution has a single singularity well outside the overlapping region and is smooth away from it. However, in the test examples below, the corresponding domain decompositions are already adapted to the problem in such a manner that an unbalanced distribution of the work load is avoided. The implementation of a more dynamical parallelization strategy is subject to future developments.

7.2.2 The L-shaped domain

We start with the Poisson equation in the L-shaped domain as our model problem. The exact solution and right-hand side as well as the aggregated frame is set up as presented in §7.1.2. In Figure 7.18, **AddSchw**, with a sequential solution of the local problems, for piecewise quadratic wavelets, $d = 3$, and the choice $j_0 = 4$ and $j_0 = 3$, is compared with **MultSchw**. It turns out that **AddSchw** indeed converges with

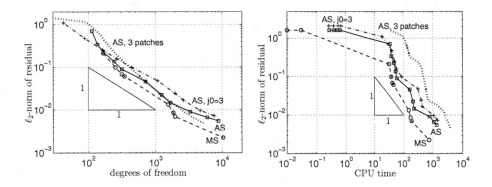

Figure 7.18: Comparison of **AddSchw** (dash-dotted line: $j_0 = 3$ and solid line: $j_0 = 4$) with **MultSchw** (dashed line: $j_0 = 4$) for the Poisson equation in the L-shaped domain with primal and dual spline orders $d = 3$, $\tilde{d} = 3$. The dotted line refers to a test of **AddSchw** with a domain decomposition consisting of three patches and $j_0 = 4$.

the optimal order $\frac{d-1}{2} = 1$ in linear time. **AddSchw** has been tested with $\alpha = 0.5$, so that $m\alpha = 1$, and for the computation of the right-hand side of a local problem as few columns of the stiffness matrix are involved as this was the case for **MultSchw**; reconsider Remark 6.4, Figure 6.1, and the discussion before Proposition 6.7. However, the performance of **MultSchw** in this experiment is slightly better. For $j_0 = 3$ the performance of **AddSchw** is also worse than for $j_0 = 4$. In Figure 7.19 it is depicted how the final approximation $\bar{u}_{2k}$ is composed of functions from $H_0^t(\Omega_0)$ and $H_0^t(\Omega_1)$, respectively. It becomes visible that **AddSchw** exhibits the same sensitivity to the maximal diameters of the supports of the primal and dual wavelets as it can be observed for **MultSchw**. In particular, for $j_0 = 3$ we observe again redundant oscillations in the local contributions that are no more present for $j_0 = 4$. For the latter case the active wavelet coefficients at termination are plotted in Figure 7.20.

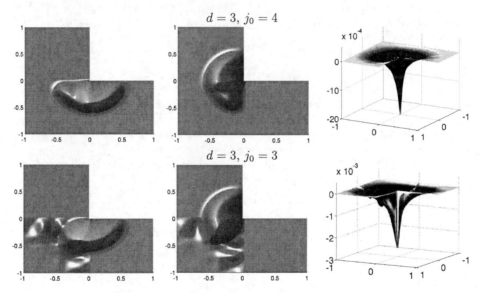

Figure 7.19: Final local approximations $\bar{u}_{2k}^{(0)}$ (left), $\bar{u}_{2k}^{(1)}$ (middle) produced by **AddSchw**, and the corresponding global pointwise error (right). From top to bottom: $d = 3$ with $j_0 = 4$, $d = 3$ with $j_0 = 3$.

Testing the adaptive additive Schwarz method for another domain decomposition consisting of three patches, $\Omega_0 = (-0.5, 1) \times (-1, 0)$, $\Omega_1 = (-1, 0) \times (-0.5, 1)$, $\Omega_2 = (-1, 0)^2$ and $d = 3$, $j_0 = 4$, one finds again that the algorithm converges with the optimal rate; see Figure 7.18. Note that this result is obtained without an implementation of the projection step involving the projector $\mathbf{P}$ from (6.2.16) as it has been proposed in the theory to obtain the verification of optimality for the case of more than two patches. With regard to the discussion in §6.2.5, we may state that in this test case the incorporation of $\mathbf{P}$ into the method is not needed.

The results of a run of **AddSchw** with $d = 3$, $j_0 = 4$, and a parallel solution of the local problems are shown in Figure 7.21. The left diagram addresses the decay of

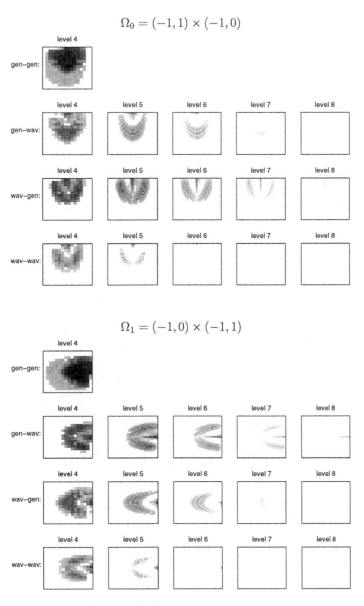

Figure 7.20: Distribution of wavelet coefficients of the final approximation $\bar{u}_{2k}$ produced by **AddSchw** on Ω_0 (upper part) and Ω_1 (lower part) for $d = 3$, $j_0 = 4$.

the ℓ_2-norm of the residuals with respect to the computing time. The right diagram shows the speedup gained by the parallelization. In order to attain the same errors, the parallel program running on two processors has indeed spent about two times less computing time.

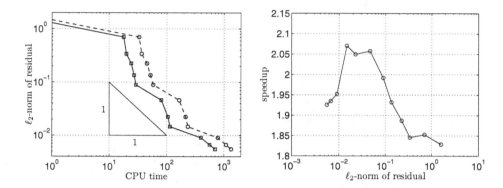

Figure 7.21: Left: ℓ_2-norms of the residuals within **AddSchw** of $\bar{u}_{2k}$ after the application of **COARSE** versus CPU time, solid line: parallel solution of local problems, dashed line: sequential solution of local problems. Right: Ratios between CPU times of parallel and sequential code.

7.2.3 A ring-shaped domain

The next model problem is the Poisson equation with prescribed homogeneous Dirichlet boundary conditions in the ring-shaped domain $\Omega = (-1,2)^2 \setminus [0,1]^2$. We decompose Ω into four overlapping subdomains $\Omega_0 = (-1,2) \times (-1,0)$, $\Omega_1 = (1,2) \times (-1,2)$, $\Omega_2 = (-1,2) \times (1,2)$, and $\Omega_3 = (-1,0) \times (-1,2)$. As usual, we construct a frame by collecting biorthogonal spline wavelet Riesz bases of order $d = 3$ with $\tilde{d} = 3$ vanishing moments and $j_0 = 4$ on all patches Ω_i, $i = 0,1,2,3$, into the global system Ψ. As the exact solution the sum of four singularity functions (4.4.4) situated around each of the re-entrant corners is chosen; see Figure 7.22 for a visualization of the exact solution and the corresponding right-hand side.

Again we compare **AddSchw** and **MultSchw** with regard to the number of degrees of freedom and computing time spent; see Figure 7.23. For the relaxation parameter, $\alpha = 0.5$ has been chosen. On the one hand, one may conclude from this result that the multiplicative Schwarz method has been slightly more efficient compared to **AddSchw** with a sequential execution of the local solution steps. On the other hand, when these tasks are distributed over 4 processors, we gain about a factor 4 in the computing time, and with this modification **AddSchw** happens to be faster than **MultSchw**; see Figure 7.23 (right) and Figure 7.24.

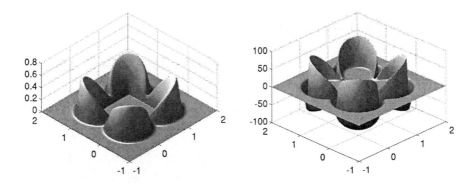

Figure 7.22: Exact solution (left) and right-hand side (right) for the Poisson equation in the ring-domain.

Summarizing, we may state that the theoretical findings from §6.2 concerning convergence and optimality of **AddSchw** for the case of two overlapping patches have been verified for nontrivial test cases. Moreover, the algorithm may reproduce optimal convergence rates also when the domain is composed of more than two patches without including projection strategies as mentioned in §6.2.3. Finally, the potential of parallel versions of **AddSchw** has been demonstrated. We learn from the above results that, for a fixed domain covering, the treatment of a problem with m instead of only one processor may lead to a reduction of the computational time by a factor m. This gain can be sufficient to outperform **MultSchw** with respect to the CPU time. Thus, the conjecture formulated at the end of §6.2.5 is confirmed. These results show that indeed the additive algorithm represents a possibly more efficient alternative in a parallel environment.

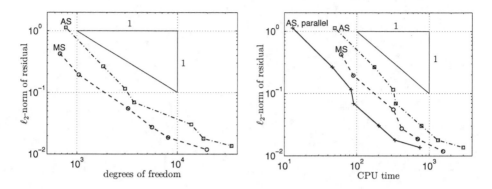

Figure 7.23: Comparison of **AddSchw** (dash-dotted line: sequential code, solid line: parallel code, both for $j_0 = 4$) with **MultSchw** (dashed line, $j_0 = 4$) for the Poisson equation in the ring-shaped domain with primal and dual spline orders $d = 3$, $\bar{d} = 3$, respectively.

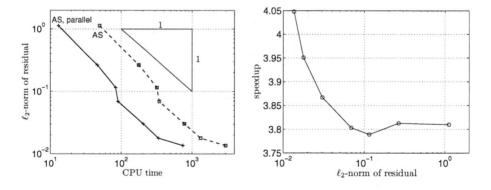

Figure 7.24: Left: ℓ_2-norms of the residuals within **AddSchw** of $\bar{\mathbf{u}}_{2k}$ after the application of **COARSE** versus CPU time, solid line: parallel solution of local problems, dashed line: sequential solution of local problems. Right: Ratios between CPU times of parallel and sequential code.

Conclusion and Outlook

For elliptic, positive order, symmetric operators, new efficient adaptive steepest descent and overlapping domain decomposition solvers subject to aggregated wavelet frame discretizations have been developed and tested. In particular, it has been verified that all essential building blocks, needed for a practical implementation of the schemes such that optimal convergence rates can be realized in linear computational time, are available. The optimal computational complexity has been theoretically proved and practically confirmed by a variety of numerical experiments for non-trivial model problems. As one important result, we have seen that the developed multiplicative Schwarz method may outperform a standard adaptive solver based on a finite element discretization with respect to the number of degrees of freedom spent. A parallel implementation of the new additive Schwarz frame method has also been developed and tested, indicating the great potential of these algorithms. Let us now summarize and discuss in detail the different achievements.

Discussion of the achieved results

Most effort was spent on the basic topics (T1) and (T2) outlined in the introduction. In Chapter 4, a fully adaptive wavelet frame method based on a steepest descent iteration has been developed and was proven to be of asymptotically optimal complexity. This finding has also been confirmed by the numerical experiments presented in §4.4. The complicated a priori determination of the optimal relaxation parameter, as needed for the damped Richardson method considered in [104], is avoided. We have discovered a really practicable method which, in addition, outperforms the damped Richardson approach in case the relaxation parameter is underestimated. After this important result, we were left with two basic deficiencies, a theoretical and a practical one. On the one hand, the verification of optimality was still based on an assumption (basically) on the canonical dual frame which is usually not known, and thus, the assumption hard to verify. On the other hand, the numerical tests have revealed the not fully convincing numerical performance, mainly induced by the rather strong redundancy of the approximation in the overlapping regions, and for spline wavelets of order $d = 4$, also by the condition number of the stiffness matrix. Fortunately, all these bottlenecks can be overcome by considering adaptive algorithms based on overlapping Schwarz domain decomposition methods as in Chapter 6. There, an adaptive multiplicative Schwarz method has been introduced for which optimality has been proved without the mentioned technical assumption made in Chapter 4. At least a sufficient overlap of the subdomains Ω_i is required (recall (6.1.16), (6.1.17)). The global elliptic problem is solved by an iterated subsequent solving of elliptic problems in Ω_i, $i = 0, \ldots, m - 1$. Removing certain degrees of freedom on Ω_i before a local solve in Ω_i, the sparsity of the approximation in the overlapping region is significantly enforced, leading, together with the natural preconditioning effect, to very good numerical performances. We have seen that for the Poisson equation in the L-shaped domain with a singular solution, the new adaptive frame method outperforms a standard adaptive solver based on a finite element discretization in terms of the number of degrees of freedom spent. Furthermore, it converges with the optimal rates in linear time also for situations where the condition on the overlap of the patches Ω_i is violated (at least to some extent). The optimality has also been confirmed for the fourth order biharmonic equation in the L-shaped domain in two space dimensions. The numerical results have revealed that the application of higher order wavelets ($d = 4$) within the domain decomposition methods really pays off. The possible worse quantitative properties of the underlying wavelet bases in this case seem to be compensated by the additional preconditioning effects. Finally, an adaptive *additive* Schwarz frame method was introduced and proven to be convergent. We have also outlined how its optimality can be verified. In the case of two patches covering the domain Ω, the local contributions on Ω_i essentially converge to the same limit functions as for the developed multiplicative solver. This is the basic property needed for the rigorous proof of optimality. For more than two subdomains, we have discussed the incorporation of a projection step to remove kernel contributions as in [104]. It was outlined how a suitable projector $\mathbf{P}$ can be implemented in practice, avoiding the approximation of L_2-inner products of primal wavelets and (certain) dual wavelets as considered in [104]. We emphasize that this consideration is is mainly motivated by theoretical reasons. Using such a technique, one may construct an optimal adaptive method without additional assumptions on the domain covering or on the unknown canonical dual frame. Yet, because of the additional amount of computational work, we have not implemented this approach. In fact, our adaptive algorithm without this modification performs optimally also for domain coverings with more than two patches, as the numerical tests have revealed.

The additive variant is of particular interest because the solution of the local auxiliary problems can be performed strictly parallel. The results obtained with a first parallel implementation have shown that *for a fixed domain covering* the treatment of a problem with m instead of only one processor may lead to a reduction of the computational time by a factor m. Then, the additive algorithm may even outperform the multiplicative variant. We may conclude that the additive scheme, although its complexity analysis appears to be slightly more complicated, represents a competitive alternative, which can be even more efficient in a parallel environment.

Moreover, this thesis has dealt with two other questions arising in the context of adaptive wavelet frame methods. The first one concerns the computation of the entries of the stiffness matrix. With this respect, we may state that, from the theoretical point of view, the results presented in Chapter 5 (and [108]) are of great importance. We have shown that suitable composite quadrature rules exist that can be used to approximate the entries of compressed versions of the discretized operators in an adaptive algorithm, so that the linear complexity will be retained. Hence, non-matching grid constellations do not prohibit the development of asymptotically optimal schemes. Nevertheless, the resulting quantitative performance might not be fully satisfactory, for which reason one should try to avoid such situations if this is possible.

In §2.3 the connection between the concept of aggregated frames and stable space splittings has been pointed out, providing a more general view on this notion compared to [37–39, 104]. In this context, the importance of a suitably smooth partition of unity $\{\sigma_i\}_{i=0}^{m-1}$, with supp $\sigma_i \subset \overline{\Omega_i}$, $\sum_{i=0}^{m-1} \sigma_i \equiv 1$, for the verification of the frame property, has become clear. Unfortunately, for many coverings that lead to matching grids, the existence of suitable partitions is not immediately clear. We have studied the two-dimensional L-shaped domain covered by two overlapping rectangles as a prototypical model case. It has turned out that, indeed, a sufficiently smooth partition can be constructed with which the existence of an aggregated frame for $H_0^t(\Omega)$, $t \in \mathbb{N}_0$, can be proved; recall §2.7. This result already involves a considerable amount of rather complicated analysis (reconsider the proof of Lemma 2.4). The determination of the convergence rates of the best N-term aggregated frame approximations in Chapter 3, for this special case, has been even more demanding. However, we have not spared the effort to deal with this complicated situation, in order to show that the developed theory really applies to constellations that typically arise in practice. We have seen that the classical results known from the basis case, most widely, carry over. Furthermore, it seems possible to generalize the presented theory to more general polygonal domains and elliptic problems.

Perspectives

A canonical continuation of the work in this thesis would be to apply the developed frame methods to problems in even more complicated bounded domains, also in *three* space dimensions. Moreover, the treatment of integral equations on a closed manifold should be studied in detail. The aggregated wavelet frames seem to be a concept with which such more demanding applications are now within reach. This hope is justified by the observation that the application of higher order wavelets seems to pay off, when domain decomposition methods are used. Thus, we may really benefit from the higher convergence rates. In this case, also possible non-matching grids in the overlapping region are much less critical, because the quantitative performance of the quadrature rules from Chapter 5 will improve. Nevertheless, one should think of the implementation (and analysis) of more sophisticated *adaptive* quadrature rules for the computation of non-diagonal blocks of the stiffness matrices. In combination with the multiplicative Schwarz method from §6.1, with which the number of columns in the non-diagonal blocks that ever have to be computed is significantly reduced, and using the integration by parts trick (5.2.5), there is some hope that good performances can also be obtained for coverings inducing non-matching grids in the case of low order spline wavelets, say, $d = t + 1$, or $d = t + 2$. We have also learned that, because of the tensor product approach to construct the reference wavelet bases, in case a face $\Gamma_{i,j}$ of the boundary of a subdomain Ω_i has at the same time a non-empty intersection with $\partial\Omega$ and Ω, then, the global smoothness of the wavelets on Ω_i near $\Gamma_{i,j} \cap \Omega$ is restricted to $C^{t-1}(\Omega)$, if at $\partial\Omega$ homogeneous boundary conditions of order $t - 1$ are prescribed. Avoiding this, though, may again lead to non-matching grids (recall, e.g., Figure 2.6). This also illustrates the importance of the quadrature aspect.

Of course, the performance of the algorithms in terms of computational time essentially depends on the implementation of all the building blocks and the data structures used. So far, the adaptive wavelet methods presented in this thesis cannot quite compete with adaptive finite element solvers with this respect, although much sparser approximations can be achieved (recall the results in §7.1.3). Investing some more man power in the scientific computing aspects, combined with a more lively exchange of experiences and ideas in this context between different research groups working in this field, would definitely help to further improve the implementations of existing adaptive wavelet methods.

For the considered adaptive algorithms, the verification of convergence is not subject to any additional conditions, whereas the proofs of optimality are quite subtle, and they may require some special assumptions. Considering the multiplicative Schwarz method, the conditions (6.1.16), (6.1.17) still restrict the class of admissible domain decompositions. For instance, for the two-patch decomposition of the L-shaped domain, (6.1.16) is violated. Resolving this problem by adding a non-convex polygonal patch around the re-entrant corner, however, seems to be practically unattractive. Because of this observation, first attempts to relax the conditions (6.1.16), (6.1.17) have been made, and this will perhaps be the subject of future investigations.

Moreover, the ideas developed in §6.2 concerning *additive* Schwarz methods definitely deserve some further study. In particular, the implementation of convenient load balancing strategies could be one stimulating mission. From the theoretical point of view, a more rigorous verification of the optimality should be worked out.

In order to further improve the quantitative performance of the adaptive methods, one may also think of imposing a tree structure onto the set of active wavelet frame elements. For the case of Riesz bases it has been shown in [29] how to modify the main procedures **COARSE** and **APPLY** to end up with a numerical scheme which realizes the tree approximation rate N^{-s} under slightly stronger regularity assumptions on the solution u. The generalization of these ideas to the setting of aggregated wavelet frames in the context of *nonlinear* problems is currently studied in [82].

Moreover, the study of other iterative solvers, for instance, domain decomposition preconditioned steepest descent or conjugate gradient methods can be considered.

List of Figures

2.1 Construction of an aggregated frame based on an overlapping domain decomposition. 43

2.2 Spaces and operators associated with a Gelfand frame. 47

2.3 Uncorrelated grids in the overlapping region of two subdomains of a domain covering used in the construction of an aggregated wavelet frame. 54

2.4 The L-shaped domain composed of two overlapping rectangles. A function $\sigma_0 : \Omega \to \mathbb{R}$, satisfying $\sigma_0|_{(-1,0)\times(0,1)} \equiv 1$ and $\sigma_0|_{(0,1)\times(-1,0)} \equiv 0$, is bound to have a discontinuity at the origin. 55

2.5 Decomposition of the boundary of Ω_0 into four segments. 57

2.6 Left: Non-overlapping decomposition of the L-shaped domain with three congruent squares. Right: Overlapping covering including the non-convex polygonal patch $\tilde{\Omega}_3$. 58

3.1 A classical DeVore/Triebel diagram for the visualization of embeddings of Besov spaces. 63

3.2 Visualization of the scales of functions spaces governing the convergence rates of linear and nonlinear wavelet approximation. 68

3.3 Classical scale of Besov spaces governing best N-term approximation in $H^t(\Omega)$ w.r.t. wavelet bases (solid line), and the relevant scale for aggregated wavelet frames in Theorem 3.6 (dashed line). 77

3.4 DeVore/Triebel diagram corresponding to the proof of Theorem 3.6. 78

4.1 Exact solution of the one-dimensional test problem (solid line) composed of a sinusoidal (dashed line) and piecewise polynomial part (dash-dotted line). 106

4.2 ℓ_2-norms of the residuals of $\bar{\mathbf{w}}_i$ versus $\#\,\mathrm{supp}\,\bar{\mathbf{w}}_i$ (left) and versus CPU time (right) for the one-dimensional test problem. 106

4.3 Comparison of **SD_SOLVE** and **R_SOLVE** for $d = 3$. **R_SOLVE** is tested with $\alpha = 0.4$ (dashed line) and with $\alpha = 0.15$ (dash-dotted line). . . . 108

4.4 Descent parameters calculated in **SD_SOLVE** for $d = 3$. 108

4.5 Distribution of wavelet coefficients of the final approximation produced by **SD_SOLVE** or **R_SOLVE** on Ω_0 (left) and Ω_1 (right) for different spline orders d. 109

4.6 Local approximations on Ω_0 (left), on Ω_1 (middle), and the corresponding global pointwise errors (right). 110

4.7 Exact solution (left) and right-hand side of the two-dimensional test problem. 112

4.8 ℓ_2-norms of the residuals of $\bar{\mathbf{w}}_i$ versus $\#\operatorname{supp}\bar{\mathbf{w}}_i$ (left) and versus CPU time (right) for the two-dimensional test problem. 112

4.9 Comparison of **SD_SOLVE** (solid line) and **R_SOLVE** for $d = 3$. **R_SOLVE** is tested with $\alpha = 0.25$ (dashed line) and with $\alpha = 0.05$ (dash-dotted line). 113

4.10 Distribution of wavelet coefficients of the final approximation produced by **SD_SOLVE** on Ω_0 (upper part) and Ω_1 (lower part) for $d = 2$. 114

4.11 Distribution of wavelet coefficients of the final approximation produced by **SD_SOLVE** on Ω_0 (upper part) and Ω_1 (lower part) for $d = 3$. 115

4.12 Distribution of wavelet coefficients of the final approximation produced by **SD_SOLVE** on Ω_0 (upper part) and Ω_1 (lower part) for $d = 4$. 116

4.13 Local approximations on Ω_0 (left), on Ω_1 (middle), and the corresponding global pointwise errors (right). 117

5.1 Left: $\operatorname{supp}\psi_\lambda \subset \overline{\Xi_{\lambda',i'}}$, i.e., an entry of $\mathbf{A}^{(\mathrm{r})}$. Right: $\operatorname{supp}\psi_\lambda \cap \operatorname{sing supp}\psi_{\lambda'} \neq \emptyset$, i.e., an entry of $\mathbf{A}^{(\mathrm{s})}$. 120

5.2 $\operatorname{supp}\psi_\lambda \not\subset \overline{\Xi_{\lambda',i'}}$ for any $1 \leq i' \leq K$, i.e., an entry of $\mathbf{A}^{(\mathrm{s})}$, but $\Xi_{\lambda,i} \subset \Xi_{\lambda',i'(i)}$ for any i. 126

5.3 The quadrature mesh on $\Xi_{i,\lambda}$. The highlighted squares represent those subcubes on which $\psi_{\lambda'}$ is not arbitrarily smooth. 128

5.4 Two simple domains made up of two overlapping patches with non-matching dyadic grids. 130

5.5 Decay of the entries in a column of $\mathbf{A}^{(\mathrm{r})}$ (upper part) and $\mathbf{A}^{(\mathrm{s})}$ (lower part) for $d = \tilde{d} = 2$ and $d = \tilde{d} = 3$. 131

5.6 Quadrature errors for single entries of $\mathbf{A}^{(\mathrm{r})}$ for different granularities N with $p = 4$ (and $d = \tilde{d} = 2$) and $p = 2$ ($d = \tilde{d} = 3$), respectively. The right pictures show the the singular supports of the wavelets involved. 132

5.7 Decay of the quadrature error in a column of $\mathbf{A}^{(\mathrm{r})}$ as function of $||\lambda| - |\lambda'||$ for fixed N, and $p = d = \tilde{d} = 2$. 133

5.8 Decay of the quadrature error for entries $\mathbf{A}^{(\mathrm{s})}_{\lambda,\lambda'}$ as function of $||\lambda| - |\lambda'||$ for fixed λ' and N, $p = d = \tilde{d} = 2$, where the singular supports of ψ_λ and $\psi_{\lambda'}$ are nested like in Figure 5.2, because $\lambda = (i, \mu)$ and $\lambda' = (i, \mu')$, i.e., they are lifted by the same κ_i. 133

5.9 Quadrature errors for single entries of $\mathbf{A}^{(\mathrm{s})}$ for different granularities N with $p = d = \tilde{d} = 2$ (top left) and $p = 4$, $d = \tilde{d} = 3$ (bottom left), respectively. The right pictures show the singular supports of the wavelets involved. The dashed line in the lower left picture refers to the case where the composite rule was applied to the right-hand side of (5.2.5). 134

5.10 Decay of the quadrature error in a column of $\mathbf{A}^{(\mathrm{s})}$ as function of $||\lambda| - |\lambda'||$ for fixed N and $p = 2$, $d = \tilde{d} = 2$, and $p = 4$, $d = \tilde{d} = 3$. 134

6.1 Location of supports of wavelets $\psi_{1,\lambda}$ being relevant for the computation of the right-hand side of a local problem in Ω_0. 149

6.2 supp $\psi_{0,\lambda}$ and supp $\tilde{\psi}_{0,\lambda}$ for $\lambda \in \Lambda_0$ with supp $\psi_{0,\lambda} \cap \Omega \not\subset \Omega_1$, and supp $u_k^{(1)}$ for some odd k assuming (6.1.17). 153

6.3 A domain covering with three subdomains. Left: The highlighted area represents the support of the contribution from Ω_0 when in **MultSchw** a local problem on Ω_2 has just been solved. Right: The same for **AddSchw**. . . 168

6.4 Schematic representation of the argument used in (6.2.9) for even k. 170

7.1 ℓ_2-norms of the residuals of $\mathbf{u}_k$ after the application of **COARSE** versus #supp $\mathbf{u}_k$ (left) and versus CPU time (right). Dash-dotted line: $d = 2$, circles: $d = 3$, $j_0 = 4$, crosses: $d = 3$, $j_0 = 3$, solid line: $d = 4$. 177

7.2 Distribution of wavelet coefficients of the final approximation u_{2k} on Ω_0 (left) and on Ω_1 (right). From top to bottom: $d = 2$ with $j_0 = 4$, $d = 3$ with $j_0 = 4$, $d = 3$ with $j_0 = 3$, the multiplicative Schwarz method without the additional removing of coefficients before a local solve, $d = 4$ with $j_0 = 5$. . . 178

7.3 Final local approximations $u_{2k}^{(0)}$ (left), $u_{2k}^{(1)}$ (middle) produced by **MultSchw**, and the corresponding global pointwise error (right). From top to bottom: $d = 2$ with $j_0 = 4$, $d = 3$ with $j_0 = 4$, $d = 3$ with $j_0 = 3$, **MultSchw** without the additional removing of coefficients before a local solve, $d = 4$ with $j_0 = 5$. 179

7.4 Left: ℓ_τ^w-norms of the iterates of **MultSchw** for the 1D Poisson equation, $d = 3$, $j_0 = 3$ (crosses), $j_0 = 4$ (circles), and for "plain DD", $j_0 = 3$ (boxes), $\tau = 2.5^{-1}$. Right: ℓ_τ^w-norms of the iterates of **MultSchw** for the 2D Poisson equation, $d = 3$, $j_0 = 3$ (crosses), $j_0 = 4$ (circles), and for "plain DD", $j_0 = 3$ (boxes), $\tau = 1.5^{-1}$. 180

7.5 Comparison of **MultSchw** (DD), **SD_SOLVE** (SD), and **MultSchw** without additional sparsening (plain DD) for the 1D Poisson equation. 181

7.6 ℓ_2-norms of the residuals of $\mathbf{u}_k$ after the application of **COARSE** versus #supp $\mathbf{u}_k$ (left) and versus CPU time (right). Dash-dotted line: $d = 2$, circles: $d = 3$, $j_0 = 4$, crosses: $d = 3$, $j_0 = 3$, solid line: $d = 4$. 182

7.7 Final local approximations $u_{2k}^{(0)}$ (left), $u_{2k}^{(1)}$ (middle) produced by **MultSchw**, and the corresponding global pointwise error (right). From top to bottom: $d - 2$ with $j_0 = 4$, $d = 3$ with $j_0 = 4$, $d = 3$ with $j_0 = 3$, **MultSchw** without the additional removing of coefficients before a local solve, $d = 4$ with $j_0 = 5$. 183

7.8 Distribution of wavelet coefficients of the final approximation produced by **MultSchw** on Ω_0 (upper part) and Ω_1 (lower part) for $d = 2$, $j_0 = 4$. 184

7.9 Distribution of wavelet coefficients of the final approximation produced by **MultSchw** on Ω_0 (upper part) and Ω_1 (lower part) for $d = 3$, $j_0 = 4$. 185

7.10 Distribution of wavelet coefficients of the final approximation produced by **MultSchw** on Ω_0 (upper part) and Ω_1 (lower part) for $d = 3$, $j_0 = 3$. 186

7.11 Distribution of wavelet coefficients of the final approximation produced by **MultSchw** without the additional removing of coefficients ("plain DD") on Ω_0 (upper part) and Ω_1 (lower part) for $d = 3$, $j_0 = 3$. 187

7.12 Distribution of wavelet coefficients of the final approximation produced by **MultSchw** on Ω_0 (upper part) and Ω_1 (lower part) for $d = 4$, $j_0 = 5$. 188

7.13 Comparison of **MultSchw** (DD), **SD_SOLVE** (SD), and **MultSchw** without additional sparsening (plain DD). 189

7.14 Comparison of **MultSchw** (solid line) with an adaptive finite element solver (dashed line). Left: Degrees of freedom vs. L_2-error. Right: Degrees of freedom vs. H^1-error. 190

7.15 Exact solution (left) and right-hand side (right) of the two-dimensional biharmonic test problem. 192

7.16 ℓ_2-norms of the residuals in **MultSchw** after the application of **COARSE** relative to the H^2-seminorm of the solution of the biharmonic equation versus #supp $\mathbf{u}_k$ (left) and versus CPU time (right). Dashed line: $d = 3$, $j_0 = 4$, solid line: $d = 4$, $j_0 = 5$. 193

7.17 Local contributions generated by **MultSchw** for the biharmonic problem in the L-shaped domain and the corresponding pointwise error for $d = 4$, $j_0 = 5$.194

7.18 Comparison of **AddSchw** (dash-dotted line: $j_0 = 3$ and solid line: $j_0 = 4$) with **MultSchw** (dashed line: $j_0 = 4$) for the Poisson equation in the L-shaped domain with primal and dual spline orders $d = 3$, $\tilde{d} = 3$. The dotted line refers to a test of **AddSchw** with a domain decomposition consisting of three patches and $j_0 = 4$. 195

7.19 Final local approximations $\bar{u}_{2k}^{(0)}$ (left), $\bar{u}_{2k}^{(1)}$ (middle) produced by **AddSchw**, and the corresponding global pointwise error (right). From top to bottom: $d = 3$ with $j_0 = 4$, $d = 3$ with $j_0 = 3$. 196

7.20 Distribution of wavelet coefficients of the final approximation $\bar{u}_{2k}$ produced by **AddSchw** on Ω_0 (upper part) and Ω_1 (lower part) for $d = 3$, $j_0 = 4$. . . . 197

7.21 Left: ℓ_2-norms of the residuals within **AddSchw** of $\bar{u}_{2k}$ after the application of **COARSE** versus CPU time, solid line: parallel solution of local problems, dashed line: sequential solution of local problems. Right: Ratios between CPU times of parallel and sequential code. 198

7.22 Exact solution (left) and right-hand side (right) for the Poisson equation in the ring-domain. 199

7.23 Comparison of **AddSchw** (dash-dotted line: sequential code, solid line: parallel code, both for $j_0 = 4$) with **MultSchw** (dashed line, $j_0 = 4$) for the Poisson equation in the ring-shaped domain with primal and dual spline orders $d = 3$, $\tilde{d} = 3$, respectively. 200

7.24 Left: ℓ_2-norms of the residuals within **AddSchw** of $\bar{u}_{2k}$ after the application of **COARSE** versus CPU time, solid line: parallel solution of local problems, dashed line: sequential solution of local problems. Right: Ratios between CPU times of parallel and sequential code. 200

List of Tables

7.1 Number of degrees of freedom and L_∞-error for **MultSchw** and the adaptive finite element solver at termination. 191

Bibliography

[1] J. Ackermann, *Adaptive Frame-Verfahren für Integralgleichungen auf der Sphäre*, Diplomarbeit, Fachbereich Mathematik und Informatik, Philipps-Universität Marburg, 2008.

[2] R. A. Adams, *Sobolev Spaces*, Academic Press, New York, 1975.

[3] B. K. Alpert, *A class of bases in L_2 for the sparse representation of integral operators*, SIAM J. Math. Anal. **24** (1993), no. 1, 246–262.

[4] L. Andersson, N. Hall, B. Jawerth, and G. Peters, *Wavelets on closed subsets of the real line*, Wavelet Analysis and its Applications, vol. 3, Academic Press, 1994, pp. 1–61.

[5] K. E. Atkinson, *The numerical solution of boundary integral equations*, Duff, I. S. (ed.) et al., The state of the art in numerical analysis. Based on the proceedings of a conference organized by the Institute of Mathematics and its Applications (IMA), University of York, York, GB, April 1–4, 1996. Oxford: Clarendon Press. Inst. Math. Appl. Conf. Ser., New Ser. 63, 223-259, 1997.

[6] W. Bangerth, R. Hartmann, and G. Kanschat, *deal.ii — a general purpose object oriented finite element library*, ACM Trans. Math. Softw. **33** (2007), no. 4, 24:1–24:27.

[7] A. Barinka, *Fast computation tools for adaptive wavelet schemes*, Ph.D. thesis, RWTH Aachen, 2005.

[8] A. Barinka, T. Barsch, P. Charton, A. Cohen, S. Dahlke, W. Dahmen, and K. Urban, *Adaptive wavelet schemes for elliptic problems: Implementation and numerical experiments*, SIAM J. Sci. Comput. **23** (2001), no. 3, 910–939.

[9] J. J. Benedetto and S. Li, *The theory of multiresolution analysis frames and applications to filter banks*, Appl. Comput. Harmon. Anal. **5** (1998), no. 4, 389–427.

[10] C. Bennett and R. Sharpley, *Interpolation of operators*, Pure and Applied Mathematics, vol. 129, Academic Press, 1988.

[11] J. Bergh and J. Löfström, *Interpolation Spaces*, Grundlehren der mathematischen Wissenschaften, vol. 223, Springer-Verlag, 1976.

[12] P. Binev, W. Dahmen, and R. DeVore, *Adaptive finite element methods with convergence rates*, Numer. Math. **97** (2004), no. 2, 219–268.

[13] K. Bittner, *Biorthogonal spline wavelets on the interval*, Wavelets and splines: Athens 2005, Mod. Methods Math., Nashboro Press, Brentwood, TN, 2006, pp. 93–104.

[14] D. Braess, *Finite Elemente*, second ed., Springer, Berlin–Heidelberg–New York, 1997.

[15] J. H. Bramble, *Multigrid methods*, Pitman Research Notes in Mathematics Series. 294. Harlow: Longman Scientific & Technical. viii, 161 p., 1993.

[16] J. H. Bramble, J. E. Pasciak, and J. Xu, *Parallel multilevel preconditioners*, Math. Comp. **55** (1990), no. 191, 1–22.

[17] S. C. Brenner and L. R. Scott, *The Mathematical Theory of Finite Element Methods*, Springer, Berlin–Heidelberg–New York, 1996.

[18] A. M. Bruaset and A. Tveito (eds.), *Numerical Solution of Partial Differential Equations on Parallel Computers*, Lecture Notes in Computational Science and Engeneering, vol. 51, Springer-Verlag, Berlin–Heidelberg, 2006.

[19] C. Canuto, A. Tabacco, and K. Urban, *The wavelet element method, part I: Construction and analysis*, Appl. Comput. Harmon. Anal. **6** (1999), no. 1, 1–52.

[20] C. Canuto and K. Urban, *Adaptive optimization of convex functionals in Banach spaces*, SIAM J. Numer. Anal. **42** (2005), no. 5, 2043–2075.

[21] J. M. Carnicer, W. Dahmen, and J. M. Peña, *Local decomposition of refinable spaces and wavelets*, Appl. Comput. Harmon. Anal. **3** (1996), no. 2, 127–153.

[22] T. F. Chan and T. P. Mathew, *Domain decomposition algorithms*, Acta Numerica (1994), 61–143.

[23] O. Christensen, *An Introduction to Frames and Riesz Bases*, Birkhäuser, Boston, 2003.

[24] C. K. Chui and E. Quak, *Wavelets on a bounded interval*, Numerical Methods of Approximation Theory (Dietrich Braess and Larry L. Schumaker, eds.), International Series of Numerical Mathematics, vol. 9, Birkhäuser, Basel, 1992, pp. 53–75.

[25] P. G. Ciarlet, *The Finite Element Method for Elliptic Problems*, Studies in Mathematics and its Applications, vol. 4, North-Holland, Amsterdam–New York–Oxford, 1978.

[26] A. Cohen, *Numerical Analysis of Wavelet Methods*, Studies in Mathematics and its Applications, vol. 32, North-Holland, Amsterdam, 2003.

[27] A. Cohen, W. Dahmen, and R. DeVore, *Adaptive wavelet methods for elliptic operator equations – Convergence rates*, Math. Comput. **70** (2001), no. 233, 27–75.

[28] _____, *Adaptive wavelet methods II: Beyond the elliptic case*, Found. Comput. Math. **2** (2002), no. 3, 203–245.

[29] _____, *Adaptive wavelet schemes for nonlinear variational problems*, SIAM J. Numer. Anal. **41** (2003), no. 5, 1785–1823.

[30] A. Cohen, I. Daubechies, and J.-C. Feauveau, *Biorthogonal bases of compactly supported wavelets*, Commun. Pure Appl. Math. **45** (1992), no. 5, 485–560.

[31] A. Cohen, I. Daubechies, and P. Vial, *Wavelets on the interval and fast wavelet transforms*, Appl. Comput. Harmon. Anal. **1** (1993), no. 1, 54–81.

[32] A. Cohen and R. Masson, *Wavelet adaptive method for second order elliptic problems: Boundary conditions and domain decomposition*, Numer. Math. **86** (2000), no. 2, 193–238.

[33] S. Dahlke, *Besov regularity for elliptic boundary value problems on polygonal domains*, Appl. Math. Lett. **12** (1999), no. 6, 31–36.

[34] S. Dahlke, W. Dahmen, R. Hochmuth, and R. Schneider, *Stable multiscale bases and local error estimation for elliptic problems*, Appl. Numer. Math. **23** (1997), no. 1, 21–48.

[35] S. Dahlke, W. Dahmen, and K. Urban, *Adaptive wavelet methods for saddle point problems — Optimal convergence rates*, SIAM J. Numer. Anal. **40** (2002), no. 4, 1230–1262.

[36] S. Dahlke, M. Fornasier, and K.-H. Gröchenig, *Optimal adaptive computations in the Jaffard algebra and localized frames*, to appear in J. Approx. Theory (2009).

[37] S. Dahlke, M. Fornasier, M. Primbs, T. Raasch, and M Werner, *Nonlinear and adaptive frame approximation schemes for elliptic PDEs: Theory and numerical experiments*, Numer. Methods Partial Differ. Equations (2008), doi:10.1002/num.20407.

[38] S. Dahlke, M. Fornasier, and T. Raasch, *Adaptive frame methods for elliptic operator equations*, Adv. Comput. Math. **27** (2007), no. 1, 27–63.

[39] S. Dahlke, M. Fornasier, T. Raasch, R. Stevenson, and M. Werner, *Adaptive frame methods for elliptic operator equations: The steepest descent approach*, IMA J. Numer. Anal. **27** (2007), no. 4, 717–740.

[40] S. Dahlke, R. Hochmuth, and K. Urban, *Adaptive wavelet methods for saddle point problems*, M2AN Math. Model. Numer. Anal. **34** (2000), no. 5, 1003–1022.

[41] W. Dahmen, *Stability of multiscale transformations*, J. Fourier Anal. Appl. **2** (1996), no. 4, 341–361.

[42] _____, *Wavelet and multiscale methods for operator equations*, Acta Numerica **6** (1997), 55–228.

[43] W. Dahmen, B. Han, R. Q. Jia, and A. Kunoth, *Biorthogonal multiwavelets on the interval: Cubic Hermite splines*, Constr. Approx. **16** (2000), no. 2, 221–259.

[44] W. Dahmen and A. Kunoth, *Multilevel preconditioning*, Numer. Math. **63** (1992), no. 3, 315–344.

[45] W. Dahmen, A. Kunoth, and K. Urban, *Biorthogonal spline-wavelets on the interval — Stability and moment conditions*, Appl. Comput. Harmon. Anal. **6** (1999), no. 2, 132–196.

[46] W. Dahmen, S. Prössdorf, and R. Schneider, *Multiscale methods for pseudo-differential equations on smooth closed manifolds*, Proceedings of the International Conference on Wavelets: Theory, Algorithms and Applications (C.K. Chui, L. Montefusco, and L. Puccio, eds.), Academic Press, San Diego, CA, 1994, pp. 385–424.

[47] W. Dahmen and R. Schneider, *Wavelets with complementary boundary conditions — Function spaces on the cube*, Result. Math. **34** (1998), no. 3–4, 255–293.

[48] _____, *Composite wavelet bases for operator equations*, Math. Comput. **68** (1999), no. 228, 1533–1567.

[49] _____, *Wavelets on manifolds I. Construction and domain decomposition*, SIAM J. Math. Anal. **31** (1999), no. 1, 184–230.

[50] I. Daubechies, *The wavelet transform, time-frequency localization and signal analysis*, IEEE Trans. Inf. Theory **36** (1990), no. 5, 961–1005.

[51] _____, *Ten lectures on wavelets*, CBMS–NSF Regional Conference Series in Applied Math., vol. 61, SIAM, Philadelphia, 1992.

[52] I. Daubechies, A. Grossmann, and Y. Meyer, *Painless nonorthogonal expansions*, J. Math. Phys. **27** (1986), 1271–1283.

[53] M. Dauge, *Singularities of corner problems and problems of corner singularities*, ESAIM, Proc. **6** (1999), 19–40.

[54] C. de Boor, *A practical guide to splines*, revised ed., Applied Mathematical Sciences, vol. 27, Springer, New York, 2001.

[55] R. DeVore, *Nonlinear approximation*, Acta Numerica **7** (1998), 51–150.

[56] R. DeVore and G.G. Lorentz, *Constructive Approximation*, Springer, Berlin, 1998.

[57] R. DeVore and V. Popov, *Interpolation of Besov spaces*, Trans. Am. Math. Soc. **305** (1988), no. 1, 397–414.

[58] M. Dobrowolski, *Angewandte Funktionalanalysis*, Springer, Berlin Heidelberg, 2006.

[59] G. C. Donovan, J. S. Geronimo, and D. P. Hardin, *Orthogonal polynomials and the construction of piecewise polynomial smooth wavelets*, SIAM J. Appl. Math. **30** (1999), no. 5, 1029–1056.

[60] M. Ehler, *On the Construction of Compactly Supported Nonseparable Bi–Frames and Their Best n-Term Approximation Properties*, Ph.D. thesis, Fachbereich Mathematik und Informatik, Philipps-Universität Marburg, 2007.

[61] H. G. Feichtinger and K.-H. Gröchenig, *Banach spaces related to integrable group representations and their atomic decomposition I*, J. Funct. Anal. **86** (1989), no. 2, 307–340.

[62] ———, *Banach spaces related to integrable group representations and their atomic decomposition II*, Monatsh. Math. **108** (1989), no. 2–3, 129–148.

[63] M. Fornasier and K. Gröchenig, *Intrinsic localization of frames*, Constr. Approx. **22** (2005), no. 3, 395–415.

[64] T. Gantumur, *An optimal adaptive wavelet method for nonsymmetric and indefinite elliptic problems*, J. Comput. Appl. Math. **211** (2008), no. 1, 90–102.

[65] T. Gantumur, H. Harbrecht, and R. Stevenson, *An optimal adaptive wavelet method without coarsening of the iterands*, Math. Comput. **76** (2007), no. 258, 615–629.

[66] T. Gantumur and R. Stevenson, *Computation of differential operators in wavelet coordinates*, Math. Comput. **75** (2006), no. 254, 697–709.

[67] ———, *Computation of singular integral operators in wavelet coordinates*, Computing **76** (2006), no. 1–2, 77–107.

[68] M. Griebel and P. Oswald, *On the abstract theory of additive and multiplicative Schwarz algorithms*, Numer. Math. **70** (1995), no. 2, 163–180.

[69] P. Grisvard, *Elliptic Problems in Nonsmooth Domains*, Pitman Publishing, Boston-London-Melbourne, 1985.

[70] ———, *Singularities in Boundary Value Problems*, Research Notes in Applied Mathematics, Springer, Berlin Heidelberg, 1992.

[71] K.-H. Gröchenig, *Describing functions: Atomic decompositions versus frames*, Monatsh. Math. **112** (1991), no. 1, 1–42.

[72] W. Hackbusch, *Multi-Grid Methods and Applications*, Springer, New York, 1985.

[73] ———, *Integralgleichungen. Theorie und Numerik*, Teubner, Stuttgart, 1989.

[74] ———, *Elliptic Differential Equations: Theory and Numerical Treatment*, Springer, Berlin, 1992.

[75] ———, *Iterative Lösung Großer Schwachbesetzter Gleichungssysteme*, Teubner, Stuttgart, 1993.

[76] B. Han, *Compactly supported tight wavelet frames and orthonormal wavelets of exponential decay with a general dilation matrix*, J. Comput. Appl. Math. **155** (2003), no. 1, 43–67.

[77] B. Han and Q. Mo, *Multiwavelet frames from refinable function vectors*, Adv. Comput. Math. **18** (2003), no. 2–4, 211–245.

[78] H. Harbrecht, *Wavelet Galerkin Schemes for the Boundary Element Method in Three Dimensions*, Ph.D. thesis, Techn. Univ. Chemnitz, 2001.

[79] H. Harbrecht and R. Stevenson, *Wavelets with patchwise cancellation properties*, Math. Comput. **75** (2006), no. 256, 1871–1889.

[80] C. Heil and D. F. Walnut, *Continuous and discrete wavelet transforms*, SIAM Rev. **31** (1989), no. 4, 628–666.

[81] A. Jonnson and H. Wallin, *Function spaces on subsets of $\mathbb{R}^n$*, Mathematical Reports, vol. 2, Harwood Academic Publishers, 1984.

[82] J. Kappei, *Adaptive Frame Methods for Nonlinear Elliptic Problems*, Ph.D. thesis, Fachbereich Mathematik und Informatik, Philipps-Universität Marburg, in preparation.

[83] D. W. Kelly, J. R. Gago, O. C. Zienkiewicz, and I. Babuška, *A posteriori error analysis and adaptive processes in the finite element method*, J. Numer. Methods Engrg. **19** (1983), 1593–1619.

[84] K. Koch, *Interpolating Scaling Vectors and Multiwavelets in* $\mathbb{R}^d$, Ph.D. thesis, Fachbereich Mathematik und Informatik, Philipps-Universität Marburg, 2006.

[85] M. Konik, *A Fully Discrete Wavelet Galerkin Boundary Element Method in Three Dimensions*, Ph.D. thesis, Techn. Univ. Chemnitz, 2002.

[86] V. A. Kozlov, V. G.Maz'ya, and J. Rossmann, *Elliptic Boundary Value Problems in Domains with Point Singularities*, Mathematical Surveys and Monographs, vol. 52, American Mathematical Society, Providence, Rhode Island, 1997.

[87] A. Kunoth and J. Sahner, *Wavelets on manifolds: An optimized construction*, Math. Comput. **75** (2006), no. 255, 1319–1349.

[88] S. Mallat, *Multiresolution approximation and wavelet orthonormal bases of* $L_2(\mathbb{R}^d)$, Trans. Amer. Math. Soc. **315** (1989), no. 1, 69–87.

[89] ———, *A Wavelet Tour of Signal Processing*, 2. ed., Academic Press, San Diego, 1999.

[90] A. A. R. Metselaar, *Handling Wavelet Expansions in Numerical Analysis*, Ph.D. thesis, University of Twente, The Netherlands, 2002.

[91] Y. Meyer, *Ondelettes, fonctions splines et analyses graduées*, Rend. Semin. Mat. Univ. Politec. Torino **45** (1987), no. 1, 1–42.

[92] P. Morin, R. H. Nochetto, and K. G. Siebert, *Data oscillation and convergence of adaptive FEM*, SIAM J. Numer. Anal. **38** (2000), no. 2, 466–488.

[93] P. Oswald, *Multilevel Finite Element Approximation. Theory and Applications*, Teubner Skripten zur Numerik, Teubner, Stuttgart, 1994.

[94] ———, *Frames and space splittings in Hilbert spaces*, Survey lectures on multilevel schemes for elliptic problems in Sobolev spaces (1997), http://www.faculty.iu-bremen.de/poswald/bonn1.pdf.

[95] M. Primbs, *Stabile Biorthogonale Spline-Waveletbasen auf dem Intervall*, Ph.D. thesis, Universität Duisburg-Essen, 2006.

[96] A. Quarteroni and A. Valli, *Domain Decomposition Methods for Partial Differential Equations*, Oxford Science Publications, 1999.

[97] T. Raasch, *Adaptive Wavelet and Frame Schemes for Elliptic and Parabolic Equations*, Ph.D. thesis, Philipps-Universität Marburg, 2007.

[98] A. Ron and Z. Shen, *Affine systems in $L_2(\mathbb{R}^d)$: The analysis of the analysis operator*, J. Funct. Anal. **148** (1997), no. 2, 408–447.

[99] T. Runst and R. Sickel, *Sobolev Spaces of Fractional Order, Nemytskij Operators and Nonlinear Partial Differential Equations*, De Gruyter, Berlin, 1996.

[100] A. Schneider, *Konstruktion von Multiwavelets und Anwendungen bei Adaptiven Numerischen Verfahren*, Diplomarbeit, Philipps-Universität Marburg, 2007.

[101] R. Schneider, *Multiskalen- und Wavelet-Matrixkompression. Analysisbasierte Methoden zur Effizienten Lösung Großer Vollbesetzter Gleichungssysteme*, Habilitationsschrift, TH Darmstadt, 1995.

[102] C. Schwab and R. Stevenson, *Space-time adaptive wavelet methods for parabolic evolution problems*, Math. Comput. (2008), DOI: 10.1090/S0025-5718-08-02205-9.

[103] H. A. Schwarz, *Gesammelte Mathematische Abhandlungen. 2 Bände*, Berlin. Springer. Bd. I. XI u. 338 S., Bd. II. VII u. 370 S. gr 8° , 1890.

[104] R. Stevenson, *Adaptive solution of operator equations using wavelet frames*, SIAM J. Numer. Anal. **41** (2003), no. 3, 1074–1100.

[105] ———, *On the compressibility of operators in wavelet coordinates*, SIAM J. Math. Anal. **35** (2004), no. 5, 1110–1132.

[106] ———, *Composite wavelet bases with extended stability and cancellation properties*, SIAM J. Numer. Anal. **45** (2007), no. 1, 133–162.

[107] ———, *Optimality of a standard adaptive finite element method*, Found. Comput. Math. **7** (2007), no. 2, 245–269.

[108] R. Stevenson and M. Werner, *Computation of differential operators in aggregated wavelet frame coordinates*, IMA J. Numer. Anal. **28** (2008), no. 2, 354–381.

[109] ———, *A multiplicative Schwarz adaptive wavelet method for elliptic boundary value problems*, Math. Comput. **78** (2009), no. 266, 619–644.

[110] A. Toselli and O. Widlund, *Domain Decomposition Methods – Algorithms and Theory*, Springer Series in Computational Mathematics, vol. 34, Springer, Berlin, 2005.

[111] H. Triebel, *Interpolation Theory, Function Spaces, Differential Operators*, North-Holland, Amsterdam, 1978.

[112] ———, *Theory of Function Spaces*, Birkhäuser, Basel/Boston/Stuttgart, 1983.

[113] ———, *Theory of Function Spaces II*, Birkhäuser, Basel, 1992.

[114] _____ , *The Structure of Functions*, Birhäuser, Basel/Boston/Berlin, 2001.

[115] _____ , *Theory of Function Spaces III*, Birkhäuser, Basel, 2006.

[116] E. Völker, *Adaptive Wavelet-Verfahren für Integralgleichungen auf dem Torus*, Diplomarbeit, Fachbereich Mathematik und Informatik, Philipps-Universität Marburg, 2004.

[117] M. Werner, *Adaptive Frame-Verfahren für Elliptische Randwertprobleme*, Diplomarbeit, Fachbereich Mathematik und Informatik, Philipps-Universität Marburg, 2005.

[118] J. Xu, *Iterative methods by space decomposition and subspace correction*, SIAM Rev. **34** (1992), no. 4, 581–613.

[119] _____ , *An introduction to multigrid convergence theory.*, Chan, Raymond H. (ed.) et al., Iterative methods in scientific computing. Papers from the winter school on Iterative methods in scientific computing and their applications held in Hong Kong, December 14-20, 1995. Singapore: Springer. 109-241 (1997)., 1997.

[120] H. Yserentant, *On the multi-level splitting of finite element spaces*, Numer. Math. **49** (1986), 379–412.

[121] _____ , *Two preconditioners based on the multi-level splitting of finite element spaces*, Numer. Math. **58** (1990), no. 2, 163–184.

[122] _____ , *Old and new convergence proofs for multigrid methods*, Acta Numerica **2** (1993), 285–326.

[123] X. Zhang, *Multilevel Schwarz methods*, Numer. Math. **63** (1992), no. 4, 521–539.